Ice Ages and Astronomical Causes

Data, spectral analysis and mechanisms

Springer
London
Berlin
Heidelberg
New York
Barcelona
Hong Kong
Milan
Paris
Santa Clara
Singapore
Tokyo

Richard A. Muller and Gordon J. MacDonald

Ice Ages and Astronomical Causes

Data, spectral analysis and mechanisms

Springer

Published in association with
Praxis Publishing
Chichester, UK

PRAXIS

Richard A. Muller
Professor of Physics
University of California
Berkeley
California
USA

Gordon J. MacDonald
Director
International Institute for Applied Systems Analysis
Laxenberg
Austria

SPRINGER–PRAXIS BOOKS IN ENVIRONMENTAL SCIENCES
SUBJECT *ADVISORY EDITOR*: John Mason B.Sc., Ph.D.

ISBN 1-85233-634-X Springer-Verlag Berlin Heidelberg New York

British Library Cataloguing in Publication Data
A catalogue record for this book is available from the British Library

Printed by MPG Books Ltd, Bodmin, Cornwall, UK

Cover design: Jim Wilkie
Cover photograph: Richard Muller
Typesetting: Originator, Gt. Yarmouth, Norfolk, UK

Printed on acid-free paper supplied by Precision Publishing Papers Ltd, UK

Contents

Preface

A physical chill settled on the 14th century at its very start, initiating the miseries to come. The Baltic Sea froze over twice, in 1303, and 1306–7; years followed of unseasonable cold, storms and rains, and a rise in the level of the Caspian Sea. Contemporaries could not know it was the onset of what has since been recognized as the Little Ice Age ...

Barbara W. Tuchman, *A Distant Mirror*, 1978

George Santayana said that those who do not study history may be condemned to repeat it. Nowhere is this more true than in climate. Truly amazing data are now available—data that show that for most of the last million years the Earth has been in a colder state than now, with ice covering much of North America and Europe. This continuous ice age has been broken every 100,000 years or so by a brief "interglacial". We are in such an interglacial period now. It started about 12,000 years ago and, shortly afterwards, agriculture began and all of civilisation developed. History shows that such warm periods in the past lasted between 10 and 20 thousand years. Based on this, we should expect our present interglacial to end any millennium now.

That is, of course, unless human-caused global warming preempts it. But don't interpret this as an argument in favour of global warming. Many scientists believe that the enhanced greenhouse effect from fossil fuel emissions will yield a rise in temperature of 2–6°C in the first 30 years of the twenty-first century. Such rapid changes have never been experienced by humans, not even during the rapid terminations of the past ice ages. As fertile regions become arid, and arid regions fertile, our survival may depend on the maturity of our political institutions to handle the world-wide disruptions that, as history shows, result from climate shifts. Nobody who has studied history could be optimistic.

Ice or heat? Or somewhere in between? What is our future? The truth is, we don't know. Enormous research programmes are now under way to guide global climate models, the large computer-based programs that simulate the behaviour of the atmosphere, oceans and glaciers. It is hard to imagine a topic more important to the economy of the world. But the models are only as good as the information we base them on. Much of the best information is in the paleoclimate records of the past million years.

Fortunately, we are blessed with truly superb data. The history of climate has been preserved in ice, rock and in layers of sediment on the sea floor. In the last quarter of the twentieth century, these amazing records have been collected at great expense. The enormous Deep Sea Drilling Project, and its successor Ocean Drilling Program, have brought back thousands of cores of sediment from around the world, with data going back millions of years. The glaciers of Greenland and Antartica have been penetrated, as have mountain glaciers near the equator, and have given us truly detailed information on the last few hundred thousand years. Analysis of sedimentary rock shows not only the most recent ice ages, but patterns of glaciation that took place hundreds of millions of years ago.

In the 1970s and 1980s, scientists attempting to understand the phenomena of the ice ages reached a tentative conclusion: the driving mechanism for the cycles of glaciation was fundamentally understood. The key force, according to the theory, was insolation, the average amount of sunlight hitting the Earth. James Croll had proposed this idea in the 1860s, and Milutin Milankovitch had modified and extended it in the early and mid 1900s. He proposed that the insolation in the northern hemisphere, with its predominance of the land mass, was dominant. The theory did not account for all the details of the data, but was believed to explain the fundamental behaviour. It appeared to be time to accept the theory and use it as a framework for future analysis.

Now, at the beginning of the twenty-first century, it is a good time to take a new look at the subject. The enormity of new and excellent data has clarified many of the questions, if not the answers. Let us reexamine the data of the glacial cycles without prejudice—or, if that is impossible, at least minimum prejudice. The Croll/Milankovitch theory has had remarkable success, but it also has remarkable problems. Can there be phenomena other than insolation that play central roles in climate? What are they? The key to understanding present climate may, indeed, lie in history.

The data are full of tantalising clues. If only we were smart enough, the complete mathematical theory of climate would be evident to us all. The goal of this book is to present the relevant techniques for analysing the data, and maybe to inspire the reader to help untangle the mystery. But there are also many established "facts" that mislead us. Mark Twain was once reputed to have said, "The trouble with most folks isn't so much their ignorance—it's knowing so many things that ain't so" (... including the authorship of this aphorism. It wasn't Mark Twain, but Josh Billings, another nineteenth century humorist.). Indeed, the published paleoclimate record is full of misinformation and error, and just one misleading or incorrect clue can derail an entire investigation. It is hard to solve a jigsaw puzzle when some pieces have previously been forced into wrong locations. The scientist must be adept at recognising false analyses and premature conclusions. More than most, this book will draw attention to the kinds of mistakes that permeate the field.

There is a problem. Many of the key breakthroughs in our understanding of climate depend on certain arcane areas of mathematics and physics that are difficult to master. These include solar system perturbation theory and spectral analysis. Not only are most geologists and climate scientists unfamiliar with these fields, but so are most physicists and mathematicians. Spectral analysis is particularly important, since it has played a key role in the elucidation of mechanisms. But it is full of traps and potential biases. There is a bad history of otherwise good scientists reporting cycles that later were shown not to be present, ranging from non-existent pulsars to claimed phenomena linked to the sunspot cycle. The whole field of finding and identifying cycles in geophysical data has a mixed reputation.

It is easy to do a Fourier transform, but trickier to know what to make of the results. There are numerous excellent books on spectral analysis, including *Spectral Analysis for Physical Applications*, (Percival and Walden, 1993) or *Spectral Analysis and Time Series* (Priestley, 1981). But most of these are books that assume a love of mathematics and prior expertise in statistics; they can be very difficult for the beginner. At the other extreme, there are books that emphasise practical approaches with minimum theoretical background. One of the best of these is *Numerical Recipes* (Press et al., 1993). But these books are too general to give details relevant to paleoclimate, and they are often misleading since the spectra available in paleoclimate are sufficiently unusual that they benefit from specialised methods. Our tone will be kept closer to that of *Numerical Recipes* than to Percival and Walden or Priestley. We will emphasise those techniques that are most applicable to paleoclimate data, but skip most of the proofs.

There are many important issues in the spectral analysis of paleoclimate that have not received direct treatment in the classic texts. Primary among these is the uncertainty in the time scale—not a problem for typical physics or engineering data. The paleoclimate signal is often measured vs. depth, but all the theories address the signal vs. time. To translate from depth to time requires assumptions that can have significant effects on the interpretation of the data. Another issue that is barely mentioned, if at all, in many standard texts is analysis of spectral shape. Rules, such as the Cramer-Rao limit, are designed to describe the regions when unbiased measurements are possible. But what can you do when the limit is not satisfied? Do you ignore the information contained in the spectral shape, or do you live with the bias and pull out useful information? Another topic frequently absent in the standard discussions of climate is the value of Monte Carlo simulations. This method has proved to be the most useful technique ever devised for understanding complex experiments in physics, and yet it has been little used in paleoclimate.

The study of the ice ages is remarkably interdisciplinary. To be an expert in this field, a scientist must have a comprehensive knowledge of geology, physics, glaciology, oceanography, atmospheric science, celestial mechanics, solar system astrophysics, cosmic rays, solar physics, interplanetary dust, climate modelling, spectral analysis and statistics. As a consequence, there are no real experts. Moreover, the subject matter is too diverse to allow a comprehensive treatment of the subject in any one text. This book, therefore, attempts to cover those aspects of the ice age that are critical in trying to deduce the origin of these climate events. A key issue is understanding how orbital changes of the earth relate to past climate.

Our emphasis will be on the information that we can glean from spectral analysis, and not on understanding the interplay of the oceans and atmosphere that turn a celestial force into a climate variation.

In fact, the evidence for the role of astronomy comes almost exclusively from spectral analysis. The seminal paper was published in 1976, titled, "Variations of the earth's orbit: pacemaker of the ice ages" (Hays et al., 1976). Their conclusions depended on spectral analysis, and yet spectral analysis is a discipline that makes many scientists uncomfortable. The strange result is the appearance of excellent books on paleoclimate, such as the recent (1999) book by Raymond Bradley, which are used as the basis for courses on paleoclimate—and yet spectral methods are not discussed at all in the text. The term "spectral analysis" is not listed in the index.

We will not try to cover the wealth of data that Bradley's book covers. Nor will we attempt to survey oceanography or its methods; for that see a text such as *The Sea Floor* (Seibold and Berger, 1996). Our emphasis instead will be to try to fill the gap, to try to give a working approach to the use of spectral analysis in paleoclimate data. Although we will touch on other climate indicators, our emphasis will be on oxygen isotopes, since those appear to give the most direct indication of the phenomenon known as the ice ages.

We believe that the current paradigms in climate theory owe too much to historical accident. Some features in early data that seemed to point the way to a theoretical understanding were not verified when subsequent higher-quality data became available. One example of this is the 19 kyr peak observed in the 1976 Pacemaker article. In subsequent oxygen isotope measurements, it has been very weak and, in many cases, completely absent. We will argue that its presence in the Pacemaker paper may have been no more than a statistical fluctuation.

As the century turns over, the authors of this book are known for taking an iconoclastic view of much climate research. We have proposed that the 100 kyr cycle of the ice ages is driven by orbital inclination, and not by eccentricity, the prevailing paradigm. The purpose of this book is not to present our viewpoint—although it certainly will creep through at various places, and will appear in more strength in Chapter 7—but we hope it will prove useful even to those who completely disagree with this theory. Our goal is to present the methods of analysis so that they can be used by anybody, particularly by those who have little prior experience in the spectral analysis of paleoclimate data.

If Sherlock Holmes were around today, and if he had an interest in science (according to Arthur Conan Doyle, he did not), would the wealth of clues we have available now point him clearly to the answer? Would he say, "It's elementary, Dr Watson," as he often did in the movie (but not the book) versions of his work? We don't know. But we think it is time to look at the data again, with open minds, as if for the first time.

Foreword

This is an interdisciplinary book, lying at the intersection of geology and paleoclimatology with physics, mathematics and astronomy. Great advances in science often come from interdisciplinary frontiers, but this work is difficult because it requires that specialists in one field learned the concepts, accumulated knowledge and ways of thinking of a different speciality. The study of the climate of the glacial Earth over the last couple of million years is one of the most active interdisciplinary frontiers of the present time, with discoveries and data pouring forth. Yet the significance of these discoveries is not always clear, particularly their implications for global climate change in the near future.

In order for the implications of paleoclimatology to be clearly understood, the emerging data must be carefully analysed with sophisticated mathematical techniques, properly applied, and interpreted in terms of the underlying physics and the astronomical driving forces. This requires that geologists and paleoclimatologists have or acquire an understanding of these very different but relevant fields of science. In this book, Richard Muller and Gordon MacDonald set out to provide the understanding of mathematics and solar-system astronomy that paleoclimatologists must have.

In a widely held if rather unsophisticated view, the sciences should be arranged in a hierarchical order of merit, depending on the degree to which their subject matter can be treated mathematically. However, it is more useful to see the sciences as occupying a spectrum, with the simple but quantitative at one end and the complex but qualitative at the other. The difference results from the nature of the subject matter, not the merit of the approach. Rigorous mathematical treatment is possible only in the absence of historical complications. In a physics experiment, it makes no different what the electron of interest was doing the day before yesterday; it always behaves as an electron should. At the other extreme, in sociology, history is almost everything; people behave in complex ways conditioned by millions of years of contingent, unpredictable and unrepeatable events.

By dealing with individual particles, idealised objects and experiments sensitive to a single variable, physicists have come to the extraordinary realisation that Nature behaves according to mathematical laws of the most exquisite elegance. Are we seeing deep into the very heart of Nature? Or is mathematics simply the best tool we have so far for approximating a still more subtle and sophisticated reality?

Scientists long believed that the physical laws of Nature controlled everything in the Universe so completely that the future is predetermined and free will is an illusion. This concept has been overturned in the last decade or so by the emerging understanding of chaos, complexity and non-linear dynamics. It is appropriate that this new development began in ecology, a discipline well along towards the complex, qualitative, historical end of the spectrum.

Midway along the spectrum is a group of disciplines, with no unifying name, but sharing the characteristic that their subject matter has been shaped by history, but not to the extent that rigorous mathematical treatment is impossible. Geology, geophysics, geochemistry, astronomy, astrophysics and cosmology thus bridge the gap which makes it almost impossible for a psychologist and a physicist to have a fruitful conversation.

This book falls squarely in this bridging area of the sciences. Muller and MacDonald may be a uniquely qualified team for this effort. MacDonald is the coauthor (with Walter Munk) of the classic book *The Rotation of the Earth, A Geophysical Discussion*. He has been the editor of the Journal of Atmospheric Sciences, the Journal of Geophysical Research, and Science. Muller is a physicist whose interests have always extended into the study of the Earth and the heavens. This is shown by some of his major contributions—the discovery of the motion of the Solar System relative to the cosmic background radiation, the invention of accelerator mass spectrometry now routinely used in radiocarbon dating, the investigation of comet showers and periodicity in mass extinctions, and the recent determination of the history of cosmic bombardment by dating lunar impact-glass spherules. Each of these contributions clarifies some aspect of the history of Earth or the cosmos.

What kind of history has that been? Has history been characterised primarily by unidirectional trends, or mostly by repetitive oscillations? Most geologists and astronomers would agree that both are involved.

It may be more useful to add a third category, and see the history of the Earth as made up of sudden events, slow trends and cycles. In astronomy and cosmology we seen events such as the Big Bang and supernovas, trends like the cosmic expansion and the progressive conversion of light elements into heavier ones by fusion, and cydles like the rotation of galaxies and solar systems and the cycling of material through exploding and condensing stars. In geology, events include impacts, volcanic episodes, and jumps in plate configuration; trends include the extraction of crustal material from the mantle and the gradual production of an oxygen- rich atmosphere by photosynthesis; and cycles range from the daily rotation of the Earth to the repeated opening and closing of the Atlantic Ocean on a 500 million-year time scale.

How can these features of history be quantified? Events like impacts and volcanic eruptions can be characterised by the statistics of recurrent intervals and magnitudes, and their physics is best understood in mathematical terms. Trends lend themselves to the techniques of the calculus of rates. Cycles are, of course, best studied by the mathematical techniques of time-series analysis, going back to Fourier. The paleoclimatological literature is thus full of spectral plots, showing the importance of time cycles of different lengths observed in paleoclimate proxy data. The widely accepted Croll-Milankovitch theory that fluctuating climate conditions during the Quaternary glaciation have been driven by astronomical cycles is based entirely on time-series analysis of paleoclimatic and orbital data.

And yet, time-series analysis is subtle and full of traps. Considerable mathematical sophistication is required in order to apply it without reaching incorrect conclusions. This new book is exactly what is needed in paleoclimatology for the interpretations of data on ancient climates to be really valid. Combining a theoretical presentation of the techniques with practical instructions and many useful Matlab recipes, and augmented with a new hypothesis for an astronomical climate driver, the book should be of the greatest use to practising geologists and paleoclimatologists.

Walter Alvarez

Professor of Geology and Geophysics, University of California, Berkeley May 2000

Acknowledgements

We are thankful to the many people who shared their insights with us, and who helped educate us on the many different areas encompassed by this book. We are particularly grateful to those who gave the manuscript a careful reading, and found many errors, especially Jonathan Levine, Daniel Karner, Ike Winograd, Richard Garwin and Walter Alvarez. This work was supported in part by the US Department of Energy Division of Environmental Sciences under contract number DE-AC03-7BSF00098. Richard Muller is particularly grateful to the Ann and Gordon Getty Foundation for continuing to support his work as it wandered in an unpredictable path from *Nemesis* to the ice ages.

1

Brief introductions to . . .

1.1 . . . THE HISTORY OF CLIMATE

Beginning in the early 1900s, the climate of the world began to warm. This is evident in Fig. 1.1, which shows the average Earth surface temperature from 1880 to 1999. The temperature is an area-weighted average over the land and ocean compiled by the National Oceanographic and Atmospheric Administration, using an averaging technique devised by Quayle et al. (1999; see also Jones et al., 1999). In the plot, "zero" temperature is defined as the temperature in 1950. The fine line shows the monthly temperatures; the thicker line shows the 12-month yearly averages.

The figure shows that the twentieth century had a temperature rise of nearly one degree Celsius. That may not sound like a lot, but its effects are quite noticeable. In Europe, the great glaciers of the Alps, such as the Mer de Glace near Chamonix, have been in retreat, and the canals of Holland almost never freeze over, as they did in an earlier era to allow Hans Brinker to silver skate into legend. The effects elsewhere on the globe are more severe, with large areas of Central Africa, once fertile, becoming arid and no longer capable of supporting a large population. Although the reason for this warming is not fully understood, many climate scientists think it is the result of the addition of carbon dioxide and other greenhouse gases into the atmosphere by humans.[1]

As we go back in time in search of earlier records, the historical record becomes less reliable. Fortunately, Nature has provided its own recording mechanism. As we will explain in Chapter 4, measurements of oxygen isotopes yield an estimate of ancient temperatures combined with total global ice volume—a combination which is just as interesting as temperature alone, if not more so. Data from a kilometre-long core taken from the Greenland glacier, as part of the Greenland Ice Sheet Project

[1] Although this warming is clearly seen in the surface measurements, satellite measurements taken between 1979 and 1998 do not show a similar increase in the temperature of the lower troposphere. According to most climate models, the global warming caused by CO_2 should contribute to both surface and troposphere temperatures. The discrepancy is not understood; see, for example, Santer et al. (2000).

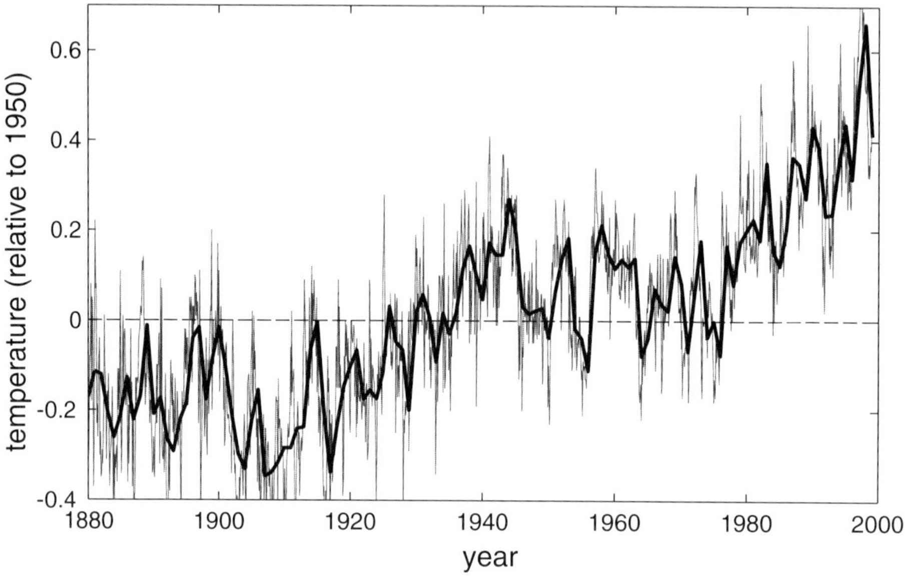

Fig. 1.1. Global warming.

"GISP2" (Grootes et al., 1993), are shown in Fig. 1.2. For comparison purposes, the zero of temperature scale for this plot was set to match that of the previous plot. For historical interest, we marked some events from European history.

The cool period preceding the twentieth century warming is now seen as a dip that lasted 700 years. This period is now referred to as "the little ice age". (The coldest periods, near 1400 and 1700, are sometimes called the *two* little ice ages.) In her popular account of the history of the fourteenth century, historian Barbara W. Tuchman argues that the low temperatures triggered social conflict and poor food production, and was thus responsible for hunger, war, and possibly even pestilence (Tuchman, 1978). Just a few centuries before, at the beginning of the second millennium, Europe had experienced the "medieval warm period" (Lamb, 1965). It was a time when civilisation emerged from the Dark Ages, art and painting flourished, and the wealth and new productivity of Europe allowed it to build the great cathedrals. Some historians will attribute this flowering to great leaders, or to great ideas, or to great inventions, but it is foolish to ignore the changes in climate. Just prior to that, in the 900s, the Vikings were invading France, possibly driven from the more northern latitudes by the cold temperatures of that century. The height of the Roman republic and empire was reached during another time of unusual warmth—even higher than the warm period of today, if the ice-reckoned temperature scale is accurate.

The next plot (Fig. 1.3) shows the data from the Greenland ice core back to 10,000 BC. Near the right-hand side of this plot, the dip of the little ice age is clear. Some scientists argue that global warming is not human-caused, but is simply a natural return to the normal temperature of the previous 8000 years. In fact, no

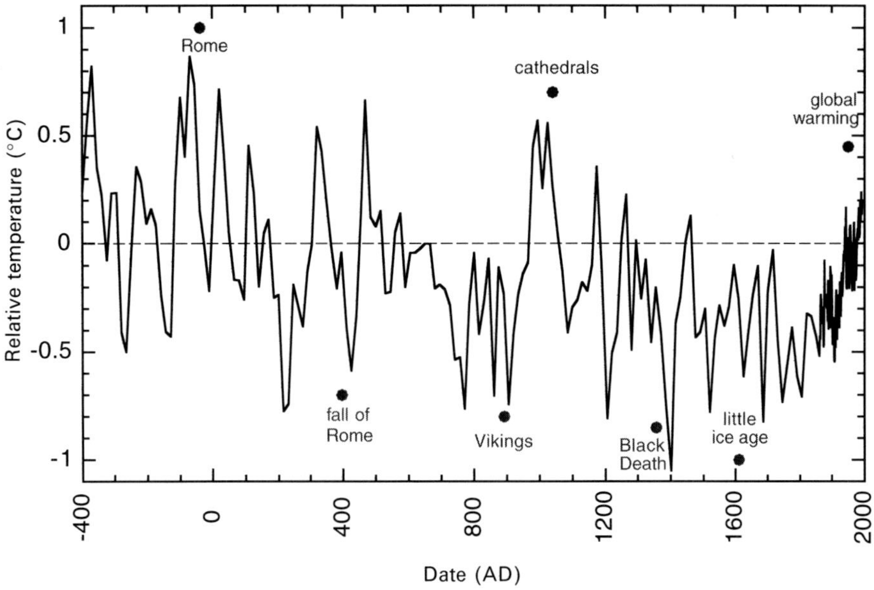

Fig. 1.2. Climate of the last 2400 years.

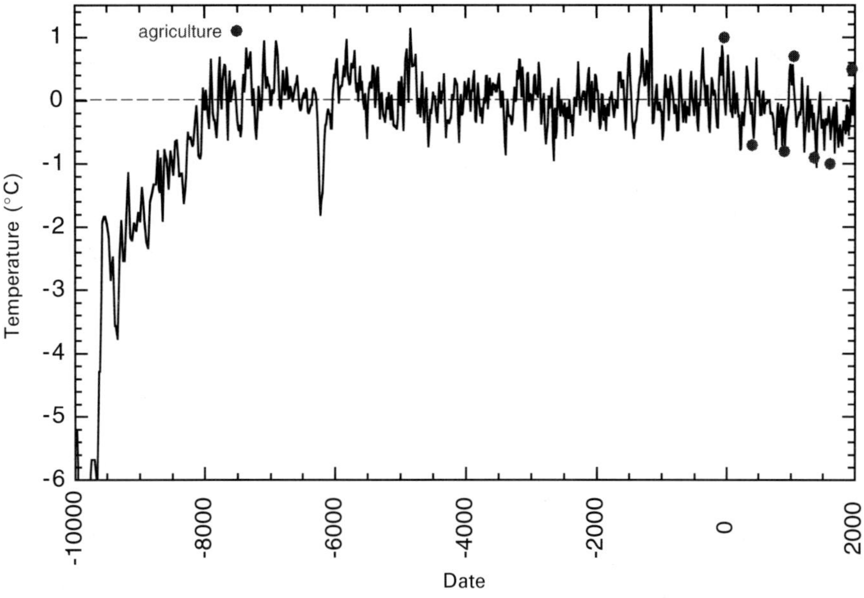

Fig. 1.3. Climate of the last 12,000 years.

one knows for sure if this is right or not. But the foundation for thinking that human effects will cause warming is substantial. Even if the recent rise in temperature is natural, human-caused effects have a high probability of dominating in the near future, and within our lifetimes the temperature of the Earth could go higher than ever seen previously by *Homo sapiens*.

The dip near 6000 BC is not understood. It actually appears to be coincident with a short-term *increase* in temperature that took place in Antarctica! So we can't easily interpret everything in these plots, at least not without studying other records. Fluctuations are evident all over the plot, and crying out to be understood.

Agriculture began about 7000 BC, as marked on the plot. All of civilisation was based on this development. Agriculture allows large groups of people to live in the same location. It allows a small number of people to feed others, so that the others can become craftsmen, artists, historians, inventors and scientists.

The sudden rise at the left side of the plot, at about 9000 BC (i.e. 11,000 years ago), was the end of the last ice age. The abruptness of the termination is startling. Agriculture, and all of our civilisation, developed since this termination. The enormous glacier, several kilometres thick, covering much of North America and Eurasia, rapidly melted. Only small parts of this glacier survived in Greenland and Antarctica, where they exist to this day. The melting caused a series of worldwide floods unlike anything previously experienced by *Homo sapiens*. (There had been a previous flood at about 120 kyr, but that was before *Homo sapiens* had moved to Europe or North America.) The flood dumped enough water into the oceans to cause the average sea level to rise 110 metres, enough to inundate the coastal areas, and to cover the Bering Isthmus and turn it into the Bering Strait. The water from melting ice probably flooded down over land in pulses, as ice-dammed lakes formed and then catastrophically released their water. These floods left many records, including remnant puddles now known as the Great Lakes, and possibly gave rise to legends that persisted for many years. As the glacier retreated, it left a piles of debris at its extremum. One such pile is now known as New York's Long Island.

In the next plot, Fig. 1.4, we show the Greenland ice data for the last 100,000 years. The very unusual nature of the last 11,000 years stands out in striking contrast to the 90,000 years of cold that preceded it. We now refer to such an unusual warm period as an interglacial. The long preceding period of ice is a glacial. During the last glacial, humans developed elaborate tools, and *Homo sapiens* migrated from Africa to Europe, but they did not develop civilisation until the ice age ended.

During the glacial, not only was the temperature lower by 8° Celsius (and some estimates put it at more than 12° Celsius—the record is a superposition of ice volume and temperature), but the climate was extremely irregular. The irregularities in temperature during the glacials, the wild bumps and wiggles that cover much of Fig. 1.4, are real, not an artifact of poor measurement. The same bumps and wiggles are seen in two separate cores in Greenland, and in data taken from sea floor records found off the California coast. The ability to adapt quickly during this wild climate ride may have given a substantial advantage to adaptable animals, such as humans, and made it difficult for other large fauna to survive. Maybe it was these rapid

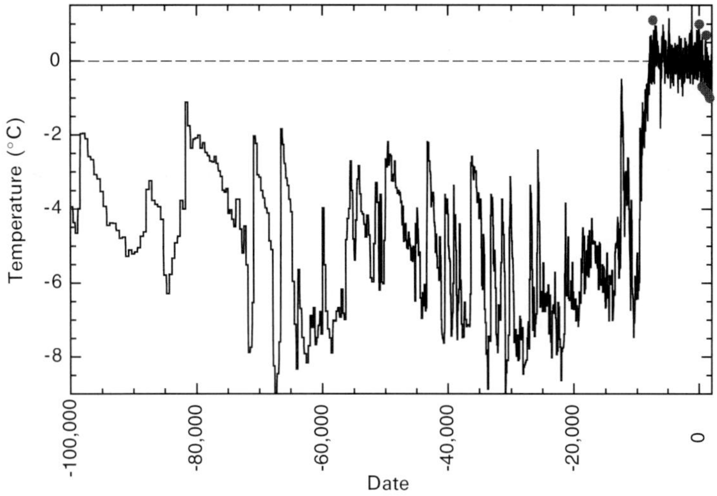

Fig. 1.4. Climate of the last 100,000 years.

changes, and not the rapaciousness of humans, that drove the mammoths, camels, giant ground sloths and giant beavers (the size of bears) of North America to extinction. Recent global warming appears negligible on this plot. However, if predictions of climate modellers are correct, global warming temperature changes will be comparable to those during the ice age.

The reliable data from Greenland go back only as far as shown in Fig. 1.4. We can continue the climate plot further back by using the records from Vostok, the Russian base in Antarctica, where another ice core was drilled. The last 420,000 years of a deuterium measurement at Vostok is shown in Fig. 1.5, with the most recent 100 kyr appended from the Greenland record, which is more detailed. The temperature scale was adjusted to agree with the scale on the Greenland record.

From this plot, it is clear that most of the last 420,000 years (420 kyr) was spent in ice age. The brief periods when the record peaks above the zero line, the interglacials, typically lasted from a few thousand to perhaps 20,000 years.

These data should frighten you. All of civilisation developed during the last interglacial, and the data show that such interglacials are very brief. Our time is about up. Data such as these are what led us to state in the Preface that the next ice age is about to hit us, any millennium now. It does not take a detailed theory to make this prediction. We don't necessarily know *why* the next ice age is imminent (at least on a geological time scale), but the pattern is unmistakable.

The real reason to be frightened is that we really don't understand what causes the pattern. We don't know why the ice ages are broken by the short interglacials. We do know something—that the driving force is astronomical. We'll describe how we know that in Chapter 2. We have models that relate the astronomical mechanisms to changes in climate, but we don't know which of our models are right, or if any of them are. We will discuss these models in some detail in this book. Much of the work of understanding lies in the future. It is a great field for a young student to enter.

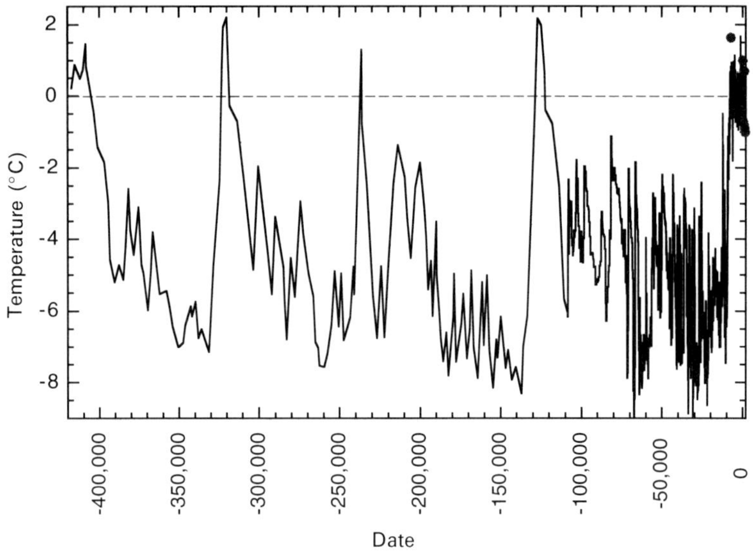

Fig. 1.5. Climate for the last 420 kyr, from Vostok ice.

The ice records take us back only to 420,000 years in the past. However, oxygen isotope records in sea floor cores allow us to go further. One of the best sets of data comes from a location in the northern Atlantic Ocean known as the Ocean Drilling Project Site 607 (Raymo, 1992). This site has climate data going back three million years, and is shown in Fig. 1.6. But before you look at the figure, let us warn you. In the paleoclimate community, there is a convention that time is shown *backwards*. That is, the present is plotted on the left-hand edge, and the past is towards the right. We are going to use this opportunity to change our convention, for the remainder of the book, so that you will have less trouble reading the literature. (The literature of "global warming" scientists, in contrast, follows the other convention, which we have used up until now.) We apologise for this change in convention, but we do not take blame for it.

In Fig. 1.6, the 10 kyr years of agriculture and civilisation appear as a sudden, barely visible, rise in temperature squeezed against the left-hand axis of the plot. The temperature of 1950 is indicated by the horizontal line. As is evident from the data, civilisation was created in an unusual time.

There are several important features to notice in these data, all of which will be discussed further in the remainder of the book. For the last million years or so (the left-most third of the plot) the oscillations have had a cycle of about 100 kyr. That is, the enduring period of ice is broken roughly every 100 kyr, by a brief interglacial. During this time, the terminations of the ice ages appear to be particularly abrupt, as you can see from the sudden jumps that took place near 0, 120, 320, 450, and 650 thousand years ago. This has led scientists to characterise the data as shaped like a "sawtooth", although the pattern is not perfectly regular.

But as we look back beyond 1000 kyr (1 million years), the character changes

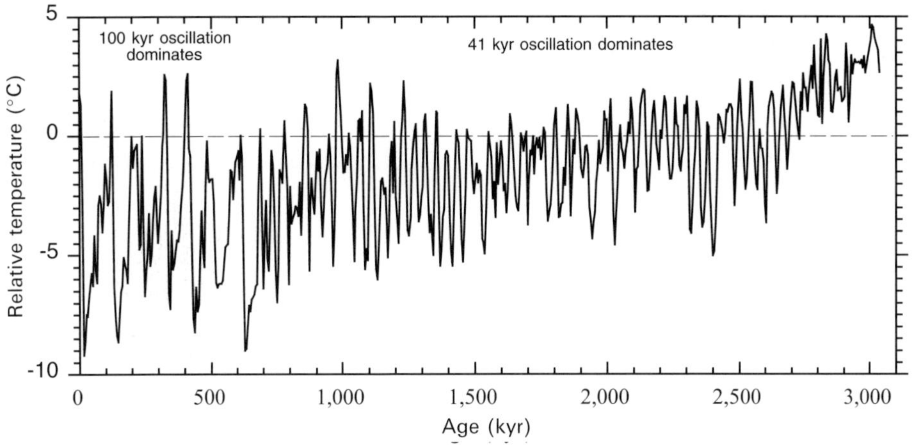

Fig. 1.6. Climate of the last 3 million years.

completely. The cycle is much shorter (it averages 41 kyr), the amplitude is reduced, the average value is higher (indicating that the ice ages were not as intense) and there is no evidence for the sawtooth shape. These are the features that ice age theories endeavour to explain. Why did the transition take place? What are the meanings of the frequencies? (We will show that they are well-known astronomical frequencies.) In the period immediately preceding the data shown here, older than 3 million years, the temperature didn't drop below the 1950 value, and we believe that large glaciers didn't form—perhaps only small ones, such as we have today in Greenland and Antarctica.

As we end this brief introduction to the history of the ice ages, let's again look to the future. As soon as the cycle of the ice ages was known, scientists realised that the ice age would eventually return. Some of them enjoyed scaring the public about the impending catastrophe. In Fig. 1.7 we show the cover from a magazine of the 1940s showing the consequences of the return of the ice age to New York City. (One of the authors of the present book saw this image as a child, and it made a lasting impression.) Unfortunately, in the public forum, the art genre of returning ice has been superceded by paintings of asteroids about to hit the Earth, usually with a curious dinosaur momentarily distracted by the unusual scene. But, as we mentioned earlier, the more likely scenario for the early twenty-first century is the continued gradual growth of global warming.

1.2 ... ICE AGE THEORIES

[The objective of geology is] to confirm the evidences of natural religion; and to show that the facts developed by it are consistent with the accounts of the creation and deluge recorded in the Mosaic writings.

William Buckland, Oxford Professor of Mineralogy and Geology
The Connection of Geology with Religion Explained (1820)

Fig. 1.7. The Ice Age returns to New York City.

To the modern reader, it may seem surprising that astronomical forces drive global climate change. In the 1950s, changes in the weather were attributed by many people to nuclear testing; in the 1990s, they were attributed to emission of carbon dioxide and other gases that enhance the greenhouse effect. Yet in the 1840s, to many people the obvious cause of climate change was astronomy. It was understood at that time that the seasons were driven by astronomical causes, as is the 24-hour cycle of day and night. So when the existence of long-term changes of climate was discovered, it was natural to see if they could be attributed to astronomical forcing. This was long before the 100 kyr and 41 kyr cycles of the ice ages had been discovered.

By the 1840s, astronomers had already shown that the orbit of the Earth undergoes slow changes. It is not the same ellipse this year that it was last year. According to Newton's physics, the orbit would be an unchanging ellipse if the Sun were the only source of attraction. The deviations come about because of the presence of the Moon and other planets. Gravity from these objects means that the force on the Earth is not a simple inverse square centred on the sun, and the natural result is that the orbit is not a simple repeating ellipse.

However the effects of the planets are relatively small. Jupiter has 10^{-3} of the mass of the Sun and, on average, it is about five times further away. Venus, although having a mass 390 times smaller than Jupiter, comes much closer, within 0.28 AU (where the Astronomical Unit, or AU, is the average Earth–Sun distance) vs. 4.2 AU for Jupiter. Since gravity varies as 1/(distance squared), and tidal gradients vary as 1/(distance cubed), the effect of Venus is often more important than that of Jupiter. But all these effects are small enough that they can be treated as perturbations—small changes—on the classical elliptical orbit. The nice consequence of this is that the Earth's orbit is always *approximately* an ellipse, and we can treat the perturbations of the planets as extra forces that gradually alter the parameters of that ellipse. For example, the axis of the Earth slowly precesses or rotates relative to the "fixed" stars. This effect is big enough that it was discovered experimentally in 120 BC by the astronomer Hipparchus, who found differences between his own measurements and those of earlier astronomical Babylonian records.

By 1749, Alexis Claude Clairaut had shown, using Newton's laws, that the north pole of the Earth precesses with a period of 25,800 years. So, for example, 13,000 years from now, the North Pole of the Earth will not be pointing towards the "North Star", but will be pointing away by an angle of about 47 degrees, close to the star Vega. This happens because the Sun and Moon exert a torque on the equatorial bulge of the Earth. This causes the axis of rotation of the Earth to wobble, an effect completely analogous to the wobble of a tilted top under the torque of gravity.

Precession implies that the signs of the zodiac change, or "advance". When astrology was defined, about 2000 years ago, a person born in January was said to be under the sign of Capricorn, since the sun was in the constellation Capricorn at that time. Since then, the precession of the axis of the Earth has changed by $2000/26{,}000 = 1/13$ of a cycle; this corresponds to a change by about one sign of the zodiac. This means that a person born in January any time in the last few hundred years, was born when the Sun was actually in Sagittarius, not Capricorn. Nevertheless, following tradition, such a person is still said to be "born under the sign of Capricorn". The more educated astrologers are aware of this change, and claim that they compensate for it (Carlson, 1985).

Another consequence of the Earth's precession is that the location of the sun at the spring equinox also changes. It is presently leaving the constellation Pisces and entering Aquarius. Astrologers say that this change could have a profound effect on our lives, and it is why they talk about the future (and sometimes the present; it depends on exactly where you draw the constellation boundaries) as "the Age of Aquarius".

Evidence for ancient glaciation, including polished bedrock and erratic boulders, is extensive. But in the early 1800s, the prevailing paradigm to explain these was diluvianism—belief that they were artifacts of the great flood of the Bible. According to William Buckland, one of the most widely-respected geologists of the time, the goal of geology was "to confirm the evidences of natural religion; and to show that the facts developed by it are consistent with the accounts of the creation and deluge recorded in the Mosaic writings" (quoted in Imbrie and Imbrie, 1979, p. 35). A detailed scientific case for ancient glaciation was first synthesised in detail by Jean de Charpentier in the early 1830s, and the great geologist Louis Agassiz became an early convert. By 1841 Charles Lyell, and even Buckland, had been won over, and over the next 20 years the theory of ancient periods of extensive ice became generally accepted.

Soon after the existence of the ice ages was postulated, they were attributed to these orbital changes. The first person to create a detailed theory was Joseph Adhémar, who published a book on the subject in 1842 called *Revolutions de la Mer, Deluges Periodics*. He believed that the 26,000 year precession cycle was the cause, and he suggested that it was the direct gravitational attraction of the Sun and Moon on the ice caps, a result that many scientists at the time correctly rejected as absurd.

One person who was inspired, in part, by Adhémar's book, was James Croll. In many ways, Croll is the true hero of the astronomical theory of the ice ages. For a moving account of his life and tribulations, we urge you to read the delightful and informative book, *Ice Ages, Solving the Mystery*, by John Imbrie and his daughter Katherine Palmer Imbrie (Imbrie and Imbrie, 1979). Croll was a carpenter who became disabled from an accident, and was forced to take a job as a janitor. But his janitorial duties were at the Andersonian College and Museum in Scotland, and there he discovered a wonderful library. He found that he could finish his mopping and polishing work early, and spend the rest of the night reading books on physics in the library. Among these was the book by Adhémar, and new calculations of the Earth's orbit by the astronomer Urbain Leverrier (one of the discovers of Neptune). Croll taught himself advanced physics, and set to work on the origins of the ice age. He decided that the most plausible driving force changing climate was variations in insolation, the sunlight hitting the Earth. He managed to get his papers published, and his theories received favourable attention. Eventually he accepted a job at the Geological Survey. In 1876, a year after his own book was published, Croll was given the high honour of being named a Fellow of the Royal Society of London.

Croll contributed many insights that are still recognised as important. He realised that the presence of large glaciers would reflect sunlight, and that this would enhance further the chill of the ice ages. He recognised the importance of ocean currents in climate, and incorporated them into his theories. He took into account not only precession, but the changes in the eccentricity of the Earth's orbit—an effect that can contribute a 100 kyr cycle to the glaciation (which, of course, had not yet been discovered).

But Croll published several predictions that proved to be very wrong, and that eventually caused his theory to be abandoned. Since the insolation in the Northern

Hemisphere is out of phase with the insolation on the Southern Hemisphere, he thought that the ice ages in the two hemispheres would alternate, and we now know they are synchronous. He also estimated that the time since the last ice age to be 80,000 years, much older than the true value; we now know it ended between 14,000 and 10,000 years ago.

The insolation theory of Croll was revived in the early 1900s by Milutin Milankovitch, a Serbian originally employed as an engineer, but who had become a professor at the University of Belgrade where he taught physics, mathematics, and astronomy. He took on the riddle of the ice ages as a challenge. Ludwig Pilgrim had recently completed new and more detailed orbital calculations, and Milankovitch made use of these. He had the critical insight that insolation on the Northern Hemisphere might completely dominate, since that is the location of two thirds of the Earth's land area. The ice ages in both the Northern and Southern Hemispheres are in synchrony because they are both driven by the same force: insolation on the Northern Hemisphere land. With this Gordian slice, Milankovitch solved the problem of alternating hemispheric ice ages.

Milankovitch then set out on the heroic task (at that time) of doing detailed insolation calculations based on Pilgrim's orbital work. Today these calculations are an interesting task for an undergraduate to do over the course of a summer, using a desktop computer. But Milankovitch had to do all the calculations by hand, and it took him many years (interrupted by war and imprisonment—again we urge you read the book by Imbrie and Imbrie). Milankovitch's calculations showed (correctly) that the insolation was dominated by a 23,000 year cycle; we have modern data for this plotted on page 137. Milankovitch concluded that the ice ages would be most intense when the insolation dropped below a certain threshold. Since the envelope of the insolation curve has an approximately 100 kyr cyclicity, his theory has implicit in it a prediction that such a cycle might be seen in the ice ages.

Milankovitch concluded, somewhat prematurely, that the problem was completely solved, and he devoted much of his time in later years to writing popular accounts of the ice ages, including a series of letters to a (presumably fictional) young girl. This effort did much to increase the public interest in the ice ages, and undoubtedly led to the artwork shown in Fig. 1.7.

But Milankovitch's theory was abandoned when precise age estimates, made possible by Willard Libby's invention of radiocarbon dating, appeared to show that the timing of the ice ages was in conflict with Milankovitch's detailed calculations. In retrospect, this was unfair. The Milankovitch theory actually explains many of the phenomena that we now see in the data. Do we throw out the astronomical theory of the seasons, simply because the first day of spring is not always spring-like? The warm weather of spring can be delayed by a month, or it can come early by a month; the important fact is that it always comes. We demand too much of a theory if we require it to predict all the details in addition to the major behaviour.

In fact, it was the observations of the regularity of the ice age cycles that led to the revival of the insolation theory. Science and technology advanced, thanks to the work of many people. Harold Urey, Cesare Emiliani and others developed and promoted the use of isotopes as proxies for changes in the Earth. The technology for obtaining sea floor cores rapidly improved. In 1970, Wally Broecker and Jan van Donk published a seminal paper (Broecker and van Donk, 1970) that showed for the first time that the dominant variation in the ice ages was a repeating cycle of 100,000 years. This was a frequency that appeared in the insolation theory. The use of geomagnetic reversals in sea floor cores allowed a vastly improved time scale. In 1976, James D. Hays, John Imbrie and Nicholas Shackleton published their paper (Hays et al., 1976) showing the presence of both a 41 and 23 kyr cycle in data derived from sea floor sediments. The same frequencies were dominant in spectral analysis of the insolation. Even if the details of the theory were wrong, the presence of the same frequencies as those present in the orbits of the planets was a strong reason to revive the astronomical theory.

Although it had been resuscitated, the insolation theory continued to have problems. In the glacial data, the 100 kyr cycle dominates, with the 41 kyr cycle weaker, and the 23 kyr precession signal weakest of all. Yet in insolation theory, it is the 23 kyr cycle that dominates, with a weaker 41, and an extremely weak 100. Why are these strengths reversed? A possible answer came from the old threshold idea of Milankovitch, which had been revived by Kenneth Mesolella and George Kukla, and put into an elegant mathematical form by Imbrie and Imbrie. They showed that a nonlinear response of the climate to insolation could greatly strengthen the 100 kyr cycle, through a mechanism that we discuss on page 195. This was both physically plausible, and seemed to solve the amplitude problem.

But there was another issue: the nonlinear ice model strengthened even more an additional cycle with a 400 kyr period that was not observed. In fact, the 400 kyr cycle should have been the strongest cycle of all, according to that theory. One way to get rid of this was to make the ad hoc assumption that the 400 kyr cycle was so long that it was suppressed by the more rapid time constants that are natural in ice production and destruction processes. At least one such assumption should certainly be allowed in any complicated theory.

Other problems continued to nag the insolation theory. In 1992, measurements of climate with precise dates became available from a water-filled cave in Nevada called Devils Hole. The sudden rise in temperature at this location appeared to precede the increase in insolation that was supposed to trigger it. This was a "causality problem", since the presumed effect appeared to precede the cause. This problem persists, and has recently become critical with the vindication of the Devils Hole chronology by radiometric dates from coral records. We will discuss the causality problem at length in Section 8.3.

Another problem with the insolation theory is called the "Stage-11 problem". Stage-11, in the jargon of paleoclimate, refers to a period about 400,000 years ago when variations in insolation were very weak, and yet the cyclical behaviour of the ice ages was very strong. A similar problem exists at present: insolation variations are weak, and yet there was a great termination just 10–14 thousand years ago; this might be called the "Stage-1 problem". Various solutions have been proposed to address these problems, but they all involve ad hoc assumptions and the introduction of arbitrary parameters.

The set of problems continues to grow. It has led some to abandon the astronomical theory altogether, and postulate that the cycles of the ice ages are driven by natural oscillations of the ocean/continent/atmosphere system. Yet it is hard to dispute that the astronomical theory can account for the values of the observed frequencies. This agreement, more than anything else, gives life to the astronomical theory. It is also the reason for the central role played by spectral analysis in this book. Moreover, it is possible to make a general argument, independent of the frequency agreement, that the ice ages are astronomically driven. This was first done by the authors of this book in a paper in *Science Magazine* (Muller and MacDonald, 1997a). The argument is based on the narrow structure seen in the spectrum of the data, and is explained in Section 2.1.

The attention given to spectrum shape has created another serious problem for the insolation theory. A high-resolution analysis of the 100 kyr cycle shows that the insolation theory, and its variants, all predict that the peak will have a split structure: it will be resolved into a 95 kyr line and a 125 kyr line. (An exception to this general rule is a model recently published by W. Berger, and explained in Section 6.4.8.) The bulk of the data shows that this prediction is contradicted. The 100 kyr cycle is a single narrow line. Ad hoc mechanisms that were plausible for eliminating the 400 kyr line are not plausible for turning the predicted doublet into a singlet. It is remarkable that this problem was not noticed until 1994. The seriousness of the problem was emphasised in an article in *Science Magazine* in 1997 (Muller and MacDonald, 1997a). Although many theories had been published to account for the ice ages, none of them predicted the narrowness of this peak. A review of the theories published in 1993 (Imbrie et al., 1993a) included a "short list" of nine different "groups" of models, reproduced on page 183, and yet every one of these theories was contradicted by the simple observation that the 100 kyr peak was narrow.

It is dangerous to dismiss a theory based on any one contradiction, since minor modifications of the theory can often solve apparently intractable difficulties. But the growing number of problems with the insolation theory is cause for serious concern. It may be the fact that the insolation theory predicted the correct values of the frequencies that leads to its tenacity in holding the minds of paleoclimatologists. But alternatives are now appearing. In 1993, it was discovered that there is another astronomical oscillation (orbital inclination) that could contribute to climate which has a spectrum that is an excellent match to the narrow 100 kyr peak. The theory based on this does not have a Stage-11 problem (or a Stage-1 problem), nor a causality problem. It does not predict a nonexistent 400 kyr peak, and it accounts in a natural way (no adjustable parameters) for the shift of frequencies that took place about one million years ago, when the dominant frequency of ice age oscillation changed from 41 kyr to the present value of 100 kyr. This is, however, a new theory, and although we like it (it is our theory), it is not yet widely accepted, so we relegate detailed discussion of it to near the end of this book, Chapter 7.

1.3 ... SPECTRA

A spectre is haunting Europe ...

Karl Marx and Friedrich Engels (1848)

During the Renaissance, when people looked through prisms at each other, they saw what appeared to be ghosts, or *spectra*, multicoloured and transparent images floating mysteriously near each other. What they were seeing, of course, were the colours of the image separating as a result of optical dispersion of the glass. The first scientific use of this phenomena was made by Newton, whose own theory of spectra turned out to be one of the few mistakes he published in physics.[2] The true explanation had to wait for the contributions of W. H. Wollaston and J. Fraunhofer in the early 1800s, who based their results on the assumption that light was a wave, not a particle. (We now know it is both—but Newton was still wrong.) They improved their experiments by using a slit instead of a pinhole to collimate their light. This produced a line rather than a spot and, as a legacy, we still refer to spectral "lines", even when the data are handled digitally, and the frequency structure is shown as a peak in a graph rather than as a brightly coloured line on a white sheet. A group of spectral peaks is still called a "band", which is what a group of adjacent lines looked like. The band-width was the width in the optical spectrum of a group of lines; we still use this term today (without the hyphen) to denote the range of frequencies covered by a system.

Today, spectral analysis refers to any process that accounts for data as a sum of individual oscillations. When a musician listens to a chord played on a piano, and then goes to the piano and duplicates it, he[3] has successfully performed spectral analysis. He has taken a complex sound and recognised it as a sum of individual oscillations. This is done using the hair cells of the inner ear, which are separated in the cochlear duct by their frequency sensitivity. That is what spectral analysis is—separation of a signal into different frequency components. The concept is made more complicated by the fact that every sound, even the roar of a waterfall, can be constructed from pure tones, although it takes a lot of them. But it is easier to reproduce the sound of a waterfall by recreating the amplitude of the sound as a function of time. This is called working in the "time domain". On the other hand, when we analyse a signal by breaking it up into a sum of pure frequencies, we say we are working in the "frequency domain".

Paleoclimate signals usually consist of a superposition of a few pure frequencies (also called lines, or tones) and background. Some of the analysis is best done in the time domain. This includes the precise timing of the glacial terminations, and the history of ice-rafting phenomena known as Heinrich events. A sudden event in the

[2] Newton believed that the refraction of light was caused by an attractive force between the light particle and the surface of glass as it approached that surface. This force would be different for light of different colours, and that would account for the different refraction angle for different colours.

[3] To avoid awkwardness, when we say "he" we mean "he or she". Please accept our apology for using this ancient and perhaps out-of-date convention.

time domain, when described in the frequency domain, appears spread out over many frequencies, so it is difficult to study that way. Likewise, a single peak appearing in the frequency domain (e.g. at 41 kyr), appears as a relatively complex object in the time domain: a sinusoidal variation that covers all time. Spectral analysis and time domain analysis are best done together, since certain phenomena stand out in one representation, and others in the other. Non-periodic events are just as interesting as periodic ones; after all, it was a single dramatic event that appears to have killed the dinosaurs. But the frequency domain is extremely useful for many other astronomical forces (such as insolation) because astronomical signals are often extremely regular in their repetitiveness. They tend to stay in tune for a long time, and that gives rise to very strong and narrow peaks in the frequency domain.

There is interesting physics behind the fact that the astronomical signals are so regular. The first is the remarkable fact that the orbits of the planets happen to be nearly perfect ellipses, as mentioned earlier. This arises because the force of gravity is nearly a perfect $1/r^2$ force. It didn't have to be this way. If the force of gravity were $1/r$, or $1/\sqrt{r}$, or $1/r^3$, we would not have closed orbits. In fact, the only other force law that gives a closed orbit is the linear spring law, $F = \kappa r$. (See Landau and Lifshitz, 1976, p. 32.) But because gravity varies with the inverse square of the separation, the orbits repeat. If friction were present (and it is for really small particles, see Section 7.2), then the orbits would not close. General relativity theory is also a slight departure from the inverse square law, and a consequence of this is that the orbits of the planets do not completely close. One consequence is the "advance of the perihelion of Mercury". When astronomers measure this, they are studying the departure from the Newtonian inverse-square law. Perturbations from the other planets also cause departures from pure single-frequency periodicity, for the same reason. Although they are individually inverse-square, each planet's contribution to the force comes from a different direction in space than that of the Sun, so when added to the Sun's gravity, the net resulting force is not inverse square.

The most natural way to do spectral analysis of data is to try to fit it to a mathematical function containing sine waves. For example, if we have data points y_m, each having an age t_m then we could try to find the best parameters a_0, a_k, f_k, and ϕ_k, such that we minimise the difference between the real data and the function:

$$y(t) = a_0 + \sum_{k=1}^{n} a_k \sin(2\pi f_k t + \phi_k)$$

is minimised. (For an illustration of this process, you can look ahead to our description of eccentricity in Section 2.2.2; in particular, Fig. 2.7 on page 33 shows an orbital function called eccentricity closely approximated by the sum of three sine waves.) Many computer programs have been written for just this minimisation process. (In the Matlab^{TM} language, for example, the program is called fmins.m.) This approach is excellent, in principle; it is closely related to a method we will describe (Section 3.7) as the "Maximum Likelihood Method". Its main disadvantage is that it is very time consuming. A much faster method is to use the Fourier transform, for which very rapid methods of computation have been devised. But if part of the signal is periodic, and part is background, and if the time scale is uncertain, then performing a Fourier transform is often not the optimum approach. The discipline called spectral analysis has developed largely to determine the best approach to analysing a periodic signal in the presence of background and uncertainty. The choice of technique depends on the nature of the signal (is it broad? a sine wave? a square wave?), the nature of the noise (Gaussian? broad or narrow band?), the accuracy of the data (error in time scale?), and the nature of the variable you want to measure (amplitude of the periodic component? frequency? phase? spectral shape?). Not only is such optimisation important for science, but it can be extremely important in a competitive industry such as communications. The result is that the field of spectral analysis has become very sophisticated, and very arcane. Sometimes it feels too arcane—so full of traps, and so difficult to master, that otherwise excellent scientists try to avoid it.

To make the optimum choice of spectral analysis technique requires a detailed understanding of the background, i.e. the part of the signal that we *don't* want to study. But we usually don't have that. The result is that there is no "correct" way to do the analysis, only ways that have different advantages and disadvantages, and different traps. The field of paleoclimate has been dominated for a long time by the use of the "Blackman-Tukey method", presented in detail in the classic book *The Measurement of Power Spectra* (Blackman and Tukey, 1958). It is also reportedly the method that Tukey himself suggested be used for paleoclimate work. It is, in fact, particularly well suited for data with an uncertain time scale, as has been the case for much paleoclimate data.

We'll give a simple example of spectral analysis here, while leaving the details to a later chapter. In an Antarctic ice core from Vostok, the relative fraction of the hydrogen isotope deuterium was measured as a function of depth. Using a model for the ice flow, this could be plotted as a function of time. The data were part of the basis for Fig. 1.5. We show these data covering the entire period 0–420 kyr, in Fig. 1.8.

There appears to be a periodicity of approximately 100 kyr. This is evident in the spectral power plot shown in Fig. 1.9. The plot was made using the periodogram method of Section 3.2.4.

From this plot, we see that in addition to the 100 kyr cycle (the peak near the frequency $f = 0.01$ cycles/kyr), there is another prominent peak near 0.024 cycles/kyr. This is identified with an orbital parameter called obliquity, which is described in Section 2.2.4. It is not evident in the time domain because the eye is distracted by the strong 100 kyr cycle.

Our real purpose in showing the spectra here is to demonstrate how phenomena that are well hidden in the time domain can stand out in the frequency domain. And it works both ways. The "sudden terminations" in the time plot are the nearly vertical lines that appear at the end of each glaciation. There is no evidence for these in the spectral plot. Such sharp changes require a conspiracy of a large number of small components at many different frequencies. These small contributions are well hidden in the frequency domain. A comprehensive analysis of data requires

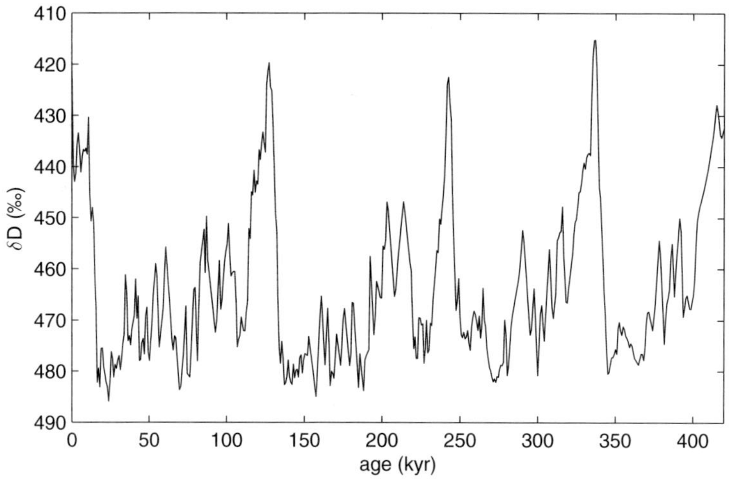

Fig. 1.8. Vostok deuterium.

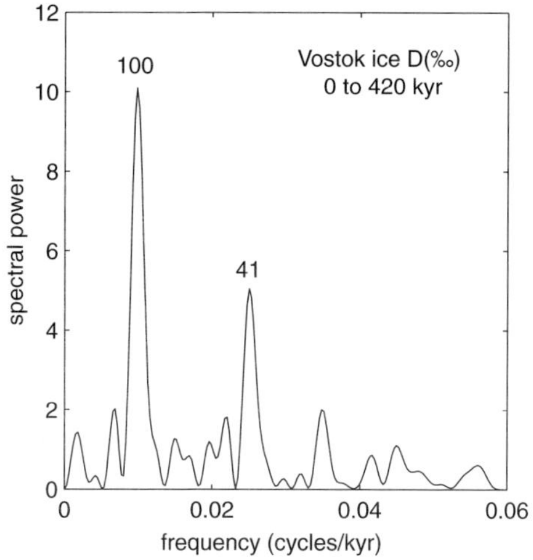

Fig. 1.9. Spectrum of Vostok deuterium.

analysis in both the time and frequency domains. Often it is necessary to do both, by breaking the data into segments of different time intervals, and doing the spectral analysis on each segment. This is important when the dominant cycles change with time, as they often do in paleoclimate data.

2

Astronomy

"Probably another reason why many Europeans consider the Chinese such barbarians is on account of the support they give to their astonomers, people regarded by our cultured Western mortals as completely useless. Yet there they rank with heads of Departments and Secretaries of State. What frightful barbarism."

Franz Kuhnert, 1888[1]

2.1 WHY ASTRONOMICAL FORCING?

In the 1840s, James Croll assumed that the ice ages were driven by changes in the orbit of the Earth. The astronomical origin was likewise taken as natural by Milutin Milankovitch, when he revived and improved the Croll theory in the early 1900s. But neither of these theories has proven successful in accounting for the dominant 100 kyr glacial cycle of the last million years, and so there have sprung up theories that ignore astronomy and attribute this cycle to processes confined to the Earth. These are sometimes called "free oscillation" theories, and they argue that the climate system of the Earth is undergoing a natural resonance, like the tone heard when you put your ear to a seashell. Yet in this book, even in the title, the astronomical origins are taken for granted. Why is this?

There are several persuasive reasons to think that astronomy is responsible. The first is the coincidence of astronomical frequencies with those found in the ice age data. In our discussion of the data, we will show that the strongest frequencies that show up often have periods 100, 41 and 23 kyr, and these are readily identified with astronomical periods. The 100 kyr period has historically been identified with the changes in the eccentricity of the Earth's orbit, although it also matches the period of another parameter called orbital inclination. The 41 kyr period is the dominant oscillation of the tilt of the Earth's poles towards the sun. And the 23 kyr period

[1] Quoted by F. R. Stephenson and D. H. Clark, *Sky & Telescope*, Feb. 1977, p. 86.

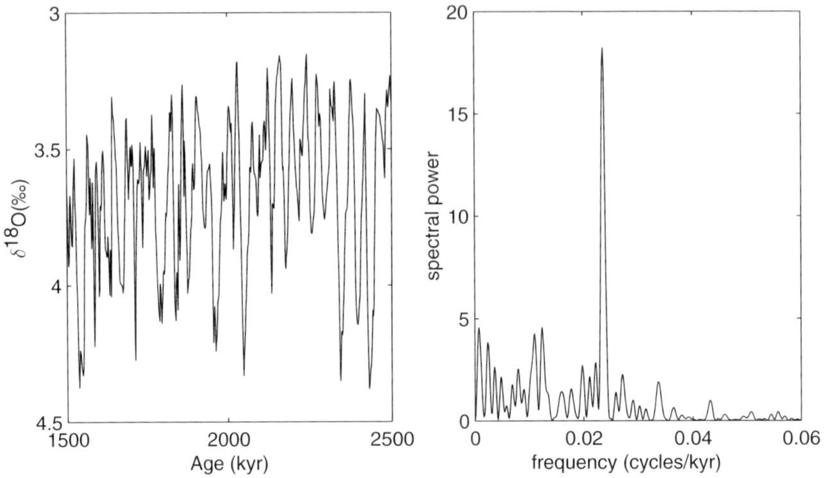

Fig. 2.1. Site 607 from 1.5 to 2.5 kyr: δ^{18}O data and spectrum.

is the dominant oscillation in the parameter that tells us how close the Earth is to the Sun in mid-summer.

A second, and independent, reason to think that the 100 and 41 kyr lines are driven by astronomy can be found in the nature of the oscillations. Over long periods (700 to 1000 kyr), these oscillations remain coherent, i.e. they keep a constant phase. This can be seen by doing a spectral analysis and looking at the widths of the spectral peaks: the peaks are exceedingly narrow. As an example, we show the variations in oxygen isotopes from a sea floor core taken in the North Atlantic in Fig. 2.1. The time scale for the data was derived by assuming that the sedimentation rate was constant. The data and the ages for the end points were taken from Ruddiman et al. (1989), but the ages for points in between were calculated by linear interpolation.

The variations in the signal are believed to represent variations in the global ice; we will discuss the reasons for this in Section 4.1. Accept this for now, and look at the rapid variations. They appear too regular to be random, and this conclusion is confirmed in the spectrum of the data, shown on the right side of the figure. The spectrum is dominated by a single narrow line that appears at the frequency $f = 0.024 \pm 0.001$ cycles/kyr, corresponding to period 42 ± 2 kyr.

The period happens to match the calculated period of the Earth's obliquity, i.e. the tilt of the North Pole towards the Sun, which varies between 22 degrees and 24.5 degrees, with a period of 41 kyr. This match is certainly no coincidence, since we expect that the tilt of the poles towards the Sun could play a major role in the continued growth of glaciation in the polar regions. But aside from that coincidence, note the narrow width of the peak. As we will show, it requires that the signal stay in phase with itself over the entire duration of the data set. If the sedimentation time scale at the end of the data set was inaccurate by as little as 20 kyr (a half cycle) then the single narrow peak would have been broadened and probably split into two adjacent peaks. Thus the sedimentation rate must have been remarkably constant during this period.

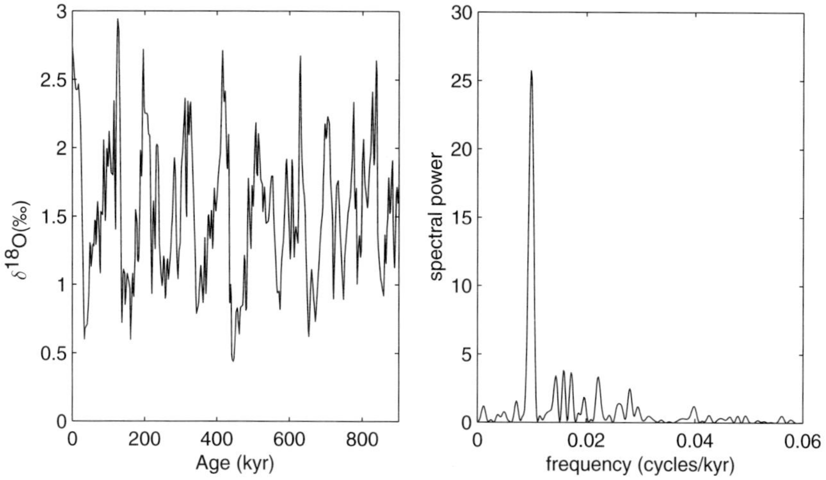

Fig. 2.2. Site 659 $\delta^{18}O$ and spectrum, 0 to 900 kyr.

We argued in 1997 (Muller and MacDonald, 1997a) that the narrowness of the peak implies that the glacial cycles are being driven by an astronomical force, regardless of what the details of the driving mechanism actually are. The reason for this is that natural processes in astronomy virtually always give rise to narrow spectral peaks, and natural processes in geology and climate do not. Narrow peaks are characteristic of processes that have low loss of energy. That means the lack of both friction and those mechanisms through which the oscillation is coupled to other processes that can drain energy away. The connection between low loss and narrow peaks is not well known to many scientists, so we will spend much of the rest of this section describing it in detail.

The argument can also be applied to the data from the last 900 kyr, which are dominated by a 100 kyr cycle. In Fig. 2.2 we show the oxygen isotope data and spectrum for Site 659 in the Atlantic off the coast of Africa. The data are plotted assuming a linear sedimentation rate, with the end point times set according to the time scale of Tiedemann et al. (1994).

Just as for the earlier data, the climate cycles appear to be very regular, but now with a period of 100 kyr. This is verified in the spectrum plot, which is dominated by a single narrow peak with $f = 0.01$ cycles/kyr. The width of this peak is the same that we would obtain with a pure sine wave 900 kyr long. Just as we argued previously that the 42 ± 2 kyr peak was astronomically caused, we argue that the 100 kyr peak must also be astronomically caused.

The subtle part of this argument is the statement that a natural climate oscillation would not show a similarly narrow peak. We will now explain the reasons in some detail. We will consider three cases: a *resonant* oscillator, a *relaxation* oscillator and, finally, a *driven* oscillator. The observed spectra are consistent only with the driven oscillator model. If this argument sounds arcane and uninteresting, then you are welcome to skip it. It is of interest primarily to those who postulate that the basic oscillations of climate are internal to the Earth, and not astronomically driven.

2.1.1 Resonant oscillator

Any oscillation of an object which maintains a single frequency is called, in physics, a *mode*. A simple pendulum has a single mode; two coupled pendula have two modes. The number of modes increases greatly if the internal vibrations of the objects are included. An object that has N molecules (where N is typically Avagadro's number, or greater), each one free to move in three dimensions, typically will have 3N modes. In quantum mechanics, only oscillations in pure modes have well-defined energy, equal to hf, where h is Planck's constant, and f is the frequency of the mode. Heat consists of the simultaneous excitation of all modes. In thermal equilibrium, for example, all modes are excited equally, and for classical systems all have average energy equal to $(1/2)kT$, where k is Boltzmann's constant, and T is the temperature. Friction in an oscillating system can be thought of as a process that takes energy from one mode and redistributes it to many others, turning it into heat. The presence of friction implies that an object will not continue to oscillate as a pure sine wave, but that the amplitude will decay with time. A spectral analysis of a decaying sine wave yields a spectrum with a broadened peak.

To discuss this mathematically, we will have to jump ahead and use some of the Fourier transform equations that will be fully discussed in the next chapter. Let the time behaviour of a decaying oscillation be given, for $t > 0$, by:

$$y(t) = \sin(2\pi f_0 t)e^{-\lambda t} = \frac{e^{2\pi i f_0 t} - e^{-2\pi i f_0 t}}{2i} e^{-\lambda t}$$

The Fourier transform of this function is obtained by multiplying it by $e^{2\pi i f t}$, and integrating (see Section 3.2):

$$\begin{aligned} Y(f) &= \int_0^{+\infty} \frac{e^{2\pi i f t} - e^{-2\pi i f t}}{2i} e^{-\lambda t} e^{2\pi i f t}\, dt \\ &= \frac{1}{2i} \int_0^{+\infty} (e^{2\pi i f t} - e^{-2\pi i f t}) e^{-\lambda t} e^{2\pi i f t}\, dt \\ &= \frac{1}{2i} \left[\frac{1}{2\pi i (f + f_0) - \lambda} + \frac{1}{2\pi i (f - f_0) - \lambda} \right] \end{aligned}$$

Note that we have taken the lower limit of the integral to be zero, to reflect the assumption that the oscillation begins at $t = 0$. The Fourier transform $Y(f)$ has a peak near $f = f_0$, the resonant frequency. The spectral power is given by the absolute square of the Fourier transform (i.e. multiply the Fourier transform by its complex conjugate). Near the peak at f_0, the square of the second term will dominate:

$$P(f) \approx \left(\frac{1}{4}\right) \frac{1}{4\pi^2 (f - f_0)^2 + \lambda^2}$$

This functional form for the spectral power is often given the name "Lorenzian". It is the spectral shape usually seen in optical spectra as well as in radioactive decay.

The spectral power will drop to half its maximum amplitude when

$$\Delta f \equiv (f - f_0) = \frac{\lambda}{2\pi}$$

Thus, when a system is highly damped (i.e. λ is large) then it will have a broad peak in the frequency domain. We can call $\Delta t = 1/\lambda$ the characteristic duration of the signal. Then the equation is

$$\Delta f \, \Delta t = \frac{1}{2\pi}$$

This equation is a special case of the *uncertainty principle*, which we will discuss further in Section 3.2.3.

The key point here is that any oscillator that has a mechanism to lose energy with a time constant Δt, will have a broadened spectral peak. The exceptions are objects that do not have coupling between modes, and thus are essentially friction free. Such modes are often called "high Q" oscillations, since the quality factor Q is defined as the number of radians for damping to reduce the energy by a factor of e:

$$Q = \frac{\pi f_0}{\lambda} = \pi \frac{\Delta t}{T}$$

where T is the period of the oscillation.

In nature, high Q systems are found only in those realms in which friction is weak, so there is little decay (i.e. λ is small). The Earth does have high Q modes of oscillation. They are the seismic modes of vibration, which have relatively low loss in the interior of the Earth, and reflect efficiently off the surface and interior layers. But internal climate modes tend to be lossy and therefore low Q. To estimate the Q of oscillatory behaviour, try to imagine how many cycles it will take before the energy is transferred into other phenomena, and then multiply your guess by 2. Q can be estimated from the width of the spectrum from the equation:

$$Q = \frac{f_0}{2\Delta f} = \frac{f_0}{FWHM}$$

where *FWHM* stands for the full width at half maximum of the resonance response. For the 41 kyr peak shown in Fig. 2.2, the $FWHM = 0.000875$, $f_0 = 1/41$, and we have $Q = 28$, assuming that it is a free oscillation. The challenge for any free oscillation model would be to explain how the lossy behaviour of the climate system is compatible with this high value of Q.

Two natural realms which have low friction, and show high Q behaviour are those of astronomy and quantum mechanics. The laws of physics were discovered in astronomy (by Newton) precisely because solar system motion occurs with very little friction.[2] Likewise, in atomic systems, electrons move without friction; states decay

[2] Galileo had to demonstrate that objects of different sizes fall with the same velocities (his famous Tower of Pisa experiment) by using two round massive objects; if one of them had been a feather then the demonstration would have failed.

only by relatively large, discrete quanta. It is important to note that a forced oscillator mimics the spectrum of the driving force. Thus, if the glacial cycles are driven by orbital oscillations, they too will have narrow spectral peaks. The peaks will be broadened only to the extent that the friction in the Earth causes a drag on the astronomical driving force.

We conclude that the high Q of the glacial oscillations system suggests that they are driven either by astronomical forcing or by quantum mechanics. Since we are aware of no mechanisms that couple quantum mechanical phenomena to climate, we conclude (by default) that astronomy is responsible.

2.1.2 Relaxation oscillator

Unlike the resonant oscillator of the previous section, a relaxation oscillator does not have a built-in frequency. Rather, it has a built-in time constant. In an electronic circuit, a relaxation oscillator works, typically, by charging a capacitor until a certain voltage threshold is reached that opens a switch and discharges the capacitor rapidly. The process then begins anew (see, for example, Horowitz and Hill, 1991.) In geology, an analogous process can take place. We might have a glacier that slowly increases in size but, when it reaches a certain level, spontaneously collapses—and then starts again. The switch that causes the collapse could be a related geophysical phenomena such as sea level. For example, we might suppose that an ice-age glacier would continue to grow until sea level has dropped to such a point that methane from clathrates was spontaneously released, causing a sudden warming that melts the glaciers completely. Sea level rises, the methane has a short lifetime in the atmosphere and disappears, and the glaciers begin to grow again. The collapse could not take place again until the clathrates recharge. (For a discussion of the possible role of clathrates on climate, see MacDonald, 1990.)

The ability of a relaxation oscillator to maintain a constant phase depends on the absence of background variability that can cause the oscillation to trigger prematurely or late. In an electronic circuit, this means that the electronic noise must be very low. In climate, it means that other variables, independent of the oscillation under question, must be constant. For example, if our relaxation oscillator has a natural time constant of 100 kyr, then any other climate signal (e.g. a 41 kyr oscillation, or a 23 kyr oscillation) can interfere with its proper triggering.

Many people would not consider that a defect, but rather a feature of the theory. The 100 kyr cycles of the ice ages do not terminate suddenly, but they seem to vary by ± 25 kyr. In fact, this is the basis for precession triggering models (discussed in Sections 6.4.4 and 6.4.8), in which the terminations are not quite free, but linked to precession cycles. Some occur after three precession cycles (i.e. 3×23 kyr $= 69$ kyr), and others after five such cycles (i.e. after 115 kyr).

The problem with this interpretation is that it requires some external force to keep the oscillations in phase. (In the precession triggering model, this is the insolation signal.) Unless there is such a timing mechanism, the phase of the oscillator tends to drift. If there is a variation of ± 25 kyr for each cycle, then after N cycles, the signal will drift out of phase by (on average) $25\sqrt{N}$ years. For $N = 9$, the approximate

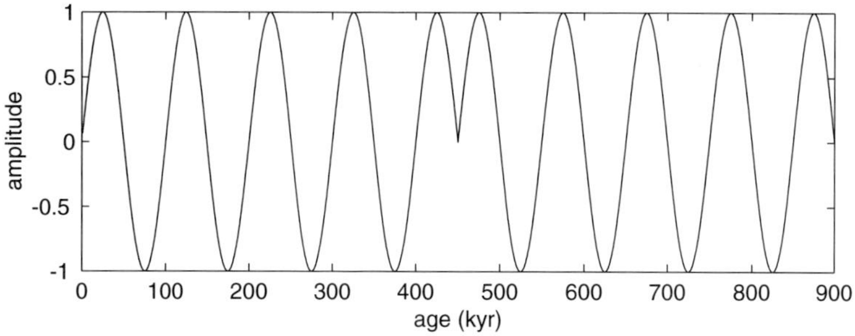

Fig. 2.3. Signal with phase shift.

number of 100 kyr oscillations observed (see Fig. 5.20), we would lose phase by 75 kyr. Yet this is not happening. (Fig. 5.20 also shows the remarkable phase stability.)

Such phase drift will also degrade the spectrum. A phase drift of 75 kyr is more than a half cycle. It means that the last half of the signal is out of phase with the first half, and that results in a broadening of the spectral peak. We illustrate this in Fig. 2.3, which shows a signal that shifts its phase by 180 degrees half-way through its 900 kyr duration. The spectral power for this signal is shown in Fig. 2.4, along with the spectral power of a pure sine wave (dashed line). Note that the width of the phase shifted line (full-width at half-maximum) is about three times that of the unshifted line. It is the presence of narrow spectral lines in the data, equivalent in width to that which we would obtain from a completely unshifted line, that leads us to conclude that the spectrum is not characteristic of a free relaxation oscillator.

We will give other examples of the broadening of the spectrum by phase instability in this book. In Fig. 3.4 we show the spectrum of a sawtooth signal that has a phase variation for each cycle of ±20 kyr. Note, however, that in this example the

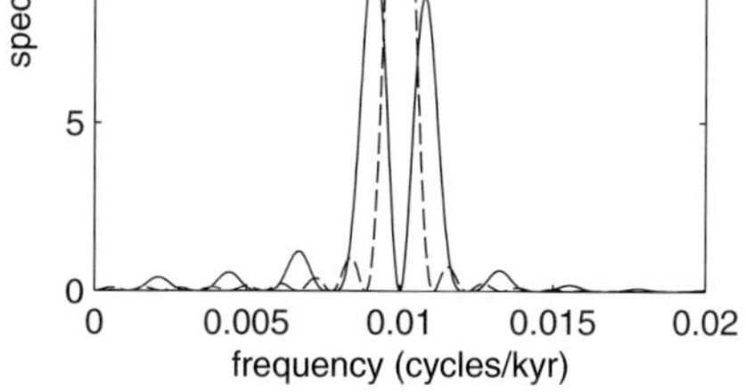

Fig. 2.4. Spectrum of phase shifted signal. The dashed line is the spectrum of a pure sine wave of the same 900 kyr duration.

phase error does not accumulate, but is equally probable on either side of a 100 kyr perfect cycle. In Section 5.6 we show the results on the spectrum of taking a narrow signal (sine wave, or orbital inclination) and putting in an artificial phase shift. In Section 5.9 we show the consequences on the spectrum of introducing a phase shift as part of the process of time-scale adjustment called *tuning*.

A detailed example of the relaxation oscillation model for the 100 kyr cycle can be found in Saltzman and Verbitsky (Saltzman and Verbitsky, 1993). This team was courageous enough to publish a detailed prediction of the signal that would be found in the Vostok CO_2 record *before* the record was measured (Saltzman and Verbitsky, 1995). Unfortunately for their model, their prediction of a maximum at age 250 kyr turned out to be completely wrong; the CO_2 at that age turned out to be a minimum. Nevertheless, they set an example for everybody else by creating a theory that made predictions, a standard that has been matched by very few other theories.

2.1.3 Driven oscillator

In contrast to the free oscillators described in the previous sections, the driven oscillator is a system that has a natural frequency f_0, but which is being driven by a force that has a difference frequency f. Tides are an example of driven oscillations. A bay may have a natural resonant frequency, the frequency of sloshing, but it is driven by the tidal gravitational force of the Moon and the Sun, which has a period of 12 hours. The key feature of driven oscillators is that they operate at the driving frequency, not at the natural frequency. Regardless of the natural period of the bay, the tides maintain a frequency of two cycles per day.

On the other hand, the amplitude and the phase of the driven oscillator depend critically on the match between the natural frequency f_0 and the driving frequency f. It is easily shown (see, for example, Marion and Thornton, 1988, p. 126), that the amplitude of the resulting motion is

$$h(t) = \frac{A\cos(2\pi ft - \delta)}{\sqrt{(f_0^2 - f^2)^2 + 4f^2\beta^2}}$$

In this equation, δ is a phase shift, and β is a damping parameter that describes the natural decay of a free oscillation. β is related to the Q of the resonance by the equation

$$Q = \frac{\pi f_0}{\beta}$$

The critical thing to note here is that $h(t)$ oscillates at frequency f, not at the natural frequency f_0. If f is close to f_0, then the denominator becomes small, and the amplitude of $h(t)$ becomes very large—but the frequency doesn't change.

If there are two frequencies in the driving force, then the situation becomes more complicated. Both frequencies appear in the response $h(t)$, but with different amplitudes. The result is that the shape of $h(t)$ is not the same as the shape of the driving force, although the frequency content is unchanged. This means the spectrum of the data will show peaks at the same locations as the spectrum of the driving force, but with different amplitudes.

An additional consequence of this fact is that if the spectrum of the driving force has a single, narrow spectral feature, then that same feature will show up in the response function $h(t)$. If there are several narrow peaks (as there will be for some of the orbital parameters we will discuss, including eccentricity and precession) then the response will also show several narrow peaks, although their amplitudes may have a different ratio.

The driving forces from astronomy have extremely narrow spectral peaks, because of the lack of damping in the motion of the planets. If these forces are responsible for glacial cycles, then we expect the spectrum of glacial cycles to show narrow spectral peaks. This is what is observed.

It is also possible, perhaps even likely, that the response of the climate to the driving force is not resonant at all, but is simply linear in the driving force. This is mathematically identical to the response that you would have if the resonant frequency f_0 is very far removed from f, and so the conclusions are the same. But with such a linear response, not only are the frequencies preserved, but so are the relative amplitudes.

If we need to enhance one frequency and suppress others, then we can do so with a resonant response. For example, we will show that the orbital parameter known as eccentricity has three frequencies inherent in it, corresponding to periods of 95, 125 and 400 kyr. However, the glacial cycles have one only frequency present, consistent with the period 95 kyr. Could eccentricity still be driving the behaviour? Yes, if a natural resonance in the climate response function is close to 95 kyr, and has very small damping (represented by β in the previous equation). This requires the resonance to have a very high Q so that the 125 is suppressed relative to the 95. If there were such a narrow resonance in the climate response, we would hardly need an astronomical driving force, since the natural oscillations of such a resonance would account for the narrow peak. The challenge in such theories is to find a physically plausible mechanism that would have such a resonance. As explained earlier, the free relaxation oscillator will not work, since it is too vulnerable to background noise.

When a driving force happens to match the resonant frequency of the response, i.e. when $f = f_0$, then there is another interesting consequence. The phase shift is given by the equation (Marion and Thornton, 1988, p. 125):

$$\delta = \tan^{-1}\left(\frac{2\omega\beta}{\omega_0^2 - \omega^2}\right) = \tan^{-1}\left(\frac{2f\beta}{\pi(f_0^2 - f^2)}\right)$$

where $\omega = 2\pi f$. Near resonance, the argument goes to infinity, and δ is $\pi/2$, i.e. there is a 1/4 cycle delay. At frequencies far below resonance, the phase delay δ is approximately zero, and at frequencies far above resonance, the phase delay is π, i.e. a half cycle. So, for example, if the 95 kyr peak is enhanced by a resonance at this same frequency, then we expect a $95/4 = 24$ kyr lag between the eccentricity cycle and the glacial cycle. The presence of this phase shift has not been widely appreciated by those who have invoked resonance models.

We conclude this section with our conclusion: that the narrow spectral features seen in glacial cycles can be understood in terms of a driven oscillator, and only astronomy provides plausible driving forces with sufficiently narrow spectral features to account for the data.

2.2 ORBITAL PARAMETERS

It may surprise some people that we *can* calculate solar system orbits back millions of years. But remember that physics was discovered when Newton recognised that the laws of gravity that affect behaviour on Earth were identical to the laws that determined the celestial motions. Newton's laws work extremely well in space, because of the relative absence of friction.

The field known as celestial mechanics, in which these calculations are done, was once one of the forefronts of scientific research. Perhaps surprisingly, it has now returned to that position. The original interest grew out of astrology. Those who believed that the locations of the stars, and in particular the planets, influenced life on Earth, saw great potential power in the possibility of predicting where those planets would be in the future. (We suspect that Franz Kuhnert, whose quote was used at the beginning of this chapter, confused astronomy with astrology.) Recently, interest in celestial mechanics has revived because of the space programme. One must calculate the orbits of spacecraft to the exquisite precision to send a mission to Mars. And interest in celestial mechanics has revived in recent years because of its value in testing General Relativity theory, and in accounting for some of the surprising discoveries made by the planetary probes. Celestial mechanics, like spectral analysis, has developed into an extremely arcane field. It is subtle and difficult if you want to calculate orbital parameters to sufficient precision to send a probe to pass a satellite of Jupiter, or to find the orbit of the Earth ten million years ago.

The basic calculations are not difficult, and can be learned in a single undergraduate course. However, just as in spectral analysis, there are many traps awaiting the incautious. In 1999, a probe to Mars failed because some calculations were mistakenly done using feet rather than metres. And it was discovered in 1991 that the published calculations of orbits older than 1.5 million years were incorrect, because of accumulating roundoff errors in the computer programs used (Quinn et al., 1991).

2.2.1 Methods of calculation

The earliest detailed calculations of changes in the Earth's orbit, those carried out by d'Alembert, Leverrier and Pilgrim, were done using a method known as perturbation theory. In this approach, the orbits of the planets were assumed to be ellipses, and the mass of the planet was distributed along the orbit of the ellipse. This approximation was based on the insight that the average effect of the planet was well described by the average mass distribution along the orbit. This insight is still extremely useful for thinking about the perturbations. The mathematics of this perturbation approach is described in detail in the text by Danby (1992).

To calculate the changes in the orbital parameters, the parameters are first expanded in a trigonometric series. For example, the eccentricity e is written

$$e = e_0 + \sum_i E_i \cos(\lambda_i t + \Phi_i)$$

and then the coefficients e_0, λ_i, and Φ_i are calculated from the equations of motion (i.e. from $F = ma$) using perturbation theory. The number of terms used in the sum depends on the precision required. In the 1970s, the most complete tables of coefficients, with up to 47 terms, were published by Berger (1978). Insolation for any date and latitude could then be easily calculated from these coefficients; for the equations see, for example, MacDonald (1990).

The issue became more complicated with the recognition that the orbits can be chaotic in character, so that a solution in terms of periodic functions misses important behaviour. This, coupled with the development of special purpose computers, led to the extensive use of "direct integration" which avoided the perturbation approach by working directly with the equations of motion. In the direct integration method, the positions and locations of the planets are put into the program. The force on each planet is then calculated from the gravitational force equation, and the acceleration is calculated from $F = ma$. A small step in time is taken (Quinn et al. (1991) used step sizes of 0.75 days to cover 3 million years) and the new location and velocity of each planet is calculated. The process is continued. Special mathematical methods are used in the calculations to avoid mathematical artifacts. These solutions are not readily described by a few equations, but the internet has come to our rescue, by allowing the detailed numeric values to be posted for all to use.

The use of direct integration led to an interesting discovery by Quinn et al. (1991), who uncovered an error in the previous perturbation calculations. They showed that the previous perturbation methods of Berger and colleagues (1978) had correctly described the main features of the data for the past 1.5 million years, but that they were quite inaccurate for older ages. Quinn et al. showed that error could be traced to roundoff error in the prior computer calculations. The ironic result is that, although Berger et al. had recommended that all of their expansion coefficients be used for accurate calculations, only the first few were accurate. Quinn et al. state that "typical [roundoff] error in the obliquity in the last 1 Myr is at least 0.001 radians, which is larger than all but 6 of the 47 terms in Berger's series". Berger and Loutre corrected the error, and their new results (Berger and Loutre, 1992) are now in agreement with those of Quinn et al. Detailed calculations going back 20 million years have now been published by Laskar et al. (1993), and 100 million years by Sussman and Wisdom (1992).

For the purposes of this book, it is not necessary to understand these calculations in detail. However, it is important to know what the relevant orbital parameters are, and it is interesting to identify the phenomena responsible for the changes in them.

The orbital parameters that are central in understanding climate are called eccentricity, obliquity, the precessional parameter (or just 'precession' for short), and the orbital inclination. The first three of these are critical in describing the insolation; the inclination is important for other processes, including extraterrestrial accretion. We will now describe each of these in turn.

2.2.2 Eccentricity

The present orbit of the Earth around the Sun is shown in Fig. 2.5, to scale. The small dot in the centre is the Sun, also drawn to scale. The Earth is too small to be seen under the line representing its orbit. The orbit appears nearly circular, but it is not. Note that the orbit is not centred on the Sun, but comes closer on the right side, and further away on the left, as can be seen by comparing it to the grid.

Eccentricity measures the departure of the Earth's orbit from that of a perfect circle. If a is the major axis of the orbit (the horizontal diameter in Fig. 2.5), and b is the minor axis (the vertical width), then the eccentricity e is

$$e = \sqrt{1 - \left(\frac{b}{a}\right)^2}$$

The present eccentricity of the Earth's orbit is 0.016722. From the equation for e, you can calculate from this that $(a/b) = 1.00014$. Thus the orbit is circular to within 0.014%. That's why it looks like a circle; the deviation is less than the width of the line. (It is a true circle only if $e = 0$.) However (and this is the important part) the

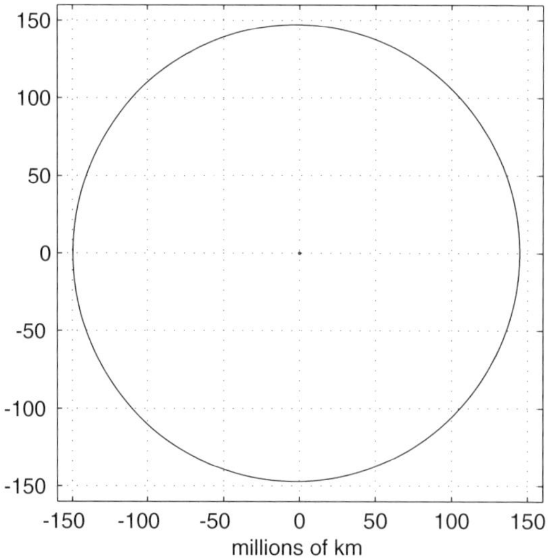

Fig. 2.5. Earth's orbit, to scale.

centre of the orbit is displaced from the location of the Sun. The distance of closest approach of the Earth to the Sun is $r_{MIN} = a(1 - e)/2$, the maximum distance is $r_{MAX} = a(1 + e)/2$, and so the fractional difference as the Earth goes around the Sun is $(r_{MAX} - r_{MIN})/r = 2e = 3.3\%$. This small variation, which can be seen in Fig. 2.5 (although just barely), is at the heart of every insolation model.

The eccentricity is not constant, but changes slowly with time. To see why this is true, it is helpful to recognise that the eccentricity is related to the angular momentum L of the Earth in its orbit. We write Newton's law of gravity in the form

$$F = -\frac{GMm}{r^2}$$

where M is the mass of the Sun, m is the mass of the Earth, and r is the distance between them. The energy of the orbit is the sum of kinetic and potential energies:

$$E = \frac{1}{2}mv^2 - \frac{GMm}{r}$$

To simplify the equations we are about to write, we define the constant k by

$$k = \frac{-2E}{G^2M^2m^3}$$

For orbits in which the planet is bound to the Sun (closed orbits), the energy E is negative, so the constant k will be positive. (The zero energy orbit, for which the planet can escape to infinity, is parabolic). Note that k is a constant only to the extent that the energy of the orbit is unaffected by perturbations. But for perturbations of the kind that occur in our solar system, this is a very good approximation. (See, for example, Danby, 1992.) With k approximately constant, it can be shown that the eccentricity is related to the orbital angular momentum L by

$$e^2 = 1 - kL^2$$

(For a derivation, see, for example, Landau and Lifshitz, 1976, p. 36. Note: in the reference, M represents angular momentum, not mass.) This is the equation that relates eccentricity to orbital angular momentum. It is the key for understanding why and how perturbations of other planets affect the eccentricity of the Earth's orbit. The equation says, for example, that in extreme eccentricity ($e = 1$, the limiting value for which the ellipse becomes a parabola), the angular momentum is zero. This makes sense; this is the value you get when the object is falling directly into the Sun, with no transverse velocity. Eccentricity reaches its minimum value of zero when L reaches its maximum value of $1/\sqrt{k}$. You can easily verify that this is for a circular orbit, i.e. an orbit for which the centrifugal force equals the centripetal acceleration:

$$\frac{mv^2}{r} = \frac{GMm}{r^2}$$

Any force which removes angular momentum from a circular orbit makes that orbit more eccentric. Any force which increases angular momentum makes the orbit more circular. The rate of change of angular momentum is related to the torque τ on the Earth–Sun system by:

$$\frac{dL}{dt} = \tau$$

Torque on the Earth–Sun system is produced by any planet that pulls on the two asymmetrically. Major causes of such torque are the planet Jupiter, because it is massive, and Venus, because it comes so close. So the torques from these planets cause the eccentricity of the Earth's orbit to change. The fact that a few planets dominate the torque leads to the relatively simple variation observed in eccentricity. A plot of the Earth's eccentricity for the last 800 thousand years is shown in Fig. 2.6.

Over this time interval, the curve is approximately fit by the sum of a constant and three sine waves:

$$e(t) = a_0 + \sum_{k=1}^{3} a_k \sin(2\pi f_k t + \theta_k)$$

The three frequencies are 0.0105, 0.0080 and 0.0025 cycles/kyr, corresponding to periods of 95, 125 and 400 kyr. The presence or absence of these frequencies in paleoclimate has been a central issue in the analysis of the data, so it is worth looking at them closely. In Fig. 2.7 we plot the sum of these waves, with amplitudes $a_0 = 0.025$, $a_k = (0.0083, -0.0069, -0.0118)$, and phases $\theta_k = (0.0561, 0.1162, 1.8108)$ radians. These three frequencies account for 90% of the variance of the eccentricity signal.

It is worthwhile to examine this plot when the 400 kyr component is removed, leaving only the 95 and 125 kyr components, as shown in Fig. 2.8. What we see here is the beating of the 95 kyr cycle with the 125 kyr cycle. They nearly cancel near 0 years, 400 kyr and 800 kyr. The near cancellation near 400 kyr occurs near a climate period known as Stage-11 when the climate modulation was large. We will discuss

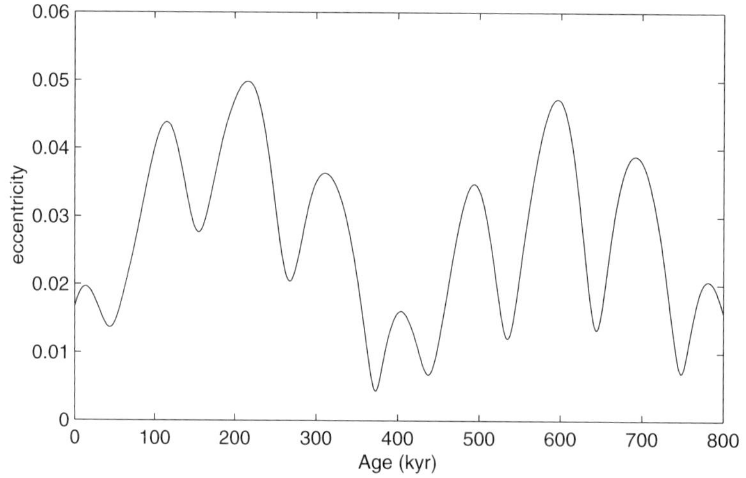

Fig. 2.6. The eccentricity of the Earth's orbit.

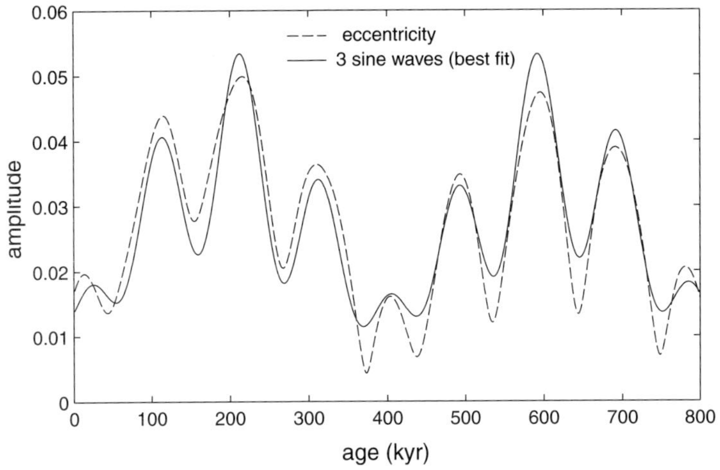

Fig. 2.7. Sum of sine waves with periods 95, 125 and 400 kyr.

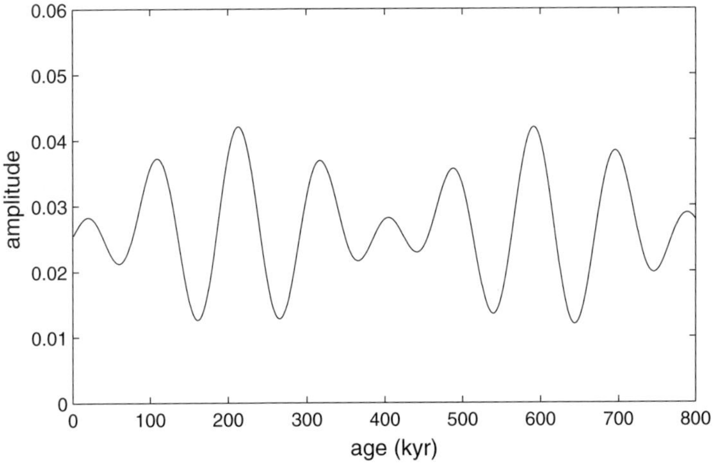

Fig. 2.8. Sum of sine waves with periods 95 and 125 kyr.

this at length in Section 8.2. The fact that the eccentricity modulation is small at this time has been considered a difficulty for the standard Milankovitch theory, and was termed by Imbrie et al. (1993a) as *The Stage-11 Problem*. Comparing Fig. 2.8, which has two frequencies, with Fig. 2.7, which has all three frequencies present, we see that the 400 kyr component takes this beat signal and raises the average value near 200 and 600 ky, lowering it near 0, 400 and 800.

The spectrum of eccentricity for two time intervals is shown in Fig. 2.9. Note that the ratio of the peaks varies with time. Because of such variations, comparisons with data should always use spectra derived from matched time intervals.

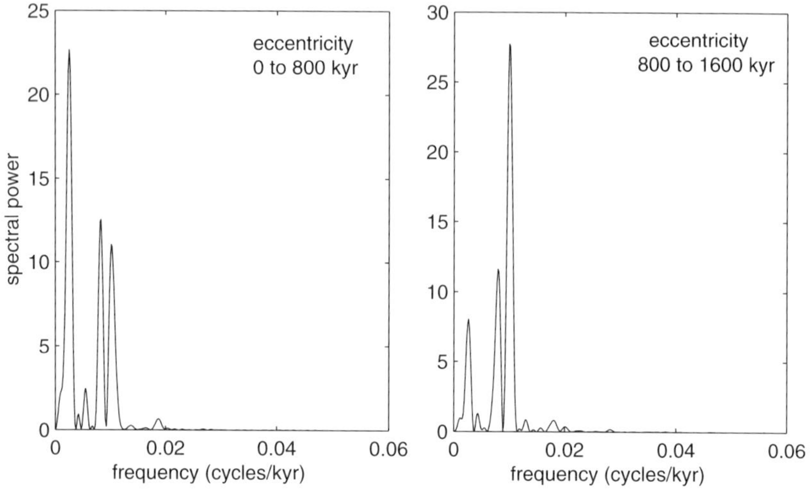

Fig. 2.9. Spectrum of eccentricity.

The three peaks correspond to periods of approximately 400, 125 and 95 kyr. These three frequencies are not really independent. They obey the relationship:

$$\frac{1}{95} - \frac{1}{125} \approx \frac{1}{400}$$

This relationship does not reflect a nonlinear process in climate. The eccentricity signal is purely astrophysical, and the relation derives from the nonlinear generation of eccentricity from the perturbations of the other planets on the orbit of the earth.

2.2.3 Precession

Precession is the slow change of the direction of the North Pole, as the Earth wobbles under the torque that the Moon and Sun exert on its equatorial bulge. The effect is completely analogous to the precession of a tilted top, whose axis rotates from the torque of gravity. The "ellipticity" of the Earth is 0.0034. (Ellipticity is mathematically similar to eccentricity, but it describes the shape of the Earth's surface rather than the shape of its orbit.) It is defined as the difference in the radius of the Earth at the equator vs. at the poles, divided by the average radius. Thus the equator is further from the centre by about 0.0034×6371 km = 21 km.[3] Because of this ellipticity, the gravitational pull of the Moon and the Sun exerts a torque on the Earth. The subsequent behaviour of the Earth is similar to that of a spinning gyroscope, which precesses because of the torque of gravity. (The torque arises in the gyroscope

[3] This makes Mt. Huascaran in Peru, which is only 11 degrees from the equator, further from the centre of the Earth than Mt. Everest, which is 28 degrees north, a fact which leads some people to argue that it is "higher". But it is not higher if the local reference of sea level is taken as the standard.

when the centre of mass is not directly above the point of support.) The net effect, as we discussed in Section 1.1, is that the direction of the Earth's axis precesses with a period of 25.8 kyr. This is the period measured with respect to the stars. In 12.9 kyr (a half period) the North Pole will be half way around the circle and will point towards Vega, 47 degrees away from the present Pole Star. In 25.8 kyr, the Earth's axis will again point towards the present Pole Star.

Precession affects climate because, when it changes, it means that the Earth's closest approach to the Sun, perihelion, occurs in different seasons. At present, perihelion occurs on 4 January, close to the Winter Solstice. In the Milankovitch theory, this is related to the fact that we are now in the midst of an interglacial. We are between ice ages because the Earth is too close to the Sun during Northern winter, and the glaciers cannot endure.

The date of perihelion also changes, however, and calculations must take this into effect. If we measure the precession of the Earth's poles with respect to this moving perihelion, the period is shorter, not 25.8 kyr, but closer to 23 kyr. Actually, because the perihelion does not move smoothly but lurches, the perihelion in this moving system is a superposition of several periods, the most important being 24, 22 and 19 kyr. The lurches in the perihelion motion are due to changes in the eccentricity of the Earth's orbit. The perihelion moves most rapidly when the eccentricity is greatest. The triplet of frequencies present in the precession parameter reflects the triplet of frequencies present in the eccentricity.

The effect of precession on climate also depends on how close the Earth comes to the Sun, and this depends on the eccentricity of the Earth's orbit. Thus the key parameter for insolation calculations turns out to be the product of the sine of the precession angle (measured with respect to the moving perihelion), multiplied by eccentricity. This combination is called the "precession parameter", often referred to as simply "precession". We give it the symbol p. We write the precession parameter p as

$$p = e \sin \omega_M$$

where e is the eccentricity, and ω_M is usually taken (by convention) as the angle between the *spring* solstice and perihelion. But it is necessary to recognise that it is not the same thing as the simple precession of the poles that we discussed in the introduction, which has a very regular period of 25.8 kyr. Because of the moving perihelion, as we described above, the period of the precession parameter is a combination of 24, 22 and 19 kyr. The amplitude of p is modulated by eccentricity, but it is not this modulation that gives rise to the triplet of peaks. For short time intervals, the peaks may not be resolved (see Section 3.15), and the precession may appear simply as a single peak near period 23 kyr.

Let's illustrate p by calculating its present value. Spring solstice typically occurs (these days) on 20 April, and perihelion occurs on 4 January. There are 106 days between these dates. This makes $\omega_M = 105$ degrees[4] $= 4.5$ radians. Using the current value of $e = 0.0167$, this gives $p = 0.0163$.

[4] Since there are 365 days per year, and 360 degrees, there is approximately one degree per day.

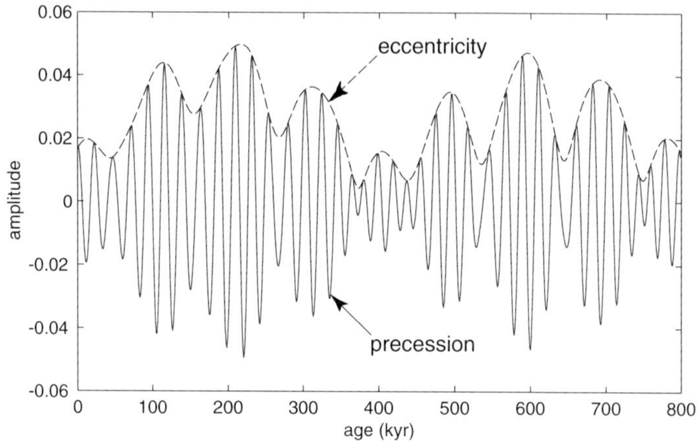

Fig. 2.10. Precession parameter p and eccentricity.

A plot of the precession parameter p for the last 800 kyr is shown in Fig. 2.10. Eccentricity is also plotted. From this diagram it is clear why we sometimes say that the *envelope* of the precession parameter is eccentricity. We can also say that the precession angle is *modulated* by eccentricity. This fact will prove very important, especially in nonlinear climate models. It is this modulation that must provide the 100 kyr oscillation, according to modern versions of the Milankovitch theory.

The spectrum of the precessional parameter is shown in Fig. 2.11 for two different time intervals, 0 to 800 kyr, and 800 kyr to 1600 kyr. The spectrum is dominated by three frequencies, corresponding to periods of 24, 22 and 19 kyr. The relative amplitudes of the three peaks vary with time. To compare a spectrum of a paleoclimate proxy with that of precession, the time interval for the two sets must be chosen to be identical.

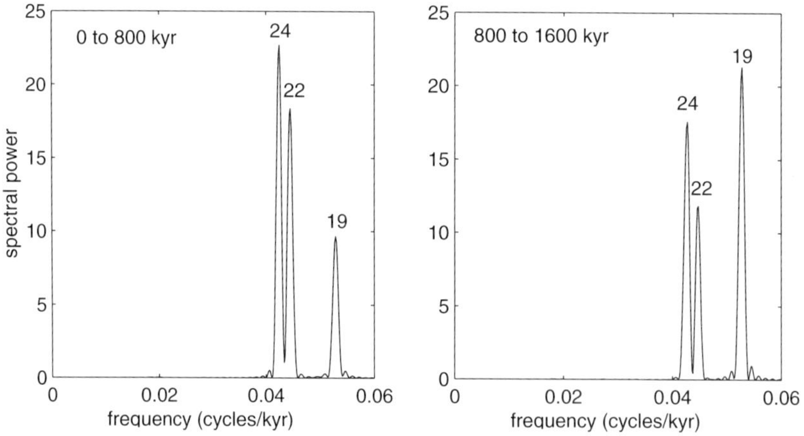

Fig. 2.11. Spectrum of precession.

2.2.4 Obliquity

Obliquity is defined as the angle of the tilt of the Earth's pole towards the Sun—i.e. how much the North Pole is tilted towards the sun in summer. If obliquity were zero, then the Sun would pass overhead at the equator every day. At the poles, half of the Sun would be visible above the horizon at all times, and the Earth would not have strong seasons. (There would be weak seasons from the fact that the distance between the Earth and the Sun varies by 3% throughout the year.) The present obliquity of the earth is 23.5 degrees. This is also the latitude of the Tropic of Cancer—the furthest north that you can go and still see the Sun directly overhead.

But obliquity is not constant: because the plane of the orbit of the Earth is constantly changing, the obliquity must be measured with respect to that plane. The Earth's polar axis precesses with a period of $1/f_P = 26$ kyr. The plane of the Earth's orbit precesses with a period $1/f_\Omega = 70$ kyr. (This variation is called Ω, and we describe it in more detail in Section 2.2.6.) The obliquity measures the differences between these two planes, and so it varies with a frequency equal to the difference of the two precession frequencies: $f_O = f_P - f_\Omega = 0.024$, corresponding to a period of $1/0.024 = 41$ kyr. Since both precession periods are quite regular, the obliquity variation is also very regular, and the 41 kyr period dominates. The current rate of change is 0.128 degrees per thousand years. That may not seem like much, but the Tropic of Cancer obelisk in Chiayi, Taiwan, erected about a century ago, is now 1.4 km distant from the correct latitude, because of the change in obliquity since it was constructed.

The obliquity for the past 800 kyr is shown in Fig. 2.12, and its spectrum is in Fig. 2.13. The variation consists of a strong 41 kyr cycle, with very small side peaks at periods 29 and 53 kyr.

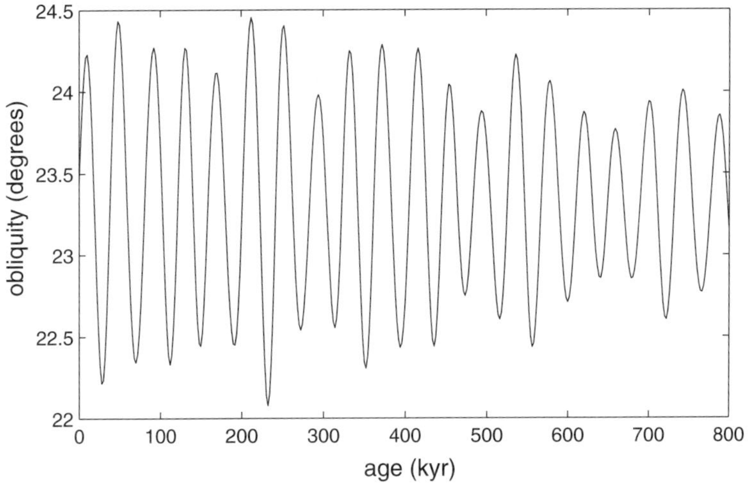

Fig. 2.12. Obliquity.

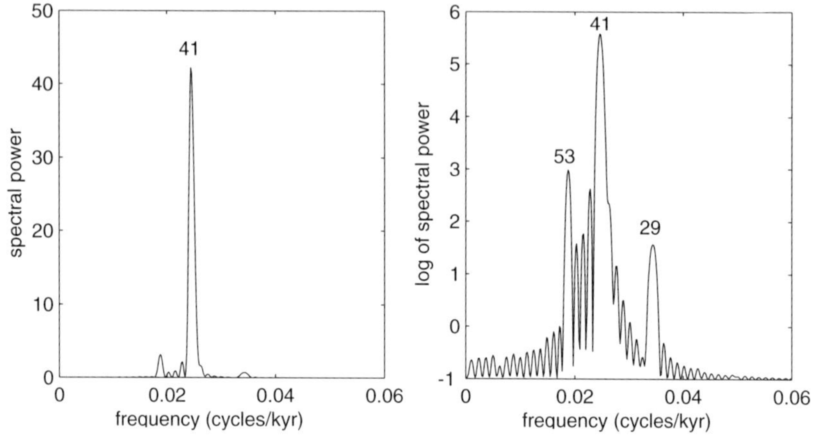

Fig. 2.13. Spectrum of obliquity, linear and log plots.

In the log plot (on the right), the additional periods at 53 and 29 kyr are most easily seen. However, they are small compared to the main peak at 41. The small rapid oscillations in the log plot are an artifact; they would be removed if we applied a taper (discussed in Section 3.4).

2.2.5 Insolation

The average flux of solar energy at the top of the Earth's atmosphere is approximately $S = 1360\,\text{Watts/m}^2$, but the energy per unit area, the insolation, depends on the tilt of that surface with respect to the incident radiation. The average daily insolation over the Earth is 1/4 of this value, $I = 340\,\text{W/m}^2$. The easiest way to see this is by recognising that the total insolation hitting the Earth is $\pi R^2 S$, where R is the radius of the Earth, but that this radiation is spread out over an area of $4\pi R^2$.

It has become standard in papers related to climate, past or present, to talk about the "equivalent radiative forcing" of various phenomena, i.e. to compare them to the insolation forcing. We show this in Table 2.1, based on a similar table of Svensmark.

Some older papers give insolation in units of calories/(cm^2 day). This unit has even been given a special name, the "Langley". We are thankful that this unit has

Table 2.1. Radiative forcing.

S	(solar constant at Earth's location)	$1360\,\text{W/m}^2$
S/4	average insolation at top of atmosphere	$340\,\text{W/m}^2$
A	Earth's albedo, the average reflected insolation	$-53.5\,\text{W/m}^2$
CO_2	estimated effect from increase 1750 to present	$1.5\,\text{W/m}^2$
CO_2	estimated effect if atmospheric CO_2 doubles	$4\,\text{W/m}^2$
Clouds	net effect	$-28\,\text{W/m}^2$

fallen into disuse (no disrespect to Langley intended). To convert to W/m^2, multiply Langleys by 0.4843.

According to the insolation theory of Milankovitch, the key driving force of the glacial cycles is the summer insolation at northern latitudes. For simplicity, most analysts pick a date and latitude, and calculate the average daily insolation at that place on that day. (The insolation is independent of longitude.) The key orbital parameters that are needed in the calculation are the eccentricity of the Earth's orbit (since the distance to the Sun is important), the obliquity at that time (since this tells how much the northern pole was pointing towards the Sun), and the precessional parameter, which depends on where in the elliptical orbit the Earth was on the date of interest. For the detailed equations, see MacDonald (1990) or Laskar et al. (1993).

The insolation for 21 July at 65N is shown in Fig. 4.38. The spectra of insolation for that date at various latitudes are shown in Fig. 2.14. There are several interesting features to note in these plots:

1. The main frequencies present are those of obliquity (0.024 cycles/kyr, period 41 kyr) and the triplet of precession (0.042, 0.045 and 0.053, corresponding to periods 24, 22 and 19 kyr).
2. Eccentricity is virtually absent. (Its frequencies are 0.0025, 0.008 and 0.0105 cycles/kyr, corresponding to periods of 400, 125 and 95.)
3. At northern latitudes the precession signal is strongest.
4. At far southern latitudes, obliquity dominates. (Note that in July, 65S is experiencing winter, not summer.)

In the insolation plots shown, only one is a good match to real δ^{18}O sea floor data. That is the last one, July insolation, 65S, which matches the spectrum of the data seen during the period between 1 and 3 million years ago, when it was dominated by obliquity (see Fig. 2.1). A similar spectrum is obtained from the January insolation at 65N, shown in Fig. 2.15.

These results suggest that it is not summer insolation that determined the cycle of the ice ages during this time interval, but winter insolation. This supports a contention that was once argued by George Kukla. He is quoted in Imbrie and Imbrie (1979) as having said, in 1968:

> *When this problem has been clarified, and the importance of winter accepted, the chance selection of summer will probably be considered the most serious mistake in Quaternary research in recent years.*
>
> (Kukla, 1968)

Note that total insolation on the Earth depends only on eccentricity. Obliquity and precession determine distribution, not the total amount. However, the *total* yearly insolation at the poles depends only on obliquity—provided that we include both poles. This is obvious, but nevertheless, it is shown in Fig. 2.16.

Thus total insolation is a contender for the driving force for the 1 to 3 Ma period ice ages. However, the absence of a 23 kyr cycle in this spectrum is the result of a

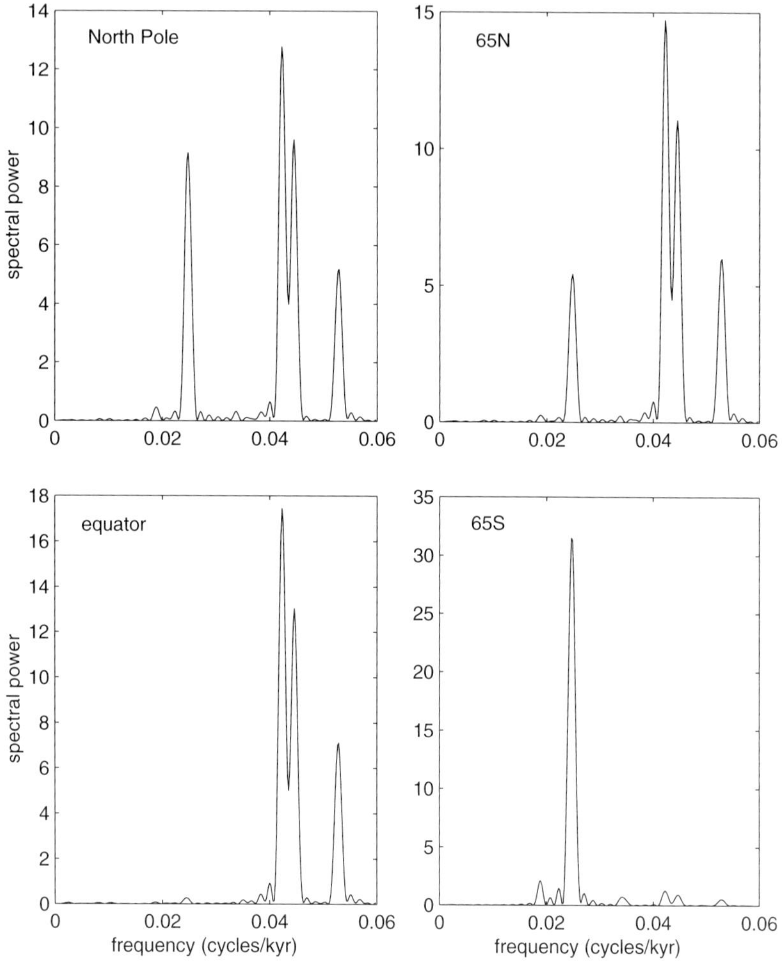

Fig. 2.14. Spectra of July insolation, at latitudes 90N, 65N, 0 and 65S.

precise cancellation of these terms in the Northern and Southern Hemispheres. Such an exact cancellation appears unlikely, so we favour the alternative explanation that the driving force is winter insolation.

Nevertheless, the insolation models for the past million years are usually based on Northern Hemisphere summer insolation. It is this insolation that Milankovitch theorists believe can best explain the features of the ice ages during that period. We will discuss these theories at length in Chapter 6.

2.2.6 Orbital inclination

The orbit of the Earth lies in a plane, but this plane is not identical to that of other planets. It differs from that of Venus by 3.4 degrees, from that of Jupiter by 1.3

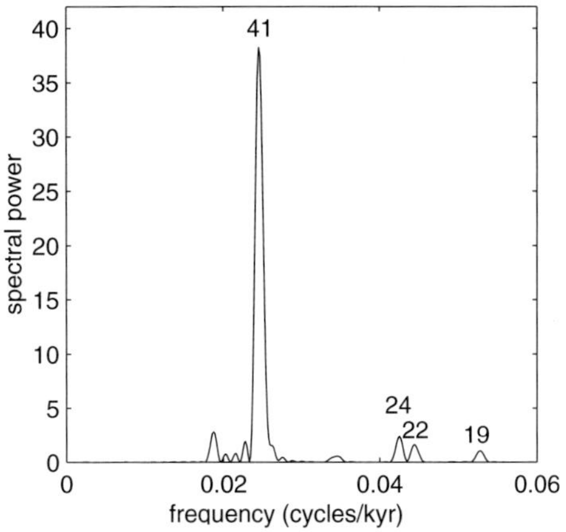

Fig. 2.15. Spectrum of January insolation, 65N.

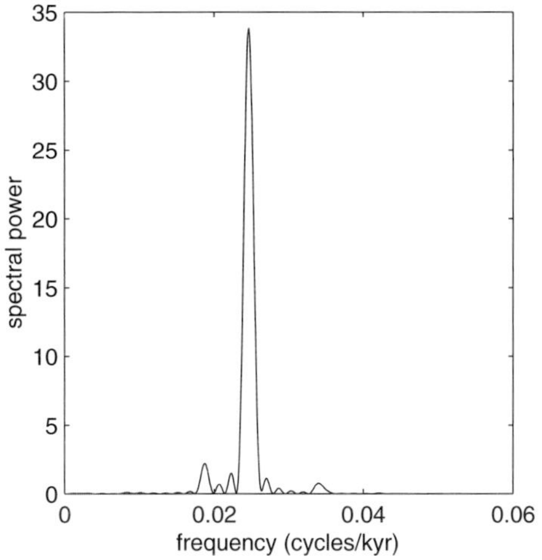

Fig. 2.16. Spectrum of total yearly insolation at poles.

degrees, and from the average of all the other planets by 1.7 degrees. In astronomy, the plane of the Earth's orbit has often been used as a convenient reference. It is called the "ecliptic". As viewed from the Earth, the Sun moves in this plane, and so it is also the plane of the signs of the Zodiac. But this is not a good reference plane for paleoclimate, because it changes with time. It is perturbed by the torques of Jupiter, Venus and Saturn, which give the Earth's motion a horizontal component of orbital angular momentum. (We distinguish between the orbital angular momentum of the Earth's motion, and that of the Earth's spin about its axis.) When we refer to the orbital plane, we must state the year that we are referring to. Croll used the plane of 1850, and present-day paleoclimate studies have generally stuck with this convention, so old calculations can be compared with newer ones. By this convention, the tilt of the Earth's present orbit is very close to, but not quite, zero.

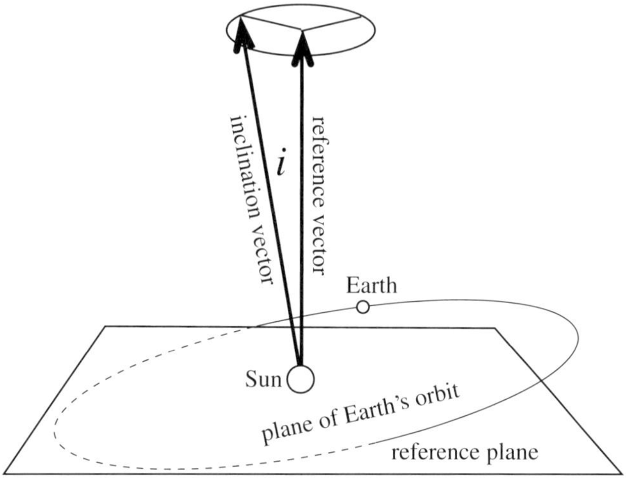

Fig. 2.17. *i* and Ω for the Earth's orbit.

The orbital inclination of the Earth is the angle between the reference plane and the plane of the Earth's orbit. It is also the angle between vectors that are perpendicular to these two planes, as illustrated in Fig. 2.17.

If our reference vector is the one that is perpendicular to the plane of 1850, then we get the simple result that the present orbital inclination is nearly zero. As we will show below, this apparently trivial choice of reference vector has led to important confusion in the field of paleoclimate. In Fig. 2.18 we plot the orbital inclination *i* for the last 600 ky, based on orbital calculations by Quinn et al. (1991). Note that it goes to zero *only* at the present (actually, in 1850). It is very unlikely that the orbital vector of the Earth was ever at this exact location before, so although the inclination approaches zero (going back in time), it never quite touches zero. Also notice that it is always positive. That, too, is a trivial result of the convention that you can always state the angle between two vectors as a positive number.

As can be seen from the plot, the inclination oscillates with a period of about 70 kyr. Little attention was paid to this variation, because Northern-Hemisphere

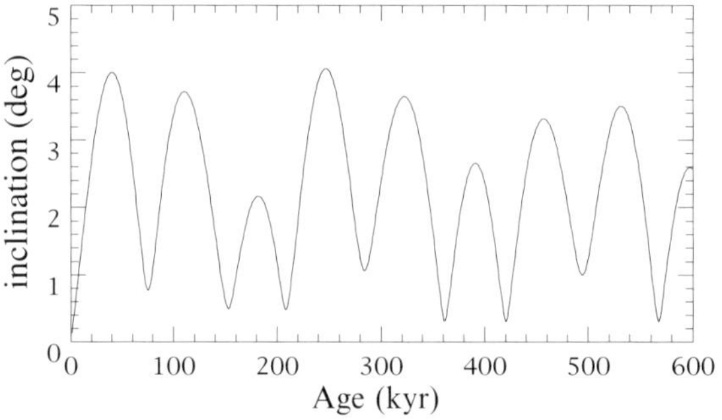

Fig. 2.18. Orbital inclination, referred to Zodiac, for the past 600 kyr.

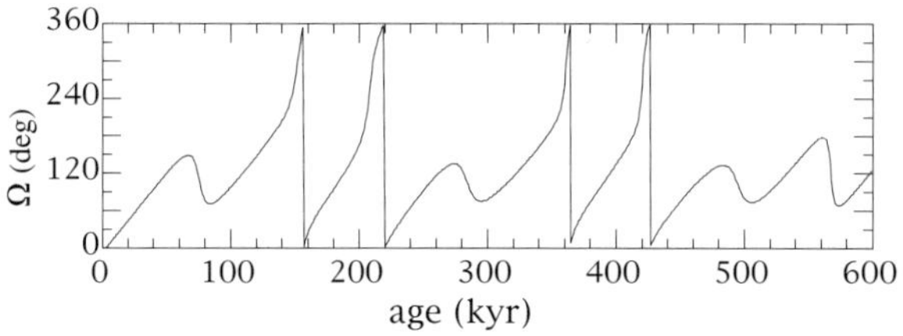

Fig. 2.19. Ω, in Zodiac frame.

insolation does not depend directly on this parameter and, according to the standard Croll/Milankovitch theory, such insolation is the only physical parameter that affects climate.

Of course, the direction of the inclination vector requires a second angle to describe its relation to the reference vector. The second angle is called Ω, and it can be defined in terms of the line of intersection of the two planes. This line is traditionally called a node, and the node on the side of the orbit for which the Earth is rising above the reference plane is called the "ascending node". This terminology will be found in the literature. Ω is then the angle between this line and a reference direction. Equivalently, we can define Ω as the azimuthal angle of the inclination vector, as shown in Fig. 2.17. The value of Ω vs time is shown Fig. 2.19.

This variation, at first look, appears irregular and chaotic, unlike many of the other plots we have seen above. However, we will show this is just an artifact of the choice of reference vector! If we plot the position of the tip of the inclination vector vs. time, we get the graph shown in Fig. 2.20.

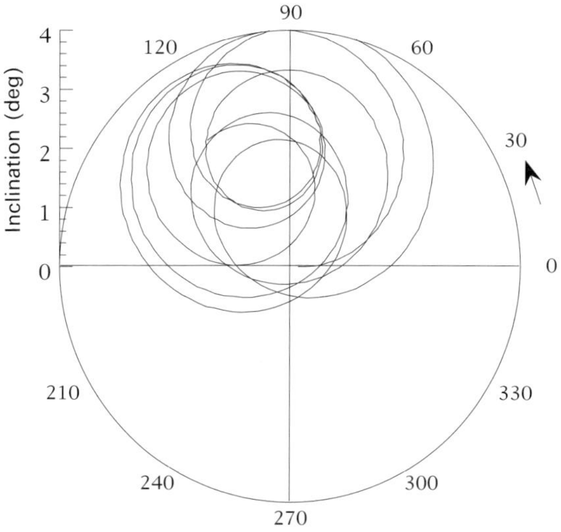

Fig. 2.20. Polar plot of inclination, referred to the Zodiac.

In this plot, i is the radial coordinate and Ω is plotted as the azimuth angle. In this plot, the behaviour of the inclination vector appears relatively simple. It goes in a circle with a slowly varying radius. The sudden jumps in Fig. 2.19 occur during those times when the vector passes close to the origin (the present inclination vector of the Earth), and thereby makes a sudden (but unphysical) jump in Ω. The inclination vector changes slowly and smoothly, and the jump is just a result of the fact that we chose the present orbit of the Earth as our reference.

A much better reference frame for paleoclimate work is the "invariable plane", defined as the plane perpendicular to the angular momentum vector of the planets. This plane is fixed in time, unlike the plane of the Earth's orbit. The invariable plane is close to the plane of Jupiter's orbit, since Jupiter dominates the total angular momentum. The invariable plane is also the average plane of the Earth's orbit. The point that the vector circles turns out to be essentially identical to the vector that defines the invariable plane. If we chose the invariable plane as the reference vector, instead of the 1850 vector, then the polar plot is simply shifted from that shown in Fig. 2.20 to that shown in Fig. 2.21.

This is a simple transformation, but it is a nonlinear transformation. The new value of i is a nonlinear combination of the former i and Ω; likewise for the new value of i. The new values of i and Ω in this new frame are plotted in Fig. 2.22, as a function of time.

Note that, with this new reference frame, the behaviour of the inclination vector is simplified. The value of i is nearly sinusoidal, with a period of about 100 kyr. This fact was first noted by us (Muller, 1994; Muller and MacDonald, 1995) and led us to

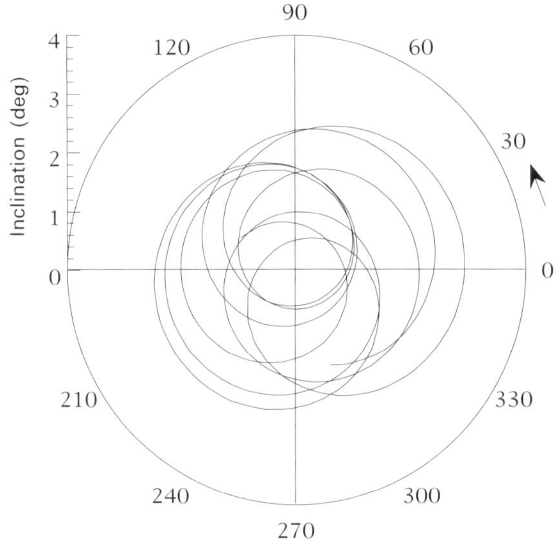

Fig. 2.21. Polar plot of inclination, referred to invariable plane.

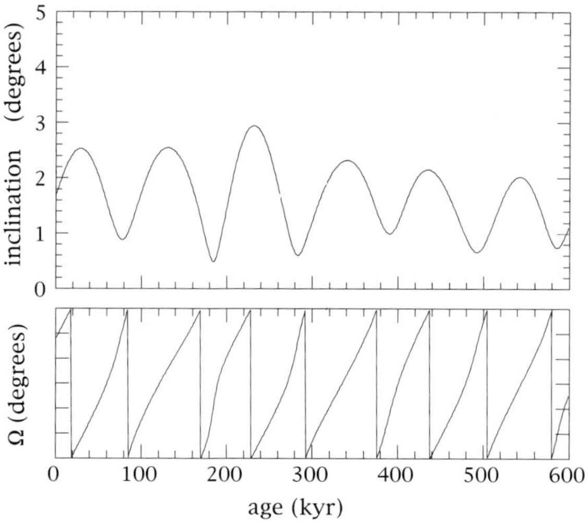

Fig. 2.22. Inclination and Ω, referred to invariable frame, for the last 600 kyr.

speculate that this variation could play a role in the 100 kyr cycle of the ice ages. The behaviour of Ω is also much more regular. It shows a nearly constant increase of 360 degrees every 70 kyr. The sudden jumps are simply the return of the azimuth angle to 0 every time it surpasses 360 degrees.

3

Spectral analysis

"If we dealt with problems involving the superposition of a few simple periodic phenomena, as do astronomers interested in binary stars and related problems, we can learn much from the periodogram. Sadly, however, almost no one else has this kind of data. As a result, the periodogram has been one of the most misleading devices I know."

John Tukey (1967)

3.1 INTRODUCTION

As mentioned in the Introduction, spectral analysis has been absolutely key in establishing the importance of astronomical forcing for climate. Yet spectral analysis is considered by many to be an arcane field to be approached only by experts. There is good reason for this: there is a long history of incorrect results published by good scientists who fell into spectral traps. Many people prefer to ignore the field, or at least not get involved. Yet that really isn't possible for scientists who want to form their own judgements about the relative contributions of proposed astronomical mechanisms.

In this chapter we will give an introduction to spectral analysis. We will attempt to describe the fundamental techniques, as well as alert the reader to traps and pitfalls. We will describe some commonly-used "tricks of the trade" such as zero padding, that make interpretation of spectra simpler and more intuitive.

We feel obliged, however, to offer a few words of caution. Although there are many ways to do an incorrect spectral analysis, there is no "correct way". The method chosen will depend on the taste and intuition of the analyst. In the literature on climate you will find proponents of the Blackman–Tukey method, of the Multi-taper method, of the Maximum Entropy method, and even proponents of the un-windowed periodogram. You will find advocates of log plots, linear plots and square root plots. Be aware that all approaches to spectral analysis entail assumptions, and that the differences in these assumptions often account for the different results of the analyses. If you know what those assumptions are, you can make a reasoned judgement of what conclusions you should draw.

In many ways, the spectral analysis of climate data is a specialised application. Many of the methods that are optimum for other kinds of data are not optimum for paleoclimate data. The traditional focus of statisticians are issues such as "leakage" (how does power at one frequency affect another), "bias" (is the estimate, on average, centred at the correct value), and "consistency" (does the precision of the estimate improve when more data are analysed). Yet these issues have been remarkably unimportant in the history of climate analysis! Far more important have been issues traditionally ignored (or at least, not emphasised) in the traditional methods. These include systematic errors, the uncertainty in the time scale, the coherence of the background across many different sites, and the need for high resolution—rather than accurate estimation of either the amplitude or the frequencies.

With a few exceptions, we will emphasise results, and not proofs. There are many textbooks that prove the theorems that we state; see, for example, those by Priestley (1981) and Percival and Walden (1993). It is more important for the paleoclimatologist to understand the limitations of the various methods rather than the derivations.

3.2 THE FOURIER TRANSFORM

Any function can be expanded in terms of polynomials. The resulting expression is known as a Taylor Series. But even if you understand the Taylor expansion, it still is not obvious (although it is true) that, instead of polynomials, you can use harmonic functions: sines and cosines. Basically, *any* function can be expressed as a sum of sine functions and cosine functions. The result is known as a Fourier series, named after the French mathematical physicist Joseph Fourier, who invented it to help solve heat flow problems.

When should you use polynomials, and when should you use harmonic functions? If the underlying forces are harmonic, then use of a harmonic series often results in the need for a much simpler and shorter expression. If upon doing the analysis you discover that a few terms dominate, then you can try to identify them with known forces that have the same periods.

Mathematicians write sums in two ways. When the sum consists of a countable number of pieces, they use the capital Greek letter Σ (which corresponds to the Roman S). This letter is now often called "the summation symbol". When the sum consists of a continuous number of terms, they use a script S, which has evolved into an integral sign $\int$. We will begin with the integral version but, since the data consist of discrete points, this will soon turn into a discrete sum, known as the digital Fourier transform.

3.2.1 Basic properties

Let the symbol $h(t)$ represent a continuous function of time, such as the insolation. Then the ordinary Fourier transform, also called the Fourier amplitude, is a function of frequency f:

$$H_T(f) \equiv \int_{-\infty}^{\infty} h(t)e^{2\pi ift}\,dt$$

The subscript on the function H_T is the letter "T" standing for "True". (We'll get rid of this in a moment, as soon as we show that we can never calculate H_T from real data, but only estimates of it.)

Everything that we are about to do can be done without the use of complex numbers. If we were to do so, we would get two Fourier transforms: a cosine transform and a sine transform. But these correspond the real and imaginary parts of the complex transform H. We will continue using complex numbers primarily because, typically, they make the equations half as long and more transparent. In the end, the spectral functions that we derive will all use real numbers.

Part of the charm of the Fourier transform comes from the fact that we can solve this equation explicitly for $h(t)$, yielding an equation that looks very similar to the previous one:

$$h(t) = \int_{-\infty}^{\infty} H_T(f)e^{-2\pi ift}\,df$$

Thus, given data $h(t)$, we can find the Fourier transform $H(f)$, and given $H(f)$, we can find $h(t)$. The two contain exactly the same information, expressed in different ways.

The spectral power P is defined as the square of the Fourier amplitude:

$$P_T(f) = |H_T(f)|^2$$

Note that the spectral power does not tell us the phases of the sine waves. In an information theory sense, it contains exactly half of the information. Of course, the phases can be obtained directly from the Fourier amplitude.

But once we start talking about real data, rather than theoretical functions, we run into a problem. Real data do not span infinite time. Even worse, if h is a climate proxy, we might have it sampled only at a few (say 300) discrete points over a time range of a few hundred thousand years. Because of this limitation, we can never calculate the true Fourier transform H_T, and never get the true spectral power P_T. The best that we can do is to get an *estimate* of P_T. The entire theory of spectral analysis is devoted to methods to obtain such an estimate.

One important estimate of H_T can be made from discrete samples of $h(t)$. Suppose that all we have available to use is the value of $h(t)$ at times t_j. We define these values as the set h_j:

$$h_j \equiv h(t_j)$$

Then an estimate of the Fourier transform can be made by the equation:

$$H(f) \equiv \sum_{j=1}^{N} h_j e^{2\pi ift_j}$$

Notice that we have turned the integral into a sum, and that H no longer has a subscript T. This H is not the true transform; it is only an "estimate" of the true H_T.

There are important theorems that can be proven for the special case that the data are equally spaced in time. If the interval between points is Δt, then define the "Nyquist frequency" as

$$f_N = \frac{1}{2\Delta t}$$

The Nyquist frequency is the highest frequency that will appear in the Fourier transform. To appreciate how important the Nyquist frequency is to your everyday life, consider the following fact. The engineers who designed music compact disks (CDs) chose the sampling period Δt to be 1/44000 sec, so that the Nyquist frequency would be 22,000 cycles/sec—above the hearing range of most people. (Some audiophiles still argue that this was not high enough, since although the human ear cannot hear individual tones at frequencies this high, inaudible signals at higher frequencies are *aliased* into the audible range. We discuss aliasing in Section 3.2.8.)

By convention, we let $f_0 = 0$ represent the average ("DC") level of the data. We will usually subtract the average value from the data before we do a spectral analysis, and this will automatically give us $H_0 = 0$. (If you ever forget to subtract the mean from the data, the thing you will notice is a very high value in the first bin of the spectral power, usually overwhelming the rest of the transform!) The lowest frequency of real interest is the one that gives one full cycle in the time interval T:

$$f_L = \frac{1}{T}$$

Now we have the highest frequency f_N and the lowest f_L. We still have to decide what other frequencies to evaluate. It turns out that a very convenient choice is to pick multiples of the low frequency f_L:

$$f_k = kf_L$$

To simplify our discussion, let's assume that N is even. If we allow k to run between 1 and $N/2$, this gives us $N/2$ frequencies going from f_L to f_N.

Why do we get only $N/2$ frequencies from N data points? The reason is that for each frequency, we obtain a complex number. The real part is the cosine coefficient, and the imaginary part is the sine coefficient. So we are obtaining $N/2 + N/2 = N$ amplitudes from N data points. Why are there at most only $N/2$ data points in the spectral power plot? The reason is that we are not plotting the phases, which account for another $N/2$ data points.

It turns out that there is some value in letting k run over a larger range, even though the frequencies that we calculate in doing so will not give us any useful information. We start with $k = 0$; that includes $f = 0$ in the calculation. If we let k run from 0 to $N - 1$, we get the following nicely symmetric pair of equations:

$$H_k \equiv \sum_{j=0}^{N-1} h_j e^{2\pi ijkf_L t_j}$$

$$h_j = \sum_{k=0}^{N-1} H_k e^{-2\pi ijkf_L t_j}$$

This set of equations for discrete data is usually called the "discrete Fourier transform". H is the complex Fourier transform of h.

These equations have another important feature. If we knew, somehow, that $h(t)$ were "band-limited" (it contains no frequencies below f_L or above f_N), then we can prove that the discrete transform above will yield the exact true function H_T! This means that there is a relationship between the continuous function $h(t)$ and the discrete values H_k. It is given by:

$$h(t) = \sum_{k=0}^{N-1} H_k e^{-2\pi k f_L t}$$

The only difference between this and the previous equations is that the function $h(t)$ here is continuous, and the subscript k has been dropped from the t in the sum. This result means that the function $h(t)$ is completely determined by its samples h_k. The band-limited condition is often met for many functions of paleoclimate interest, such as eccentricity and insolation. The condition is often not met for geologic data. Nevertheless, this equation means that the discrete Fourier transform can often serve as a excellent method for estimating the true spectrum.

3.2.2 Glacial Fourier transform

We will now discuss the numerical calculation of the Fourier transform. We begin with a method we call the "Glacial Fourier Transform". This method is extremely easy, but almost never used, and yet it is probably the best method to start with for a beginner. Most experts will tell you to use the "Fast Fourier Transform" or FFT. Don't listen to them! We think the FFT, with the burden of jargon that you must learn before you can even start to use it, serves as a barrier to keep beginners away from playing with and understanding the Fourier transform.

The equation for the discrete Fourier transform (from the previous section) can be broken into its real and imaginary parts:

$$\begin{aligned} H(f) &= \sum_{j=1}^{N} h_j(\cos(2\pi f t_j) + i\sin(2\pi f t_j)) \\ &= \sum_{j=1}^{N} h_j \cos(2\pi f t_j) + i \sum_{j=1}^{N} h_j \sin(2\pi f t_j) \\ &\equiv H_R(f) + iH_I(f) \end{aligned}$$

(Note that although the quantity $iH_I(f)$ is imaginary, the function $H_I(f)$ is real.) For paleoclimate work, these real and imaginary components of H can be evaluated from this equation on a small computer. Once the data h(k), the ages t(k), the desired frequency f, and the number of data points N have been read into the computer, the code in Basic calculates the real and imaginary components (Hr and Hi) of the

Fourier amplitude and then sums and squares them:

```
Hr = 0
Hi = 0
for j = 1 to N
        Hr = Hr + h(j) * cos(6.2831853 * f * t(j))
        Hi = Hi + h(j) * sin(6.2831853 * f * t(j))
next j
P = Hr^2 + Hr^2
```

As we mentioned earlier, you will almost never see code like this, since computer programmers consider it inefficient. But what is inefficient for the computer may be simple and clear for the scientist. The method will not be mentioned in texts on numerical methods. But that is silly. On a modern computer, the computation is so fast, that it is more important to write easily understood code than to write fast code. As an example, we applied the Glacial Fourier transform to calculate 250 frequencies for 500 data points, using the computer language Matlab™ on a laptop computer. The entire computation took 0.12 seconds.

This is a good time to look at the equation, and get a sense of what the Fourier transform does. It takes your data, multiplies it by a sine wave, and then sums the results. If the data oscillate in phase with the sine wave, so that they are positive together and negative together, then all the terms in the sum are positive and the Fourier amplitude is large. If they drift into phase and then out of phase, then half of the values in the product will be positive and half will be negative, and the sum will be close to zero. You can also see that the sum is not particularly sensitive to sharp changes in the data (e.g. sudden terminations); it is more sensitive to the bulk behaviour of the data, e.g. are most of the data points positive when the sine wave is positive?

You can calculate the phase from

$$\phi = \tan^{-1}\left(\frac{H_I}{H_R}\right)$$

Unfortunately, the inverse tangent function in many computer languages, called *atan*, returns a value limited to the range $-\pi/2 < \phi < \pi/2$, which only covers half of the angular range. This problem can be avoided in Fortran, Matlab, and in some Basic interpreters, by using the alternate function atan2:

$$\phi = \text{atan2}(H_I, H_R)$$

We like the program Matlab™ for our computations of the Fourier transform.

Matlab represents the list of values $h_1, h_2, h_3, \ldots, h_n$ by a single symbol h. The Matlab code for the Glacial Fourier transform becomes very simple:

$$\mathrm{Hr} = \mathrm{sum(h.} * \cos(2 * \mathrm{pi} * \mathrm{f} * \mathrm{t}));$$

$$\mathrm{Hi} = \mathrm{sum(h.} * \sin(2 * \mathrm{pi} * \mathrm{f} * \mathrm{t}));$$

If you think of the word "sum" in these equations as integral signs, then the code looks just like the equations defining the Fourier transform.

In paleoclimate studies, the only time you need to use the faster FFT is for Monte Carlo studies, when it is necessary to calculate thousands of spectra. Also, if you are using someone else's spectral analysis program, it probably has an FFT built in. To learn to use the FFT, of course, you must learn a few more things:

- how to interpolate your data to uniformly spaced time intervals;
- how to pad the data with zeros, so the number of data points will be an exact power of 2;
- how to interpret the mysterious order of the output vector, which will have every value appearing twice;
- how to calculate the frequencies that go along with the FFT, by using the Nyquist formula;
- how to interpolate between the FFT frequencies to get the value at the frequencies that you really want (e.g. at the known astronomical frequencies).

We give Matlab routines in Appendix 2 that do all these steps.

3.2.3 The uncertainty principle

The Fourier transform has an elegant theorem associated with it called the "uncertainty principle". We are going to say more about it than anyone in paleoclimate needs to know, just because we find its connection to quantum mechanics very interesting. If you don't, then skip this section. All we will need is the final equation.

The uncertainty theorem of Fourier analysis is directly related to the uncertainty principle of quantum mechanics. The uncertainty principle is most simply stated in the following way. Suppose the function $h(t)$ is significantly different from zero in only a limited region near $t = t_0$. Let the width of this region be Δt. Then the spectral power will also have a significant value only over a limited range of frequencies Δf. Moreover, there is a simple relationship between these widths:

$$\Delta t\,\Delta f \geq \frac{1}{4\pi}$$

This inequality can be proven mathematically if we define Δt to be the root-mean-square (RMS) width of $h(t)$, and Δf to the RMS width of $H(f)$. The equality is true only if $h(t)$ happens to have a particular shape known as a Gaussian. The relation is the basis of the famous Heisenberg Uncertainty Principle of Physics. If we multiply both sides of the equation by Planck's constant h, and then use the quantum-mechanics relationship that the energy $E = hf$, we get

$$\Delta t \, \Delta E \geq \frac{h}{4\pi}$$

We can derive a similar physics equation for momentum. If our quantum mechanical wave is a function of position x, the wavefunction can be expanded in terms of spatial frequencies f. We define the wavenumber k by $k = 2\pi f$. The uncertainty equation now reads

$$\Delta x \, \Delta k \geq \frac{1}{2}$$

Multiplying by $h/2\pi$, and substituting the quantum mechanics definition of momentum $p = hk/2\pi$, we get the equation that Heisenberg is most famous for:

$$\Delta x \, \Delta p \geq \frac{h}{4\pi}$$

But let us now return to the mathematical version of the equation. Suppose that we have a record that lasts for a time T. What is Δt? It is the root-mean-square deviation from the average time. What is that? We can make a quick guess from the fact that all the points are less than $T/2$ from the average time, so Δt must be less than $T/2$. If you calculate the RMS for a flat distribution, it is not hard to show that the correct value is $\Delta t = T/\sqrt{12} \approx 0.29T$. So according to the Uncertainty Principle,

$$\Delta t \, \Delta f = \frac{T \Delta f}{\sqrt{12}} \geq \frac{1}{4\pi}$$

$$\Delta f \geq \frac{\sqrt{3}}{2\pi} \frac{1}{T} \approx \frac{0.28}{T}$$

This, of course, is still an inequality. To get an equality, we can perform the calculation explicitly by performing a Fourier transform of a pulse of duration T. Moreover, when we do this we can conveniently pick Δf to be the full-width at half-maximum (*FWHM*) of the spectral power, rather than the RMS deviation from the average of the Fourier amplitude. From this calculation, we get the result:

$$\Delta f = \frac{0.89}{T}$$

(The numerical constant 0.89 is actually $1.3916 \times 2/\pi$, and 1.3916 is the solution to the transcendental problem $(\sin x/x)^2 = 1/2$.) This equation is often simplified even further to read:

$$\Delta f \approx \frac{1}{T}$$

and this is the way you will frequently see it stated in the literature, sometimes with the "$\approx$" symbol replaced with a "$=$" symbol. In this explanation, T was the duration of a pulse. The same equation results if instead T is the duration of data containing a sine wave. (To show this, we can represent the data as a product of an infinite sine wave and a pulse, and then use a result known as the "convolution theorem".)

Applied to paleoclimate data, this equation gives a relationship between the width of the peak in the spectral power, and the duration of the signal. So, for example, if you do an analysis of data that has a duration of 600 kyr, and the data contain a perfectly periodic wave that stays at constant frequency for that period, then the *FWHM* of the spectral peak will be $0.89/600 = 0.0015$ cycle/kyr. If the record is longer, then the width will be correspondingly narrower.

But you can do much worse. If windowing is applied to the data (see the sections to follow) then the resolution is degraded. For example (see Priestley, 1981, Sections 7.3–7.4), the Parzen window degrades the resolution by a factor of $6/\pi = 1.91$. And you can also fool yourself into thinking that you can do better, e.g. with Maximum Entropy Analysis (see Section 3.8).

3.2.4 Periodogram

The "periodogram" was invented by Arthur Schuster in 1898 for the purposes of studying climate. It was the first form of numerical spectral analysis and, although experts frequently prefer other methods, it is still the most widely used. As conceived by Schuster, the periodogram is the absolute value squared of the Fourier transform, applied to evenly-spaced data. Even though the data exist only at discrete times, the periodogram could be evaluated and plotted at any frequency—provided that you had the patience and time to do the computation. Thus the periodogram is the technical term for the method of estimating spectral power that we have been describing.

Many competitors to the periodogram have been invented subsequently. As we said in the Introduction, there is no one "correct" method for spectral estimation. One important advantage of the periodogram is that, unlike many other popular methods (e.g. maximum entropy analysis and multitaper analysis), its statistical properties are well understood. For example, if we represent our model for the data as a sum of sine waves plus error, then Whittle (1952) showed that the periodogram gives the best maximum likelihood estimate of those frequencies. If you are really interested in these formal aspects of spectral analysis, you will find them described in a readable review by MacDonald (1989).

It has become popular in statistical circles to object to the use of a simple periodogram. For example, M.B. Priestley (1981) states that

> *"the periodogram is an extremely poor (if not a useless) estimate of the spectral density function." (p. 420)*

In the otherwise insightful book, *Numerical Recipes*, W. Press et al. state (1993):

> *"... the variance of the periodogram estimate at a frequency f_k is always equal to the square of its expectation value at that frequency. In other words, the standard deviation is always 100 percent of the value, independent of N!"* (p. 422)

But these statements are highly misleading for the student of paleoclimate, since they are meant to apply to continuous spectra or resolved spectral peaks. (A resolved peak is one which is broader than the Δf resolution limit of the Uncertainty Principle.) If we have an unresolved spectral line, as is often the case in astronomy or astronomically-driven phenomena, then the periodogram is often an excellent method. John Tukey, one of the founders of modern spectral analysis, summarised this conclusion (Tukey, 1967):

> *"If we dealt with problems involving the superposition of a few simple periodic phenomena, as do astronomers interested in binary stars and related problems, we can learn much from the periodogram. Sadly, however, almost no one else has this kind of data. As a result, the periodogram has been one of the most misleading devices I know."*

Paleoclimate is precisely one of those "related problems" that is dominated by a few simple periodic phenomena. In this chapter we will show the application of many spectral methods to paleoclimate data, and it will often turn out that the periodogram, sometimes with a simple modification called a taper, is the most useful method to apply. Particularly for the study of the 100-kyr cycle, the signal to noise is huge (often 10 : 1 or greater). There is negligible leakage from adjacent peaks, and bias from a sloped background is often unimportant. For the study of such phenomena, there is little value to be obtained in reducing the resolution by the use of other methods, and there is much to be gained by the high resolution of the periodogram.

The periodogram has been maligned because of a curious feature: signal to noise of a continuous spectrum cannot be improved by taking a longer set of data. The reason for this is simple: in the periodogram, the bin size is proportional to $1/T$, and thus for long records, the frequency resolution improves proportionately—so that the power in one bin will not increase. If the signal is continuous or the spectral feature resolved, then the signal to noise level does not change. Other techniques have been developed for this purpose, including the "Blackman–Tukey method", the windowed periodogram, and the multi-taper method. These methods also have other advantages as well; they are devised to reduce systematic effects known as leakage and bias.

But you cannot gain something without losing something, and each of these methods gains its signal-to-noise advantage by throwing away frequency resolution. The loss of resolution can be extremely misleading if the key difference between models lies in the shape of the spectral peak. In the periodogram, the information that you gain about the spectrum does increase with the length of the spectrum since you measure the frequency with greater accuracy. But, historically, statisticians have been more concerned with power estimation than with frequency estimation, hence their dislike of the method. In paleoclimate, the situation is exactly the opposite. We are more concerned with the detection of narrow spectral peaks than in a precise estimate of their power levels. This makes the periodogram, in many cases, ideal.

For the unresolved spectral feature, the periodogram can be superior to many of the other standard techniques, because it puts the power into the fewest number of bins. It can be shown that, if the power at a frequency f is normalised with respect to the local noise (i.e. the local noise has value 1), then the standard deviation σ_P of the power P is given by:

$$\sigma_P = \sqrt{2P - 1}$$

This implies that the fractional uncertainty of the power is

$$\frac{\sigma_P}{P} = \frac{\sqrt{2P-1}}{P} \approx \sqrt{\frac{2}{P}}$$

Since the spectral power of an unresolved peak increases with time, the uncertainty in the estimate of the power decreases rapidly, as the square-root of time (in contradiction to the overgeneralised statement of Press et al. given above). For a pure sine wave, the power increases proportionally to time, whereas the noise (local background) remains constant, so the signal-to-background estimate improves as the square-root of time.

In paleoclimate work, the favoured approach has been the Blackman–Tukey method which we will describe in Section 3.3. In fact, the periodogram can be considered a special case of the Blackman–Tukey method, applied with the Blackman–Tukey lag parameter set to unity.

3.2.5 Sharp feature in the time domain

In paleoclimate data, there are often features in the time domain that are abrupt, such as the "sudden terminations" of the glacials. It is important to note what effect such a feature has on the spectrum. Mathematically we call a brief spike a "delta function"; an abrupt change in level (e.g. an abrupt termination) is a "step function".

The Fourier transform of a delta function is particularly easy to calculate, since it is zero except at one time:

$$H_T(f) \equiv \int_{-\infty}^{\infty} \delta(t - t_0) e^{2\pi i f t}\, dt = e^{2\pi i f t_0}$$

$$P_T(f) = |H_T|^2 = 1$$

This shows that the spectral power of a delta function is constant, i.e. it contributes equally at all frequencies. For a digital Fourier transform, the result is approximately the same. We recommend that the reader, particularly the reader new to digital transforms, try this digitally. Represent a delta-function by setting the data equal to zero except at one point. Then do a Fourier transform. The result is shown in Fig. 3.1 below for the age range 0–600 kyr, with one point at $t = 300$ kyr set equal to 1. The function is not quite constant, although most of the points have value near 1. The function goes to zero at $f = 0$ because we subtracted the average from the data points before performing the transform.

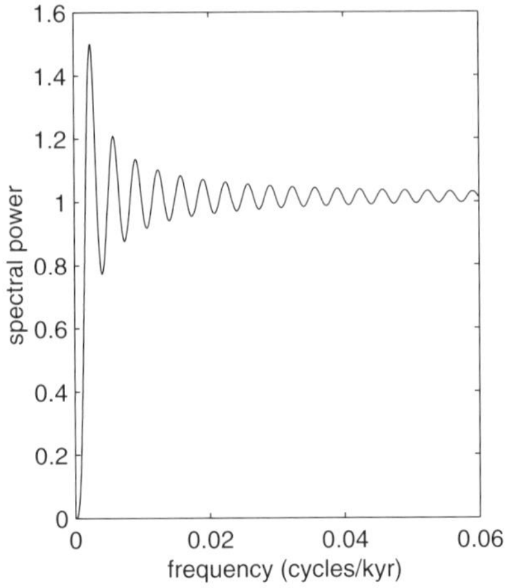

Fig. 3.1. Spectrum of delta function.

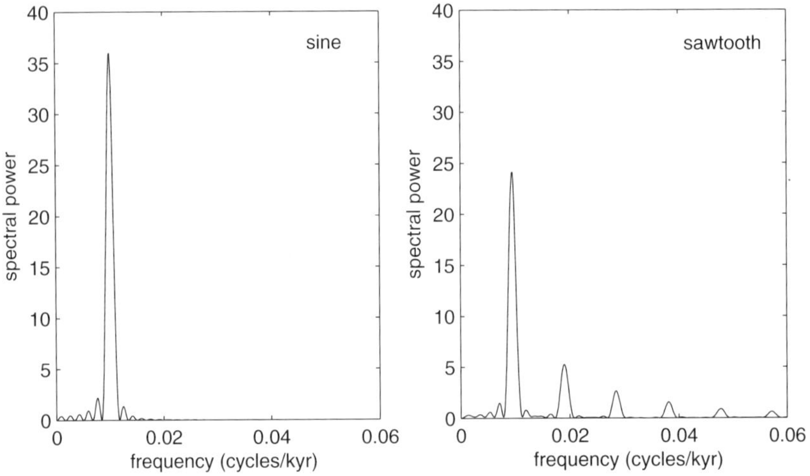

Fig. 3.2. Spectra of sine and sawtooth.

It is very interesting to see the effect that sudden terminations have on data. In Fig. 3.2 we show the spectral power of a pure sine wave and that of a 100 kyr sawtooth (i.e. triangular) wave:

You'll note that the primary difference between the two are the reduced power in the fundamental peak (at 0.01) and the presence of the harmonics at multiples of the fundamental frequency. The area under both curves is the same, by our normalisation convention (mean power = 1). The harmonic peaks are so small that they can be easily lost in, or cancelled by, background (if it were present, as it would be in real data). The net result is that the properties, or even the existence of the abrupt terminations, can be hidden or lost in the frequency domain. They are best studied in the time domain.

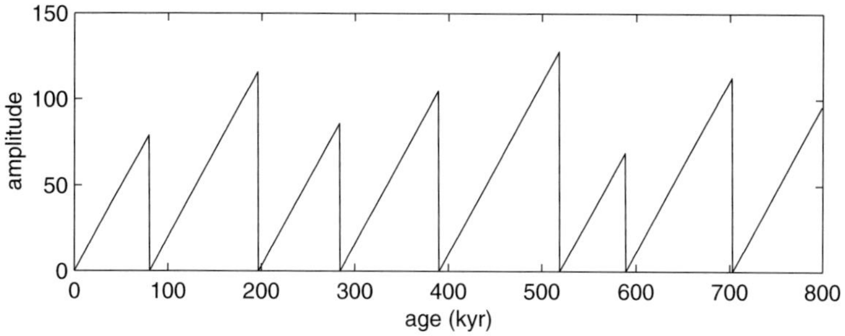

Fig. 3.3. Randomised sawtooth.

Another possibility, however, is that the glacial cycles are astronomically driven, but that the sudden terminations have a random component. The first day of snow rarely occurs on the first day of winter. The ice does not usually crack on the first day of spring. We have simulated this by generating a sawtooth in which the abrupt transition occurs every 100 kyr $\pm\sigma$ where we set σ equal to a random number between ± 20 kyr. The data generated in this way are shown in Fig. 3.3.

The spectrum for these data is shown below (Fig. 3.4). The 100 kyr peak is still strong, but the secondary peaks no longer show a simple regular behaviour. The harmonics, multiples of the fundamental 0.01 frequency (and have periods of 50 kyr and less) are gone—destroyed by the ± 20 kyr random terminations! The spectral power of these harmonics has been spread to nearby frequencies. This again illustrates that the analysis of the terminations is not readily performed in the frequency domain.

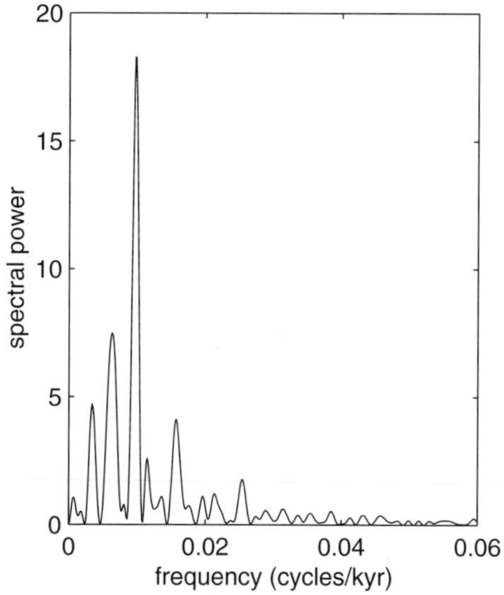

Fig. 3.4. Spectrum of randomised sawtooth.

3.2.6 Fast Fourier Transform (FFT)

The FFT is simply an extremely efficient way to calculate the Fourier transform for the case of equally-spaced data. It is such a popular technique that many analysts will extrapolate their data to evenly-spaced points just so they can use this method. An ordinary digital Fourier transform takes N data points, and calculates the real and imaginary amplitudes at $N/2$ frequencies. To do this requires N^2 transcendental calculations (i.e. computation of sines or cosines). Based on a suggestion of Richard Garwin, Cooley and Tukey developed a method of doing the identical calculation in a way that was proportional to $N \log_2 N$ rather than to N^2. For very large computations this provides an enormous saving in computer time.[1] For example, in astronomy, it is frequently necessary to compute the Fourier transforms of $N = 10^7$ data points. The ratio of computation times for the FFT vs. the Glacial Fourier Transform (GFT) is then $\log_2 N/\mathrm{N} = 2.3 \times 10^{-6}$. A computation that takes one second using the FFT would take five days using the GFT.

Many people who use the FFT find it confusing initially. The FFT requires data at equally-spaced time points, and is most efficient when the number of points is an exact power of two. The process of interpolation to evenly-spaced data introduces additional biases and systematic error. The format of the output can also be confusing. For real data consisting of N data points y_j, each taken at time t_j, the power spectrum output from the FFT appears as a set of $N + 1$ data points. The first and last data points are the same, and they represent the power at frequency zero. The second through to the $N/2 + 1$ data points represent the power at evenly-spaced frequencies up to the Nyquist frequency ($= 1/2\Delta t$, where Δt is the time between equally-spaced data points). The spectral power for a given frequency is spread out over several bins; an optimum determination of the power requires combining this information. In the standard FFT, the original data are overwritten with the Fourier transform, erasing it. This procedure dates back to the days when computer memory was a limiting expense. It is hardly ever worthwhile today.

Despite these complications (which are routinely handled by a computer program, once it is written and debugged, and forgotten by the user), the FFT is worth using whenever speed is important. Especially when doing a thousand or more Monte Carlo simulations (a procedure we highly recommend whenever you encounter a new kind of data), the speed of the FFT proves valuable.

3.2.7 Padding with zeros

The FFT calculates the amplitude for a particular set of pre-chosen frequencies. If there are N data points, then $N/2$ complex amplitudes are calculated at $N/2$

[1] Although the practical use of the FFT began with the Cooley/Tukey implementation, there is a long history of mathematicians recognising that faster computations could be done. Gauss observed that he could break a Fourier series into two parts, and reduce the computation significantly, but he never published his work. Danielson and Lanczos in 1942 derived the entire algorithm. Could a faster algorithm be invented some day? For standard computing (Turing machine), the answer is no.

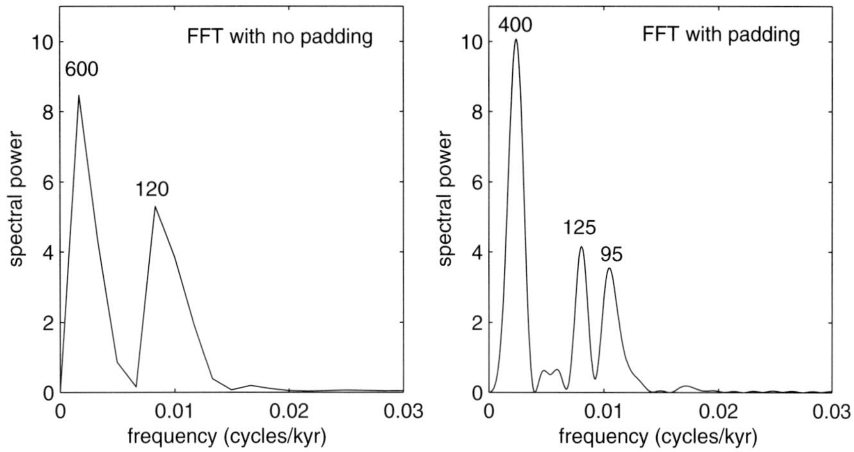

Fig. 3.5. Effect of padding on spectrum of eccentricity.

different frequencies. The lowest (non-zero) frequency is $1/T$, where T is the duration of the record. This frequency corresponds to one cycle over the entire data set. All other frequencies are multiples of this. Unfortunately, none of these may correspond to the true frequencies present in the record. For example, suppose we have 200 data points, covering a period of 0 to 500 kyr. Then the lowest frequency is $1/500 = 0.002$. If the data are dominated by a 41 kyr signal, then the frequency of that signal is $1/41 = 0.0244$. This is not an even multiple of 0.002, so the FFT will not calculate the amplitude at this frequency.

A simple and elegant way around this is to subtract the mean from the data, and then pad it with zeros, i.e. extend the data beyond T (to, say, $10 \times T$) by adding zeros to it. The FFT takes about 10 times longer, but computers are fast. The ideal way is to pad to an exact power of 2. With this padding, the minimum frequency is now 0.0002, and the FFT *will* calculate the amplitude for the frequency 0.0244. (Of course, 1/41 is not exactly 0.0244...)

We illustrate the value of padding by calculating the spectral power of eccentricity for the period 0 to 600 kyr, with and without zero padding. The results are plotted in Fig. 3.5. Note that without padding, the structure of the spectral peaks is invisible. That is because much of the information is hidden in the phase, which is not included in the plot of spectral power. With the padded data, intermediate frequencies are calculated, and the clear contributions of the peaks with periods of 400, 125 and 95 kyr are evident.

3.2.8 Aliasing

A problem arises if there are fluctuations in your data that are more rapid than your sampling. (This is the same as saying that there are frequencies in the data that are higher than the Nyquist frequency.) Suppose, for example, that you have samples of Greenland Ice that represent the temperature every 100 years. But your sampling is not precisely spaced by a year, so sometimes you measure winter ice, and other samples are summer ice. Even if there is no long-term variation in the temperature, you will see jumps up and down in your data. Unless you know the time scale exceptionally well, you may not be able to tell that the variations you see are actually seasonal. Yet, the Fourier transform will "explain" all these fluctuations as being due to cycles with periods greater than 100 years. It will do this because the transform is a mathematical identity; the data can *always* be expressed in terms of sines and cosines, even though the Fourier transform uses only a period longer than twice the sample period. The maths finds a fit, even though the real frequency is not included in the expansion series. The Fourier transform is lying to you!

As an example, suppose you measure data at 50 kyr intervals, i.e. at $t = (0, 50, 100, 150, 200, \ldots)$, and get the values $y = (1, -1, 1, -1, 1, \ldots)$. Let us assume that these data points resulted from an oscillating signal with a period of 20 kyr, i.e. they are explained by the function $y = \cos(2\pi ft)$ with $f = 1/20 = 0.05$. But if we didn't know that, we might think that they were due to frequency $f = 1/60$. We plot the data points along with both sine waves in Fig. 3.6, and you will notice that both frequencies agree with the data points. There is no way for the Fourier transform to know which of the frequencies is the correct one, since it has only the sampled data.

Which will the Fourier transform pick to represent the data? It will give a large amplitude at frequency $f = 1/60$, despite the fact that we created the signal from $f = 1/20$. This misidentification is given the name "aliasing". It occurs whenever the sampling is less frequent than the period of the true oscillation. For a further discussion of the mathematical aspects of aliasing, see Press et al. (1993).

Aliasing can be a problem in ice data, as the example illustrates. It is usually not a problem in sea floor data, since the periods of interest (from 19 kyr to 400 kyr) are usually much longer than the sample interval. Any phenomena that tend to smooth the data (such as bioturbation) tend to eliminate aliasing.

We mentioned earlier (Section 3.2.1) that some audiophiles complain that music CDs suffer from aliasing. The sample period for these CDs was chosen to be 1/44000

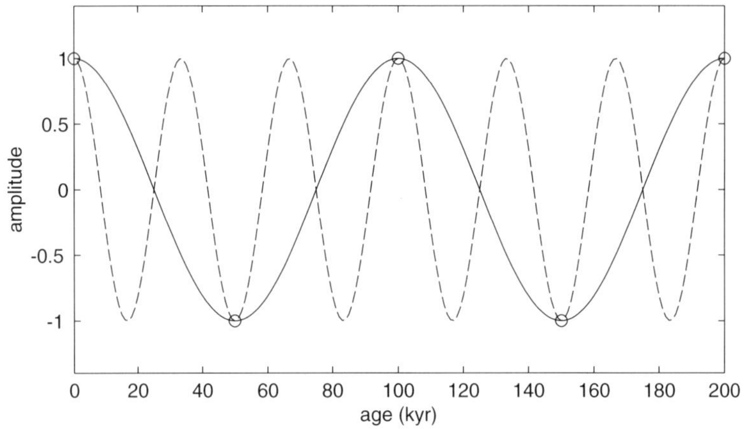

Fig. 3.6. Aliasing. The circles are the data points.

sec, so that the Nyquist frequency is 22000 cycles/sec, above the audible range. But musical instruments produce sounds at higher frequencies. When music from these instruments is recorded digitally, the sound is aliased to lower frequencies. When this happens, the CD contains sounds at audible frequencies which would not be present in the live performance. Recording engineers can eliminate such aliased sounds by putting in a filter that removes the higher frequencies prior to the music being digitised. If they are absent in the original, then they are not aliased. However, any filter applied to the signal has *some* effect at lower frequencies, particularly since the ear does not respond in a linear way to sound, and this too can introduce an objectionable distortion. Audiophiles will say that they prefer listening to music to which no filter has been applied. Of course, any recording medium (such as records) has a frequency response, and therefore acts as a filter. The question isn't whether or not we should apply a filter, but which filter has the most objectionable effect on the music.

3.3 BLACKMAN–TUKEY METHOD

The Blackman–Tukey method was the most commonly applied spectral analysis tool for paleoclimate studies in the last three decades of the twentieth century and, as such, it deserves special attention. It was introduced in a classic book (Blackman and Tukey, 1958) that introduced many scientists to fundamental techniques of spectral analysis. It was published before the rediscovery of the Fast Fourier Transform by Tukey. Blackman and Tukey were primarily concerned with the difficult case of analysing a continuous spectrum that was biased by the presence of sidelobes from strong peaks. Part of the attraction of the book is that it offered a prescription for scientists to follow. The book also offered an abundance of warnings—that the algorithm and suggested parameters should not be followed slavishly, but that it should be used only as a starting point. As an added bonus, the method made efficient use of tricks to reduce the required calculation of sines and cosines. This was important at that time since the book predated the FFT by seven years.

The Blackman–Tukey method is based on a fundamental theorem of Fourier analysis that states that the Fourier transform of a correlation is equal to the product of the Fourier transforms. The correlation of two functions A and B is defined as

$$C(\tau) \equiv g \otimes h \equiv \int_{-\infty}^{\infty} g(t+\tau)h^*(t)dt$$

The star * indicates complex conjugation. The correlation theorem states:

$$FT(g \otimes h) = FT(g)FT^*(h)$$

A special case of this, when $g = h$, is known as the Wiener–Khintchine theorem:

$$FT(g \otimes g) = |FT(g)|^2 = P$$

where P is the spectral power. In words, the Fourier transform of the autocorrelation function is identical to the spectral power. In the Blackman–Tukey method, an estimate of the spectral power is obtained by taking the Fourier transform of only part of the autocorrelation function.

The algorithm involves several steps. The first, optional, step is called prewhitening; it is designed to reduce the bias in frequency determination that happens when there is a strongly varying background under a weak peak. The next step is also optional: the data are tapered by application of a window; for further discussion of this, see Section 3.4. But the heart of the method is the calculation of the partial autocorrelation function, defined by

$$A_{1/N}(\tau) = \int_0^{T/N} f(t+\tau)f(t)dt$$

Since we are using data with real values, we have assumed that $f = f^*$. The key difference between this and the full autocorrelation function is that the integral is evaluated only to T/N, where T is the length of the data record. If $N = 1$, then this is the full autocorrelation function, and the method gives the same answer as the ordinary periodogram. A more typical value is $N = 3$; when this is chosen (and it is the value that Blackman and Tukey recommended as a starting value), then we say that "a lag of 1/3" was used. Alternatively, we can specify the number of data points that appear in the overlap integral. Thus, for example, if there are 99 data points, and $N = 3$ was used, then we might say that there were $99/N = 33$ lags used.

Note that this "partial" autocorrelation function is symmetric, i.e. $A(-\tau) = A(\tau)$. If we take the Fourier transform of this we get P_{BT}, the Blackman–Tukey estimate of the power for 1/3 lag:

$$P_{BT}(f) = \int_{-\infty}^{+\infty} e^{2\pi ift} A_{BT}(\tau)d\tau \equiv FT(A_{BT})$$

Since the autocorrelation function is symmetric, only the real part is non-zero. Therefore, in the pre-FFT days, we could save half of the computation time by computing only the symmetric part:

$$P_{BT}(f) = 2\int_0^{+\infty} \cos(2\pi f)A_{BT}(\tau)$$

If the lag were one, then this procedure gives the same estimate of the power as we get by squaring the Fourier transform:

$$P(f) = \int_{-\infty}^{+\infty} e^{2\pi if\omega\tau} A(\tau) \equiv FT(A)$$

However, if only part of the full autocorrelation function is used, the result is different. To see what the effect is, consider the pedestal function

$$B(\tau) = 1 \quad \text{for} \quad -T/3 < \tau < T/3$$

$$B(\tau) = 0 \quad \text{otherwise}$$

Then we can write

$$A_{1/3}(\tau) = A(\tau)B(\tau)$$

and the Blackman–Tukey estimate of the power is

$$\begin{aligned} P_{BT} &= FT(AB) = FT(A) \otimes FT(B) \\ &\equiv P \otimes b \end{aligned}$$

Where $b \equiv FT(B)$. This shows that the effect of the Blackman–Tukey method is to blur (or smooth) the spectral power function. The amount of smoothing is the same that you get by convolving the spectral power with the blurring function b. For the simple pedestal function $B(\tau)$ given above, b is

$$\begin{aligned} b(f) &= \int_{-\infty}^{+\infty} e^{2\pi ift} B(t)dt = \int_{-T/3}^{T/3} e^{2\pi ift} dt \\ &= \int_{-T/3}^{T/3} [\cos(2\pi ft) + i\sin(2\pi ft)]dt \\ &= \frac{\sin\left(\dfrac{2\pi fT}{3}\right)}{\pi f} \end{aligned}$$

The approximate width of this blurring function is given by the condition $\Delta f = 3/(2T)$. Recall that for the full Fourier transform, the resolution is approximately $1/T$. In using 1/3 lags, we introduced another blurring, and we have lost a factor of about 3/2 in resolution.

The kind of blurring that we get with the Blackman–Tukey method is fundamentally different than the blurring that we get when we use a taper (Section 3.4). With a taper, it is the Fourier transform that is blurred; with Blackman–Tukey, it is the spectral power. This difference is important. A spectral amplitude that is rapidly varying will be averaged to zero if the blurring is done with a taper. However, with the Blackman–Tukey method, a rapidly varying amplitude does not necessarily average to zero, since the process of squaring can make the function positive over the region of blurring.

If resolution is lost, why use the Blackman–Tukey method? One reason is because blurring is sometimes desired. Drift in the time scale behaves like phase modulation, and can introduce spurious frequencies. We will discuss this at length in Section 5.1. Blackman–Tukey blurring can average these sidelobes into the main peak, and thereby give a better estimate of the true power. Note that windowing of the data with a taper function will not have the same effect, since the sidelobes will not have the same phase, and if averaged in amplitude, they can reduce the strength of the spectral peak. We'll demonstrate this property with examples.

But this advantage only works if the variations in sedimentation that are being compensated are small enough that the width is only broadened a little. Suppose the sedimentation rate were different in the two periods 0–300 kyr, and 300–600 kyr.

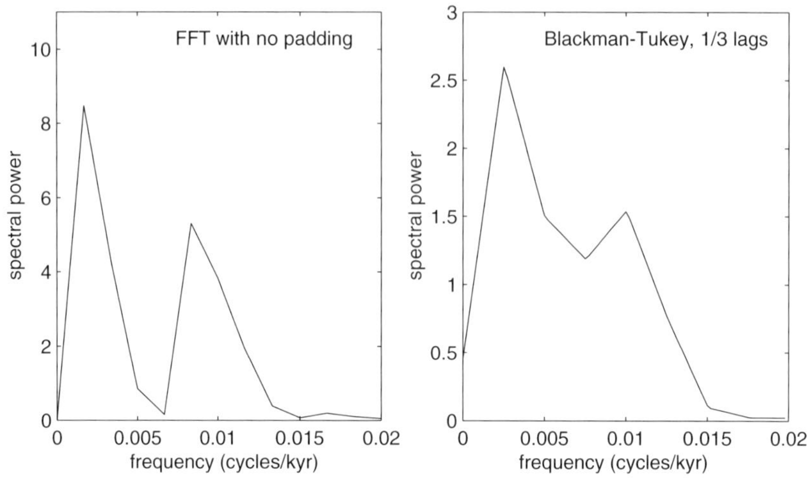

Fig. 3.7. Eccentricity spectrum, using FFT and 1/3 lags Blackman–Tukey.

Then the apparent frequencies of the 100 yr cycle would be different in those two records. The Blackman–Tukey method would average the two peaks together. Note that using large lags (e.g. 1/3) makes the least sense when tuning is used, since tuning (if done correctly) removes the low frequency drifts.

3.3.1 Example: eccentricity

As a vivid example of the loss of resolution that can result by the unthinking application of the Blackman–Tukey method with 1/3 lags, we show the spectrum of eccentricity for the period 0–600 kyr, calculated in three different ways. In Fig. 3.7 we show the spectrum from an unpadded FFT, and from the Blackman–Tukey method. In the FFT plot, the separation between the 400 kyr peak ($f = 0.0025$) and the 100 kyr peak ($f = 0.01$) is clear. In the Blackman–Tukey plot, the peaks are not even resolved (by the Rayleigh criterion, the dip in the centre should drop to half the maximum of the nearest peak). Of course, it makes no sense to use 1/3 lags for eccentricity, since the time scale is perfectly known. In Fig. 3.8 we show the spectrum calculated using the FFT with zero padding. With this method, the full triplet of eccentricity peaks is easily resolved.

3.3.2 Example: Site V12-122

We'll illustrate the value of the Blackman–Tukey method by showing several spectra from an historically interesting data set published by Broecker and Donk, for Site V12-122, for the relatively short time period of 0 to 370 kyr. First, we show the data, in Fig. 3.9.

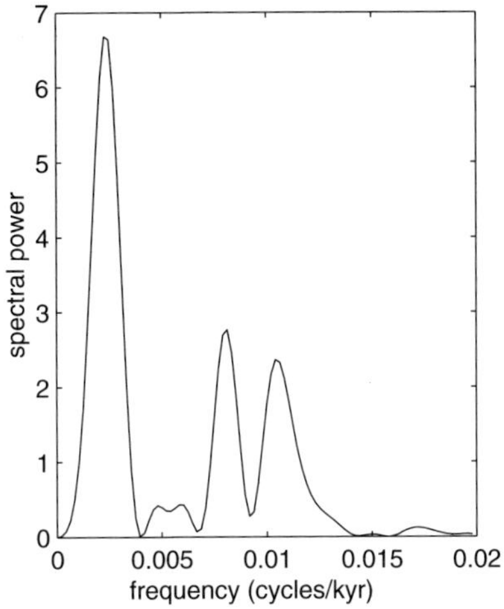

Fig. 3.8. Eccentricity spectrum using FFT with zero-padding.

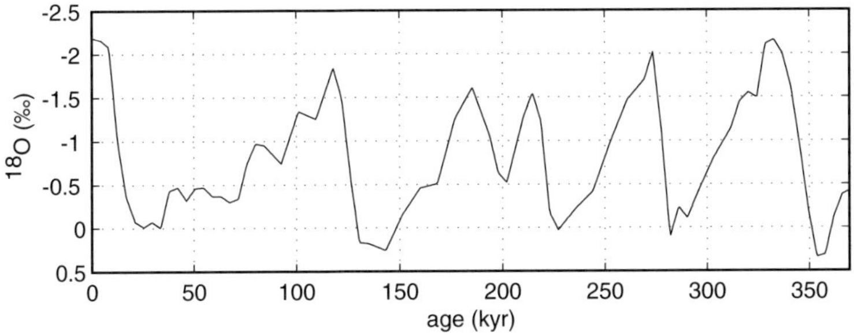

Fig. 3.9. $\delta^{18}O$ for Site V12-122.

If we plot the unwindowed periodogram for these data, we get the spectrum shown in Fig. 3.10.

The Blackman–Tukey spectrum, with lag of 2/3, is shown in Fig. 3.11. The spectrum appears smoother and more readily interpretable. There appear to be two major peaks, one near frequency 0.01, and one near 0.03 cycles/kyr.

In Fig. 3.12 we show the result of using a lag of 1/2. The pattern is becoming evident. The shorter the lag used, the greater is the blurring of the spectrum. In Fig. 3.13, we show the results of using a lag of 1/3. And finally, in Fig. 3.14, we show the Blackman–Tukey spectral estimate for a lag of 1/4.

From looking at these, we would probably pick the lag of 1/2 as the most useful for determining the frequencies and relative powers of the peaks. What would you pick? (Remember, there is no "right" answer.)

As we mentioned earlier, there is another reason for using the Blackman–Tukey method. The Blackman–Tukey book was published in 1958, several years prior to the introduction of the FFT. Prior to the FFT, the calculation of Fourier transforms

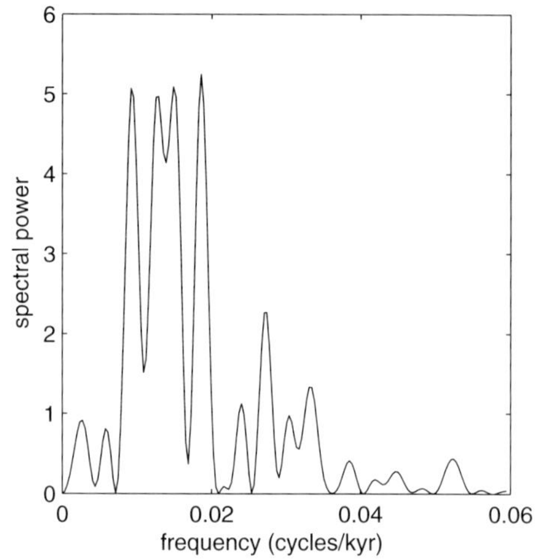

Fig. 3.10. Spectrum of V12-122, no window.

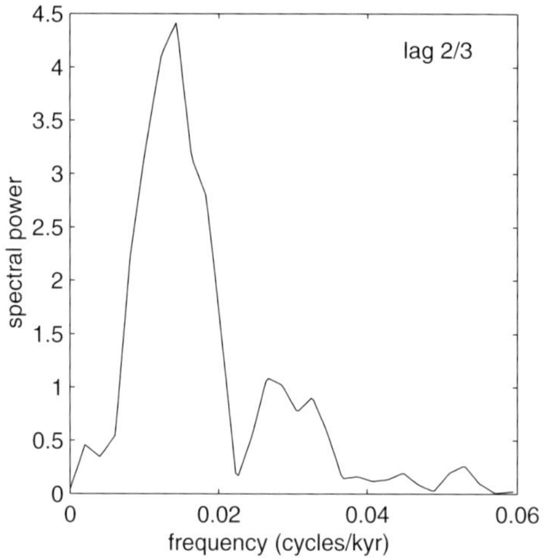

Fig. 3.11. Blackman–Tukey Spectrum V12-122, lag 2/3.

was a slow process that used a lot of computer time. A Fourier transform of N points would require the calculation of N^2 sines and cosines. For the computers of the day, even relatively small values for N (such as 1000) were a considerable expense. The Blackman–Tukey method saved substantial time by reducing the size of the data set by a factor of the lag. If the data had chatter, then there was no loss; in fact, if the sidelobes were brought back into the main peak, the estimate of spectral power was improved.

But even more time was saved by the fact that the autocorrelation function is symmetric. This meant that in taking its Fourier transform, only the cosine components need be calculated. This saved an additional factor of two. The corre-

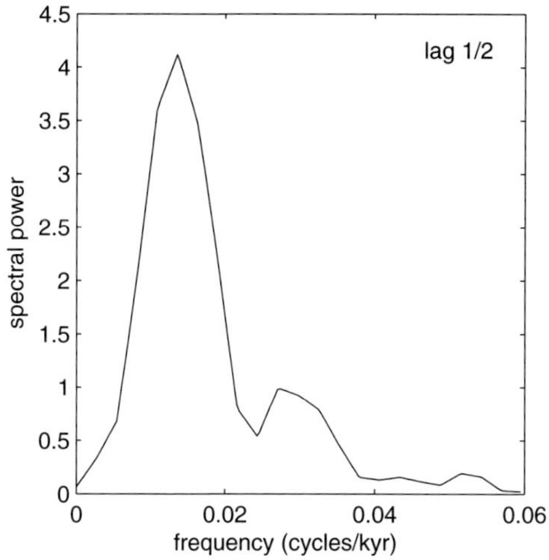

Fig. 3.12. Blackman–Tukey of V12-122, lag 1/2.

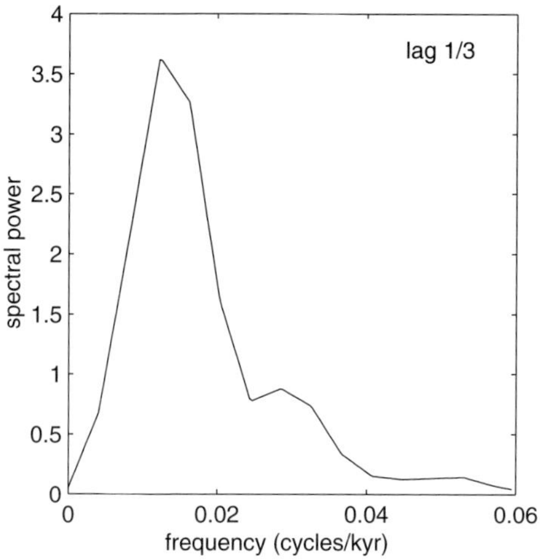

Fig. 3.13. Blackman–Tukey of V12-122, lag 1/3.

sponding disadvantage is that the method does not give any of the phases of the Fourier components. But if all you need is the power estimate, then this could be considered a worthwhile saving of time, at least in the pre-FFT days of slow computers.

A second disadvantage of the Blackman–Tukey method is that you have to estimate the proper lag to do before you do the calculation. If you then want to apply a different lag, you have to redo the entire calculation. In their book, Blackman and Tukey recommended starting with the value 1/3 for the lag, but

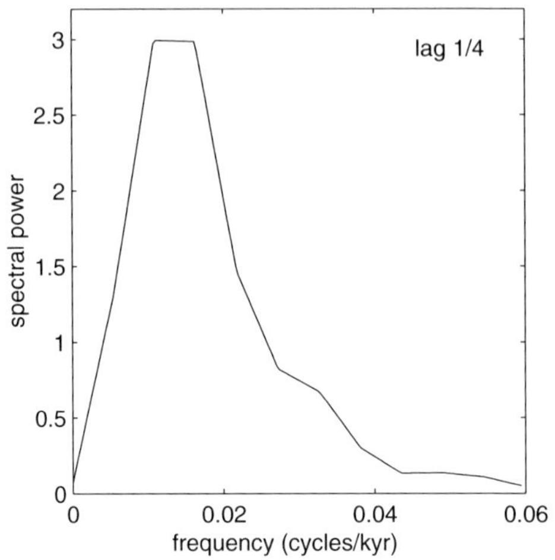

Fig. 3.14. Blackman–Tukey of V12-122 with lag 1/4.

then varying it until it gave the best performance. However, the paleoclimate community has rarely done this in practice. In the period 1976 to 1995, the vast bulk of spectral analysis carried out by the paleoclimate community used the Blackman–Tukey method with 1/3 lags. Few papers ever mention other lags. (Actually, we are not aware of *any* paper that used the Blackman–Tukey method with lag other than 1/3, although we haven't studied every paper written during that period.) Lags of 1/3 became the *de facto* standard, even in software programs distributed within the community. The result was a loss of resolution, and the missed discovery that the 100 kyr peak was single and narrow, in disagreement with the predicted structure from the Milankovitch models.

3.4 TAPERS

Tapering is a very simple, very clever mathematical technique that is extremely popular in spectral analysis. Instead of doing spectral analysis on the data $f(t)$, the analysis is carried out on the data multiplied by a taper function $g(t)$, also called the window function. The function g is usually taken to be a simple, slowly varying function, often going to zero (or close to it) near the edges. Several important tapers, for the interval $t = 0$ to T, are:

Sine taper	$g(t) = \sin(\pi t/T)$
Hanning (offset cosine) taper	$g(t) = (1/2)(1 - \cos(2\pi t/T))$
Hamming taper	$g(t) = .54 - .46\cos(2\pi t/T)$
Parzen or Bartlett (triangle) window	$g(t) = 1 - (t - T/2)/(T/2)$
Welch (parabolic) window	$g(t) = 1 - (t - T/2)^2/(T/2)^2$
Daniell (untapered or rectangular) window	$g(t) = 1$

The last of these is included only because windowing is so common that even an untapered (unwindowed) spectrum is sometimes referred to as if it were.

The effect of tapering is sometimes described as reducing the artificial presence of the high frequencies generated by fact that the data end suddenly at $t = 0$ and $t = T$. In fact, if discontinuities are present in the original signal, then the Fourier transform introduces distortions at those locations, a property known as the Gibbs phenomenon. Of course, tapering also introduces distortions. You just have to decide which kinds of distortions are more acceptable.

The ability of tapering to suppress high frequencies generated by sharp changes finds a very practical use in optics (especially radar) where the radiation pattern is the Fourier transform of the aperture function. A tapered aperture had significantly reduced sidelobes. This effect is shown in the following example. Let

$$h = \sin(2\pi t/95) + \sin(2\pi t/125)$$

represent data with two oscillations, one with period 95 kyr and the other with period 125 kyr, analogous to eccentricity. We will apply a Hanning taper to the data. (In Appendix 2, we give explicit code for calculating the spectral power with and without the taper.) The results are shown in the two plots of Fig. 3.15. The solid line is the spectrum of the untapered data; the dotted line is that of the tapered data. Both plots show the spectral power; the left one is a linear plot, and the right one is a semilog plot.

It is clear from the linear plot on the left, that application of the taper has severely degraded the resolution. It is equally clear from the right plot that the sidelobes of the tapered spectrum are much reduced. If your primary interest is in the resolution of the two peaks, then the untapered periodogram is superior. If your primary interest is to reduce the bias that sidelobes from these peaks could apply to other nearby peaks, then the tapered spectrum is superior.

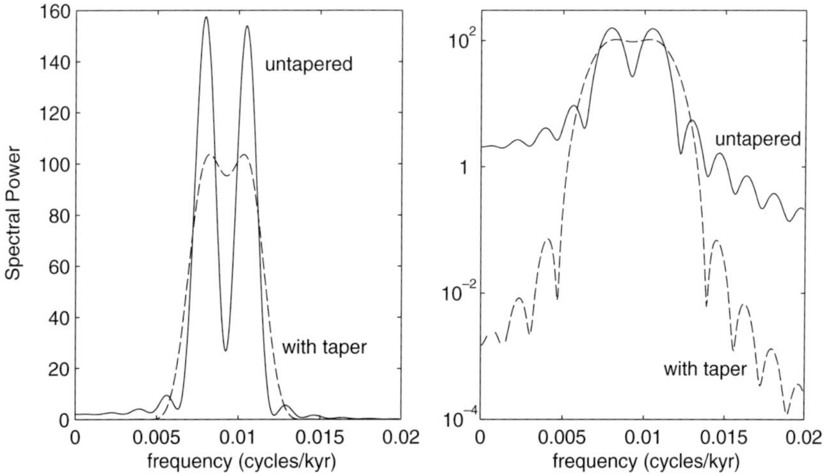

Fig. 3.15. Effect of tapering (dotted line) on a doublet peak.

The effect of the windowing can most easily be understood when the spectral analysis is done by a Fourier transform. The Fourier transform of a product is equal to the convolution of the Fourier transforms of the individual functions:

$$FT(fg) = FT(f) \circ FT(g)$$

Because the taper functions are broad and slowly varying, their Fourier transforms $FT(g)$ are narrow. The effect of tapering the data then is to convolve the Fourier transform with the narrow Fourier transform of the taper function. This amounts to smoothing. The difference between this and the Blackman–Tukey smoothing is that the smoothing is done in the complex Fourier space before the spectrum is squared.

The effect of tapering on the sidelobes can also be understood in terms of this convolution. An untapered sine wave of length T is equivalent to an infinite sine wave multiplied by a boxcar function. The Fourier transform of this is the transform of an infinite sine wave, i.e. a δ-function, convolved with the Fourier transform of the boxcar. For a tapered sine wave, the Fourier transform is the δ-function convolved with the Fourier transform of the taper function. If the taper function has lower sidelobes than the boxcar, then the spectral power of the tapered sine wave will also have lower sidelobes.

It is also possible to invent tapers that increase the resolution of the spectrum, by adding additional power at the ends, and decreasing the amplitude in the centre of the data. For example, we could multiply our data by the taper

$$g(t) = (1 + \cos(2\pi t/T))/2$$

The spectrum that we get with this taper is shown Fig. 3.16. It has better resolution than the untapered plot, and a more dramatic separation of the two peaks, but the sidelobes are getting uncomfortably large, and the fraction of the power in the main peaks is decreasing, making the spectrum more susceptible to background.

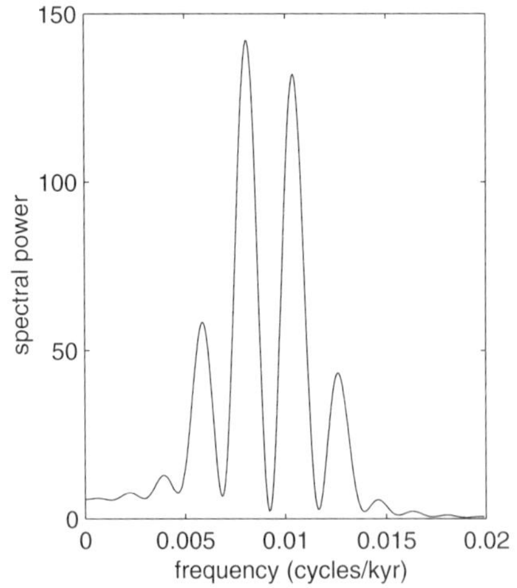

Fig. 3.16. Spectrum of the doublet using a high-resolution taper.

Some authors are adamantly opposed to tapering. Yuen has argued that tapering is "statistically unsound" (Yuen 1979, quoted in Percival and Walden, 1993, p. 215). Fougere calls tapering "tampering" (quoted in Percival and Walden, 1993, p. 215).

3.5 LOMB–SCARGLE PERIODOGRAM

Lomb (1976) and Scargle (1982) improved on the simple periodogram by a slight alteration. What they showed is that if the cosine and sine coefficients are normalised separately (we'll show explicitly what we mean by this in a moment) then the classic periodogram can be used with unevenly spaced data, and yet the statistical behaviour of the power is identical to the behaviour you would expect if you had evenly-spaced points.

To calculate the Lomb–Scargle periodogram of a data set (t_k, y_k), first calculate the usual mean and the variance:

$$\bar{y} = \frac{1}{N}\sum_{k=1}^{N} y_k$$

$$\sigma^2 = \frac{1}{N-1}\sum_{k=1}^{N}[y_k - \bar{y}]^2$$

For every frequency f, define the time constant τ by

$$\tan(4\pi\tau) = \frac{\sum \sin(4\pi f t_k)}{\sum \cos(4\pi f t_k)}$$

Then the Lomb–Scargle periodogram estimate of the spectral power $P(f)$ at frequency f is given by

$$P(f) = \frac{1}{2\sigma^2}\left\{\frac{\left[\sum_k (y_k - \bar{y})\cos 2\pi f(t_k - \tau)\right]^2}{\sum_k \cos^2 2\pi f(t_k - \tau)} + \frac{\left[\sum_k (y_k - \bar{y})\sin 2\pi f(t_k - \tau)\right]^2}{\sum_k \sin^2 2\pi f(t_k - \tau)}\right\}$$

This equation is less imposing than it looks. It has two terms, one for the cosine transform, the other for the sine transform. Each term is normalised separately. The only complication is that each frequency uses a different time offset τ. Other than these changes, the equation looks just like an ordinary digital Fourier transform.

The Lomb–Scargle method has several advantages over the classic periodogram. One, obviously, is that much paleoclimate data are not evenly spaced. Although this can be handled by interpolation, the statistical effects of such interpolation can be complicated. Secondly, there is a limit to the ordinary periodogram that comes about from a process called aliasing. What this means is that two signals of different frequencies can have identical sets of values if the samples are taken at exactly even spacing. With unevenly-spaced data, this effect can be substantially reduced. The net result is that the Lomb–Scargle periodogram can measure frequencies that would be aliased in evenly-spaced data.

One note of caution: although superior to many others, it takes a degree of courage (recklessness?) to use the Lomb–Scargle method in the paleoclimate community. We have had referees, unfamiliar with the method, urging rejection of our papers for using a method so "obscure"! (Obscure, in this context, presumably means that the referee had not heard of it.)

3.6 MULTITAPER ANALYSIS

We discussed in Section 3.4 how applying a taper (or window) to the data can reduce the sidelobes of spectral lines, and thus reduce the "leakage", i.e. the effect that strong lines can have in putting excess power at adjacent frequencies. D. Thomson recognised that the following problem had not been solved in a general way: how can you achieve the best sensitivity for finding a small signal of known frequency in the presence of another signal which had a different frequency and much greater power? The solution turned out to be a procedure in which several different tapers were applied to the data, and the resulting powers then averaged. This method is called "multitaper analysis".

The method is widely misunderstood. It has been incorrectly referred to as having "high resolution", although its basic idea is to trade resolution for sidelobe reduction. To apply the method, you must choose the number of tapers that you will apply. If you use only a few tapers, the resolution will not be degraded as much, but you will not get as great a sidelobe reduction. For the method to be optimum, the tapers should be Slepian functions. These are not easily computed, and numerical integration is usually employed. There are built-in Slepian functions available in the optional Matlab Spectral toolbox, and they are quite fast. If you don't have this, you should consider an alternative suggested by Walden et al. (1995): you can do almost as well using sine wave tapers instead of Slepian functions.

The advantages and disadvantages of the multitaper method are illustrated in Fig. 3.17. This plot shows two analyses of the signal $y = \sin(2\pi f_1 t) + \sin(2\pi f_2 t)$ where $f_1 = 1/95$ and $f_2 = 1/125$ cycles/kyr. (These, of course, are two important frequencies present in the eccentricity variation.) Shown is a periodogram of the untapered function, and multitaper analysis using four tapers. The periodogram is superior at resolving the two close peaks, but the multitaper spectrum is superior at suppressing sidelobes outside the main peak.

3.7 MAXIMUM LIKELIHOOD ANALYSIS

The maximum likelihood method is a generic approach to problems in statistical estimation. If you have a model with adjustable parameters (e.g. the amplitudes of eccentricity, obliquity and background), then you can calculate as a function of those parameters the relative probability that your model would yield data that match the real data. The parameters are then adjusted to maximise this relative probability. The relative probability that the model and a set of parameters account for the data

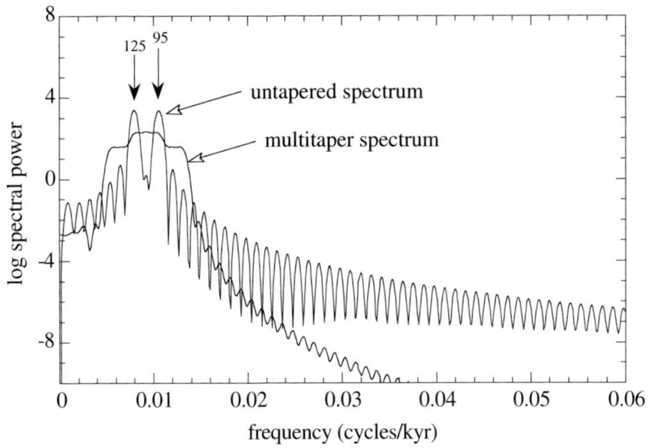

Fig. 3.17. Multitaper method compared to periodogram.

is called the likelihood. Many methods apply this approach, and differ only in the details, e.g. what specific parameters they adjust.

In this section we will discuss an approach to spectral analysis developed by MacDonald (1989). If the record is a noise process with a structured signal, this method provides the best means of signal detection in the maximum likelihood sense. We will describe the method, and then apply it to one of the historically most interesting data sets, the stack used in the 1976 article, "Pacemaker of the Ice Ages" (Hays et al., 1976). We will show that the data clearly verify the discovery of the important 41 and 23 kyr peaks, but that the claim in that paper that they saw a 19 kyr peak is questionable.

The method consists of the following steps:

1. First select an acceptable false alarm probability. The selection is a matter of judgement. If the probability is set too low, physically significant peaks in the spectrum may be missed; if it is set too high, peaks that are due to noise will be mislabelled as true peaks. A typical value might be 0.05. Then the chance of identifying a noise peak as a true peak is 5%.
2. Calculate the normalised periodogram. (Note: if the record is a noise process without any signal, this method is inappropriate.) If the data are not equally spaced, use the Lomb periodogram instead of the FFT.
3. Identify the maximum peak and test it against the false alarm probability.
4. If the maximum peak meets the false alarm test, determine by least squares the amplitude and phase of the sinusoid representing the peak.
5. Using the amplitude and phase information from the least-squares fit, subtract the sinusoidal curve from the data. (Because the record length is left unchanged, this method is superior to band-filtering out the peak.) Removal of the peak also removes the disturbing sidelobes as the analysis continues, and is the appropriate form of prewhitening. After peak removal, the variance in the

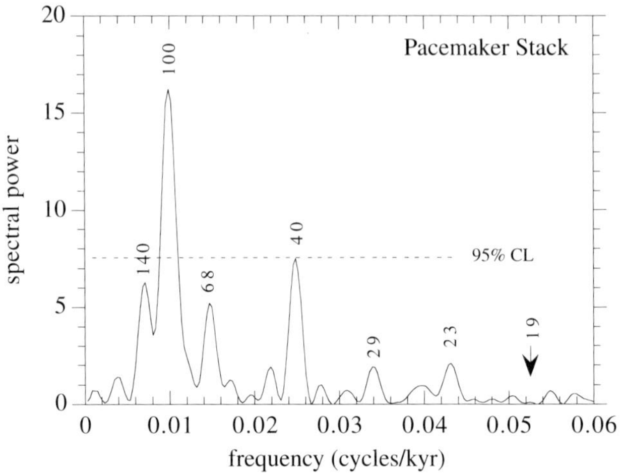

Fig. 3.18. Spectrum of Pacemaker stack.

total record is reduced. The change in variance provides guidance as to the rate of convergence of the peak removal process.

6. Recalculate the periodogram, and repeat the process of sine wave removal, until no further peaks meet the false alarm criterion. The remaining series is an estimate of the continuum part of the spectrum of the original series. The continuum residual can then be analysed to determine the nature of the noise or of the deterministic continuum.

We now apply this procedure to the "Pacemaker stack" of Hays et al. (1976). The first step is to do a spectral analysis. We did this using the Lomb–Scargle method on the $\delta^{18}O$ data of the stack. The resulting spectrum is shown in Fig. 3.18.

The maximum peak in the spectrum is at $f = 0.01$ cycles/kyr. Next, we do a least-squares fit of the Pacemaker data to a sine wave with this frequency. The results are shown in Fig. 3.19, plotted on top of the original $\delta^{18}O$ data. Next subtract this sine wave from the data. The data, with the 100 kyr cycle removed, is shown in Fig. 3.20.

A new spectrum is calculated, and it is found that the strongest remaining peak is at 41 kyr period. We obtain the best fit sine wave of this period, and subtract it from the data. The process then continues. The four strongest periods removed in this procedure turn out to be 102, 41, 140 and 69 cycles per kyr. When these have all been removed, the spectrum appears as shown in Fig. 3.21.

The next peaks that are removed are at 23 and then at 46. When these have been subtracted, the resulting plot looks as follows (Fig. 3.22).

The strongest remaining peak is at 30 kyr, followed by a pair at 20 and 19. How many of these peaks were statistically significant? That is trickier to answer than you might think. One conclusion is clear: the 19 kyr peak that was reported as significant in the original Pacemaker paper, has extremely low statistical significance. There are eight other peaks of greater statistical significance.

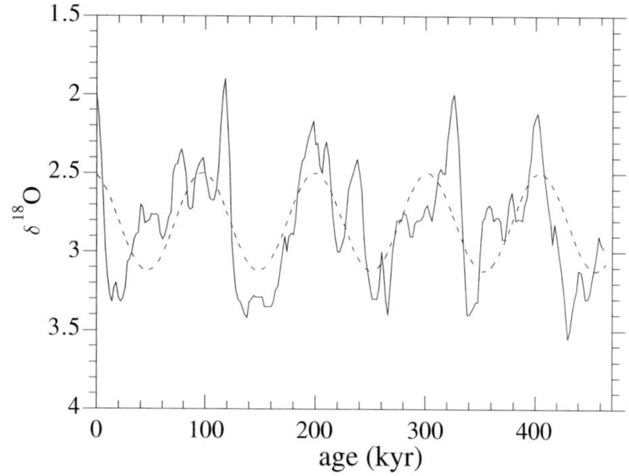

Fig. 3.19. Pacemaker stack, fit to 100 kyr cycle.

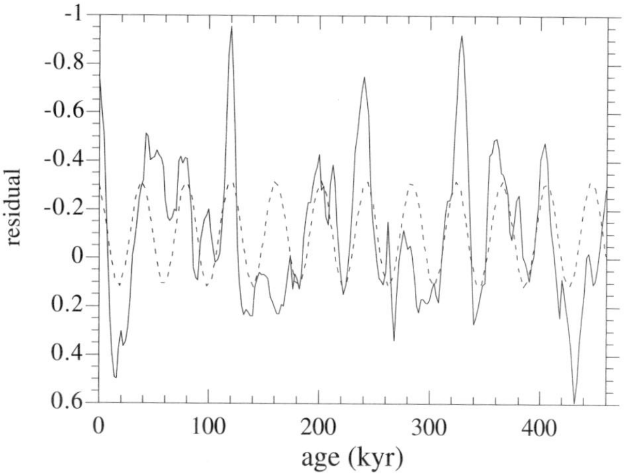

Fig. 3.20. Pacemaker residuals, 100 kyr removed, fit to 41.

The evidence presented in favour of the 19 kyr peak, in the original Pacemaker paper, was a Blackman–Tukey analysis using 1/3 lags. A plot using this method is shown in Fig. 3.23, on a log plot. (The 19 is too small to see on a linear plot.)

The small peak at 19 is what led to the reported presence of this peak. But was it real, or just a bump on top of a continuous red noise spectrum? Let us analyse the data yet another way. In Fig. 3.24 we plot the Fourier spectrum of the Pacemaker stack using a Hanning taper. A line marks the expected location of the 19 kyr peak. The location of the 19 kyr peak occurs less than half way up the side of a peak near the 20 kyr period. We conclude that the 19 kyr peak is not present in this analysis. It was not present in the Maximum Likelihood analysis. It was doubtful in the original

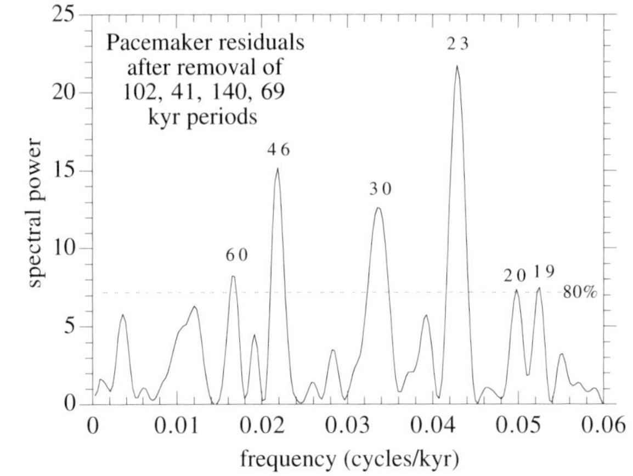

Fig. 3.21. Pacemaker stack spectrum after removal of 4 cycles.

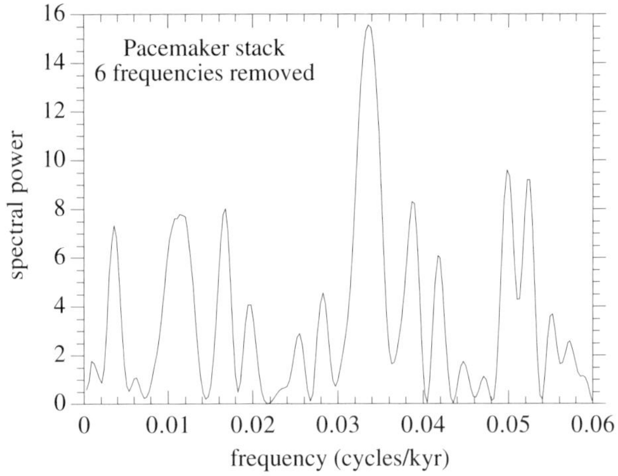

Fig. 3.22. Pacemaker stack spectrum, 6 frequencies removed.

Blackman–Tukey analysis. We conclude that its presence is not statistically significant.

But the 19 kyr peak appears to show up in other data, e.g. the Vostok atmospheric $\delta^{18}O$ (Fig. 4.19) and the terrigenous component from Site 721 (Fig. 4.23). Is it worthwhile to reanalyse old data? Yes, it is necessary, if we are truly to look at all the data as if for the first time. To untangle the mechanisms of the ice ages, it is very important to know where signals appear, and where they don't. Many people who study paleoclimate believe that the 19 kyr signal showed strongly in the Pacemaker stack—after all, that is where it was "discovered" (after having been predicted theoretically). To solve a jigsaw puzzle, it is just as important to remove pieces that were improperly placed as it is to put new pieces in their correct locations.

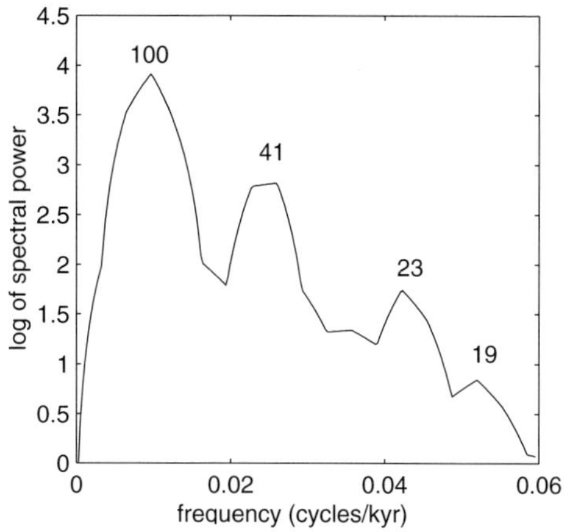

Fig. 3.23. Pacemaker spectrum, log of Blackman–Tukey with 1/3 lags.

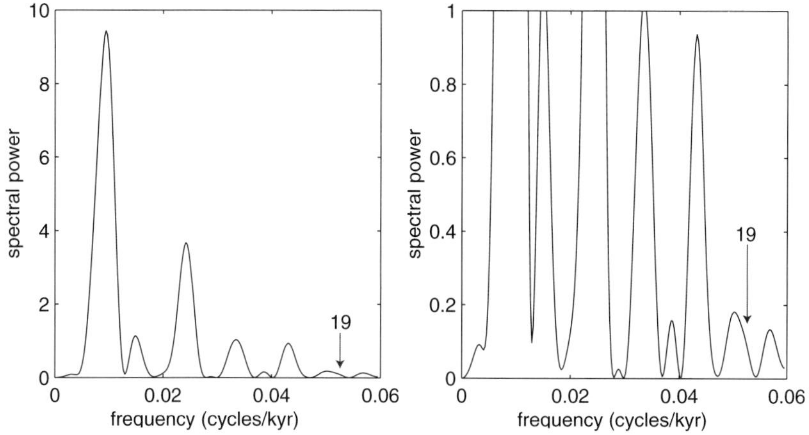

Fig. 3.24. Pacemaker spectrum, Hanning taper.

3.8 MAXIMUM ENTROPY METHOD

The maximum entropy method (MEM) applies a mathematical approach that is strikingly different from most of the other approaches described in this book. It assumes that the true power spectrum can be approximated by an equation that has a power series—but instead of making the spectrum proportional to that power series, it puts the power series in the denominator! For a readable description of why this makes some sense, see Press et al. (1993). It is full of advantages and disadvantages, just the ingredients to inspire passion among spectral analysts. The maximum entropy method finds the spectrum which is closest to white noise (i.e. it has a maximum randomness or "entropy") while still having an autocorrelation function that agrees with the measured values—in the range for which there are measured values. It has a curious feature: because it extrapolates the data to times outside the measured range, it yields spectral lines that are often much narrower than you would get by any other method. But don't be fooled by this into thinking that the method can resolve spectral lines better than an ordinary periodogram. The narrowness of these lines is one of the mathematical artifacts that can be misleading. It is worthwhile quoting Press et al. (1993) about MEM:

> *"The maximum entropy property has caused MEM to acquire a certain "cult" popularity; one sometimes hears that it gives an intrinsically "better" estimate than is given by other methods. Don't believe it. MEM has the very cute property of being able to fit sharp spectral features, but there is nothing else magical about its power spectrum estimates .... With noisy input functions, if you choose too high an order, you will find spurious peaks galore! Some experts recommend the use of this algorithm in conjunction with more conservative methods, like periodograms, to help choose the correct model order, and to avoid getting too fooled by spurious spectral features."*

Tukey is quoted (by Priestley 1981, p. 606) as saying that the maximum entropy method is suitable only for relatively "smooth" spectra. This makes it highly unsuitable for data with a few narrow lines, as we have in paleoclimate. Recall that Tukey supported the use of the ordinary periodogram for such cases.

Nevertheless, you will see the maximum entropy method used in paleoclimate data, and there will be referees of your papers who will recommend—maybe even demand—that you use it! A program to use it, based on an optional package available in Matlab, is given in Appendix 2. Programs in C, Fortran and Basic are available in Press et al. (1993). We show the spectral power calculated for the SPECMAP $\delta^{18}O$ data, with both a linear and a log plot, in Fig. 3.25. The analysis was done for the time interval 0 to 600 kyr. The maximum entropy parameters were set as follows: the number of coefficients was 250, and the number of frequencies was 800.

For comparison we show in Fig. 3.26 the maximum entropy spectral estimate for eccentricity and orbital inclination. Note how the resolution obtained appears even better than for the periodogram. However, notice that the ratio of spectral peaks is quite different than in the periodogram.

3.9 CROSS SPECTRUM AND COHERENCY

Coherency is a statistical measure that is used to measure the similarity of two records $a(t)$ and $b(t)$ in frequency space. Suppose, for example, $a(t)$ is a climate proxy, and we would like to determine whether it is dominated by a theoretical driving force $b(t)$. We can examine the spectra of the two signals, and see if their amplitudes are similar. This procedure can be made more formal by looking at the following product function of the Fourier transforms:

$$C(f) = A(f)B^*(f)$$

where A is the Fourier transform of a, and B^* is the complex conjugate of the Fourier transform of b. This simple product is given the fancy name "cross spectrum". The cross spectrum of a signal with itself is the spectral power. The cross

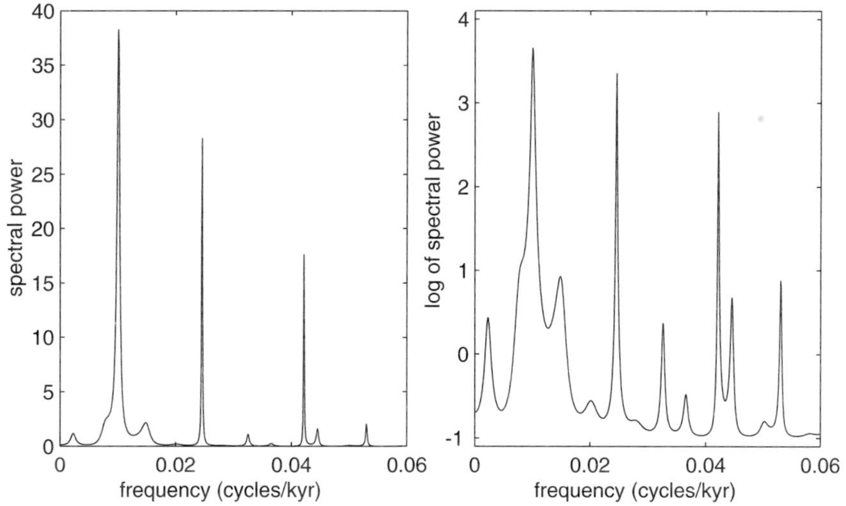

Fig. 3.25. Maximum entropy spectra of SPECMAP.

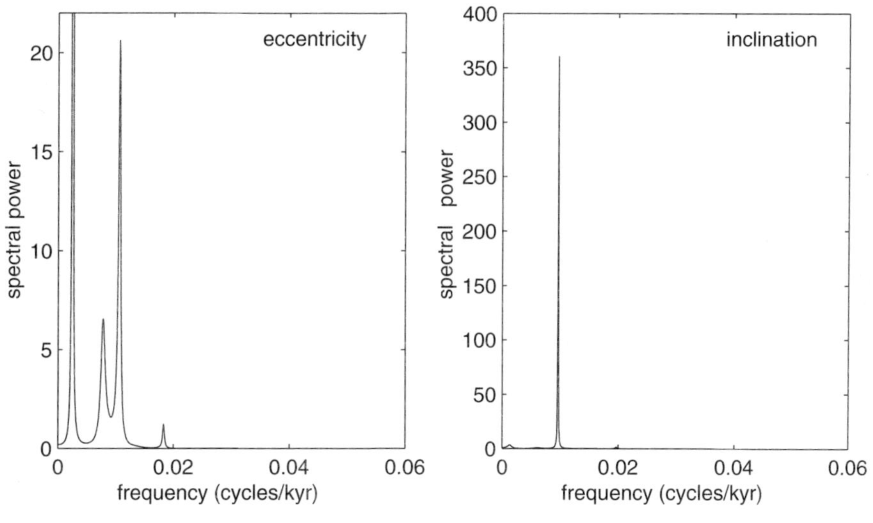

Fig. 3.26. Maximum entropy spectrum of eccentricity and inclination.

spectrum is large only at those frequencies in which there is significant spectral power in both *a* and *b*. Some books like to define the cross spectrum as the Fourier transform of the cross correlation function. But, by the correlation theorem, that definition is the same as ours.

If you are particularly interested in whether the two signals are in phase with each other, regardless of amplitude, then we can take the cross spectrum, square it, and divide by the spectral powers of the individual signals. This gives:

$$c(f) = \frac{|C(f)|^2}{P_a(f)P_b(f)}$$

There is a problem with this definition. It is identically equal to 1 at all frequencies. This is mathematically trivial, but conceptually it happens owing to the fact that a Fourier transform is a sum of perfect sine waves, and a perfect sine wave in signal $a(t)$ that has the same frequency as a sine wave in $b(t)$ will always have the same relative phase. In other words, they differ by a phase factor δ, but this is independent of time. When we take the absolute square, we get $|\exp(i\delta)|^2 = 1$. So measuring the coherency of perfect sine waves makes no sense. Coherency is valuable only if we have two signals that are varying with time, and that implies that they are not single frequencies, but (at the minimum) a band of frequencies. Then it really is interesting to see if they stay in phase, since if one is driving the other, they will.

We build this concept of a band into the definition of coherence by averaging over a frequency band. So the truly useful definition of coherency becomes:

$$\text{coherency}(f) \equiv \gamma^2(f) \equiv \frac{|\langle C(f)\rangle|^2}{\langle P_a(f)\rangle\langle P_b(f)\rangle}$$

By the brackets we mean that adjacent frequencies have been averaged. A typical way to do this is to take two spectra that have N frequencies and average them three at a time, resulting in a coherency that has $N/3$ independent values. If the coherency is calculated by the Blackman–Tukey method, and 1/3 lags are used, then an averaging of adjacent points is automatic. This averaging is essential to make the coherency into something useful.

We should mention that some authors prefer to show γ, the square-root of γ^2. Unfortunately, the function γ is also often called the coherency, so if the authors don't tell you, you may not know which convention they used (unless they use the symbol γ).

Although the coherency is extremely useful in many fields, its value is limited by the need to do the averaging. Resolution is lost, just as spectral resolution is lost in the Blackman–Tukey method. As an illustration of this, we have calculated the coherency of the Earth's eccentricity and orbital inclination for the last 400 kyr in Fig. 3.27. Note that the coherency is extremely high (>0.9) in the vicinity of 0.01, i.e. the 100 kyr peak. Because of the degraded resolution, coherency cannot distinguish the 100 kyr peak of eccentricity from the 100 kyr peak of orbital inclination, any better than the Blackman–Tukey method can.

Coherency measures only the phase relationship; it is insensitive to amplitude, since in the definition we divided by the spectral powers. The process of tuning, described in Chapter 5, adjusts a parameterised sedimentation rate to try to bring the fluctuations of the data into agreement with those of the target. Thus a tuned data set is automatically made coherent with the target model. Since coherency is insensitive to amplitude variations, it does not detect disagreements in amplitude, such as the "Stage-11" problem (see Section 8.2 on page 233). A small number of

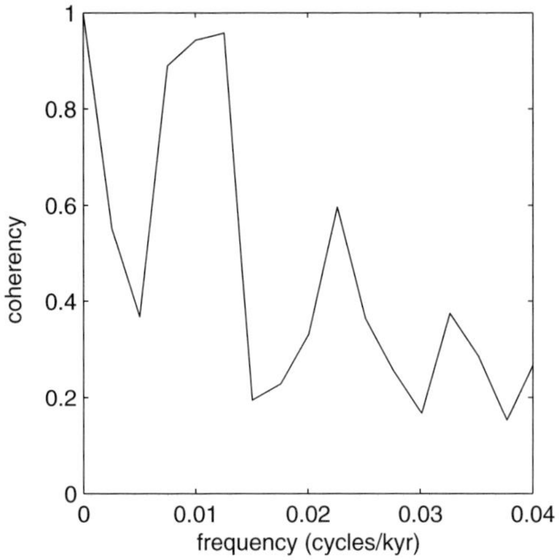

Fig. 3.27. Coherency of eccentricity and inclination.

tuning points (less than one every 100 kyr) can bring the phase of any δ^{18}O signal into agreement with that of any target (e.g. eccentricity); since amplitudes are ignored by coherency, the two signals will be in apparent complete coherency. The same procedure will yield agreement between δ^{18}O and orbital inclination. However, orbital inclination and eccentricity cannot be so adjusted, since their time scales are well known. But with one tuning adjustment, we could make an adjusted eccentricity coherent with inclination over 800 kyr! This illustrates the limitation of the use of coherency: it becomes virtually useless—maybe worse than useless (since it represents circular reasoning)—when working with tuned data.

3.10 BISPECTRA

The bispectrum of a data set is the simplest of the statistical quantities known as "higher order" spectra. Bispectra measure an aspect of data which is ignored in the usual spectral estimation (whether it be periodogram, Blackman–Tukey, or multi-taper), which is the coherency relationship between several frequencies present in the data. Bispectra have proven invaluable in the analysis of many complex phenomena, both in mechanics and in geophysics.

Briefly, a bispectrum shows a peak whenever two conditions are fulfilled: (1) a triplet of frequencies f_1, f_2, and f_3, are present in the data that obey the relationship $f_1 + f_2 = f_3$, and (2) the phase relationship between the three frequencies is coherent for at least a short averaging time for a band near these frequencies. Such frequency relationships occur frequently in astronomy. They are generated in the nonlinear processes that create the orbital changes for the Earth. The usefulness of the bispectrum is simple: if a driving force (e.g. eccentricity or inclination) has frequency

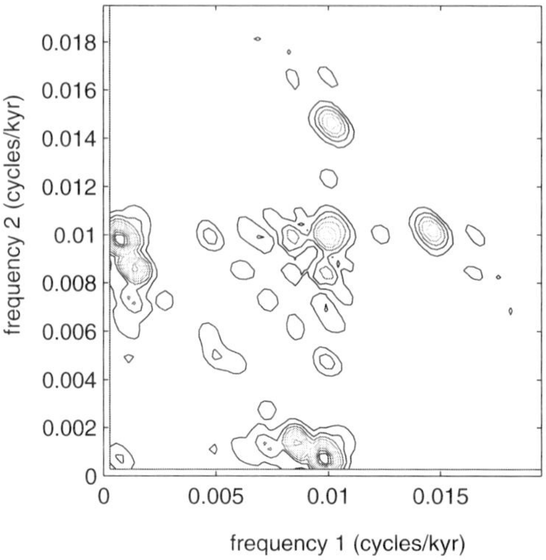

Fig. 3.28. Bispectrum of SPECMAP stack. Note that the figure is symmetric about the diagonal.

triplets which are coherent, then we expect the response (i.e. climate) to contain the same frequency triplets. So, for example, if δ^{18}O is driven by eccentricity, it should have the eccentricity triplets in it.

We have included an appendix that describes the properties of bispectra in some detail. In this chapter, we will confine ourselves to giving several examples of its use. A plot of the bispectrum has two axes, representing the frequencies f_1 and f_2. We then calculate the amplitude of a coherent signal f_3 in the data, and plot that amplitude at the location (f_1, f_2). Thus, any amplitude on the bispectral plot indicates the presence of a triplet. We make full use of these plots later, e.g. in Section 8.6.

For an example, we plot in Fig. 3.28 the bispectrum for the SPECMAP δ^{18}O data, using the corrected time scale discussed in Section 5.4. In Fig. 3.29 we show the bispectra for the same time interval of eccentricity and orbital inclination.

The way to interpret these is as follows. Nonlinear processes in a driving force, either eccentricity or inclination, result in coherent relationships between triplets of frequencies. These same relationships are likely to appear in the bispectra of the climate proxy. In addition, nonlinear processes in climate could add additional frequencies to the bispectra of the climate proxy that do not appear in the driving force. In Fig. 3.28, for example, we see several small peaks that are not present in either eccentricity or inclination, for example, at $(f_1, f_2) = (0.01, 0.01)$. This small peak implies the existence in the data of a harmonic at 0.02. This harmonic is not in either driving force, and it should not necessarily be expected to be.

It is clear from the figures that the bispectrum for inclination is a much better match to that of SPECMAP than is the bispectrum of eccentricity. We will return to this result when we discuss the bispectrum of Site 806 in Section 8.6.

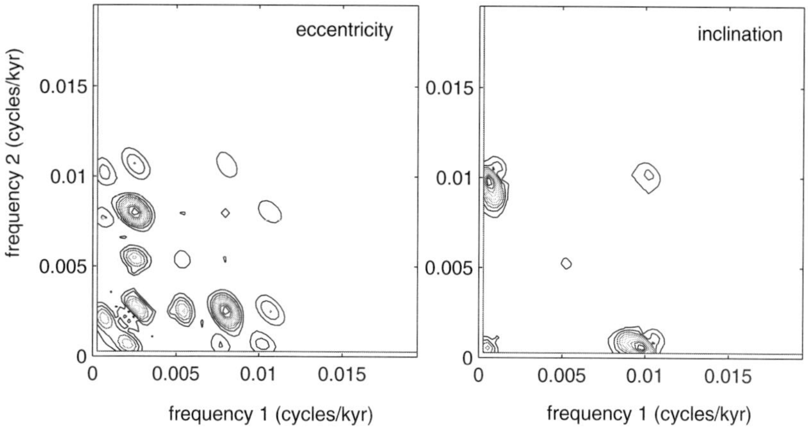

Fig. 3.29. Bispectra of eccentricity and orbital inclination.

3.11 STATISTICS OF THE FOURIER POWER SPECTRUM

The most common errors made in spectral analysis come from poor estimates of the effects of background. By background, we mean all the fluctuations in the data that come from something other than the feature (e.g. the spectral peak) under investigation. Background is not necessarily *noise*. It may be true climate variations, driven by some mechanism other than astronomy. In fact, the background may be similar in structure at different locations around the world.

An extremely important feature of the spectral power is that it does not obey the statistics of a normal distribution. We will show that the statistics, under very general conditions, are exponential rather than Gaussian. What this means is that fluctuations are much more common than would be expected from experience with other data. If the mean of the noise level is M, then the relative probability Prob of a peak fluctuating up to a power p is given by

$$\mathrm{Prob}(p) = e^{-\left(\frac{p}{M}\right)}$$

rather than by the normal (Gaussian) equation

$$\mathrm{Prob}(p) \neq \frac{1}{\sqrt{2\pi}\sigma} e^{-\frac{p^2}{2\sigma^2}}$$

The reason for this behaviour is straightforward. The Fourier amplitudes (but not the Fourier power) *do* follow normal statistics. This is because the Fourier transform is a linear transform of the data points, which themselves are assumed to obey Gaussian statistics. Both the real and imaginary parts are linear combinations of the data points h_k. By the Central Limit Theorem of statistics we expect the individual Fourier amplitudes to have Gaussian statistics for both their real and imaginary parts. So far, no surprises. If the h_i average to zero, then the expectation value for the real and imaginary parts of the Fourier amplitudes R and I is zero, and we have the normal equations

$$\text{Prob}(R) = \frac{1}{\sqrt{2\pi}\sigma} e^{-\frac{R^2}{2\sigma^2}}$$

$$\text{Prob}(I) = \frac{1}{\sqrt{2\pi}\sigma} e^{-\frac{I^2}{2\sigma^2}}$$

But the results will be very different for the Fourier power. The probability of obtaining a power p is the same as the probability of obtaining an amplitude A such that $p = A^2$. The probability of obtaining A is the joint probability of obtaining any combination of R and I such that $R^2 + I^2 = A^2$. This consists of all the points that lie in a ring in the complex plane of radius A and thickness dA. Therefore we write:

$$\begin{aligned}\text{Prob}(p)dp &= \left(\frac{1}{\sqrt{2\pi}\sigma} e^{-\frac{R^2}{2\sigma^2}} + \frac{1}{\sqrt{2\pi}\sigma} e^{-\frac{I^2}{2\sigma^2}}\right) 2\pi A dA \\ &= \frac{\sqrt{2\pi}}{\sigma} e^{-\frac{A^2}{2\sigma^2}} A dA \\ &= \sqrt{\frac{\pi}{2\sigma^2}} e^{-\frac{p}{2\sigma^2}} dp\end{aligned}$$

This is the desired result. It states that the probability of obtaining a power P is proportional to $\exp(-p/M)$, where M is the mean power ($= 2\sigma^2$), i.e. it shows that the distribution is exponential in character rather than Gaussian.

The exponential rule is extremely important but counter intuitive. It has several implications that can surprise the unwary. These include the fact that large fluctuations are very probable, particularly if you are accustomed to working with data that have a normal distribution. The probability of getting a power equal to three times the mean is $\exp(-3) = 5\%$. Thus, in random data, we expect to see peaks at levels three times the average in 5% of the frequency bins. If there are 20 independent frequencies, then we expect, on average, one of them to show a power greater than three times the mean, just from statistical fluctuations. Thus "large" fluctuations are common in the Fourier power spectrum. This contrasts with normally-distributed data, in which the probability of exceeding three standard deviations is 0.27%.

The rule also implies that the most likely power in any frequency is zero! (That is when $\exp(-P/M)$ has its maximum value.) The mean power, of course, is M. But it is about twice as likely to get bins below the mean as it is to get peaks above the mean. A similar paradox occurs in physics of any process with an exponential distribution, e.g. radioactivity. If we have one atom which has a mean life of τ, then when is the most likely time for it to decay? The answer is, at $t = 0$. That is when $\exp(-t/\tau)$ has its maximum value. This is easier to understand if you take a large number of such atoms. Indeed the maximum decay rate will be at $t = 0$, and it will decay exponentially after that. This result demonstrates that the most likely time to decay is at $t = 0$.

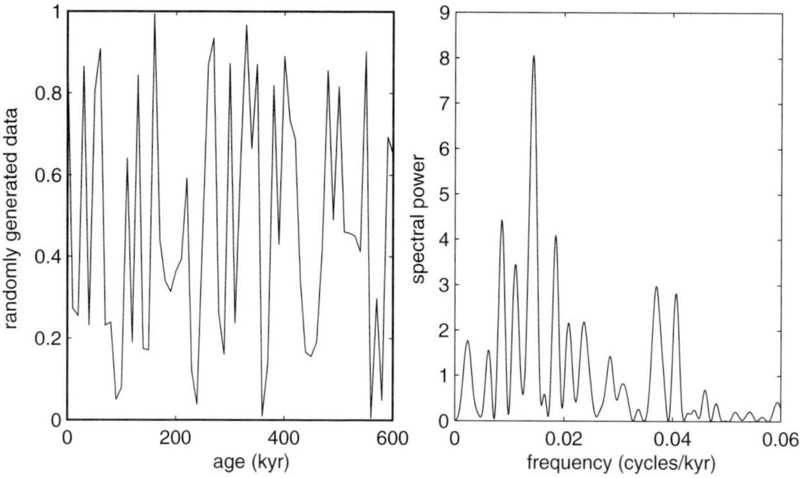

Fig. 3.30. Random signal and its spectrum.

The net effect of these results is that even random data appear, to scientists familiar with normal distributions, to have statistically significant peaks in it. This is best demonstrated with Monte Carlo simulations. In Fig. 3.30, we generated semi-random data by picking random numbers between 0 and 1 to represent the data every 10 kyr. Data for intermediate points (every kyr) was generated by interpolation. The data and spectral power for one such plot is illustrated above. This plot was the fourth one generated.

The apparent structure in the spectrum plot is misleading. It was selected over the first three because it appeared to show a structure that a paleoclimatologist might find interesting. There is a peak (maybe a split peak) near 0.01 and another set near 0.04. Could these be the eccentricity peak and the precession peak? With a little bit of time-scale adjustment, perhaps the data could be made to fit this model. And yet, the data were generated by a random-number generator!

The highest peak in the data has value 8. The spectrum has a mean of 1, so we calculate (incorrectly, it will turn out) that the probability of getting such a peak is $\exp(-8/1) = 3 \times 10^{-4}$. However, there are approximately 60 independent data points, and so 30 independent frequencies in the plot. This was the fourth one generated, so we looked at a total of 120 different frequencies to find this one. So the probability is $120 \exp(-8) = 4\%$. This still seems statistically significant! In fact, one might say "the peak is significant at the 96% confidence level". We publish. But it is nonsense! These are randomly generated data. Where did we go wrong?

The mistake we made is that we underestimated the background. To show this, we generated 1000 different spectra, using the same random procedure that we used to create Fig. 3.30, and then averaged the spectra. The result is shown in Fig. 3.31.

It is clear that the background is not the same at all frequencies. This is an example of a "red" background that we will discuss it at more length in Section

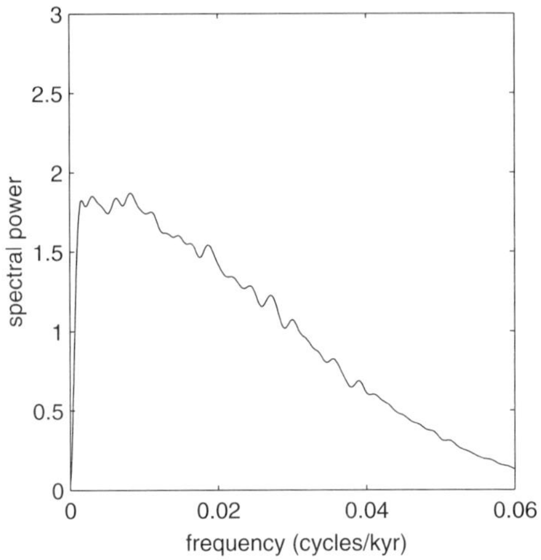

Fig. 3.31. Red spectrum from sum of many randomly-generated backgrounds.

3.13.1. Note that the background is high in the very region that we found the peak we thought was significant. The average spectral power goes to zero at $f = 0$ because we subtracted the mean before doing the Fourier transform.

It is not an accident that what appeared to be the most significant peak occurred in this region. From this plot, you can see that the background level in the vicinity of the peak is approximately 1.8, rather than 1. To estimate the statistical significance of the peak, we need to calculate the ratio of its power to that of the background at the same frequency. So the correct estimate of the probability is $100 \times \exp(-8/1.8) = 1.2 = 120\%$. So we expect to find (on average) 1.2 peaks of similar statistical significance in every four random data sets.

The above example is meant to illustrate how easy it is to overestimate the statistical significance of peaks, unless the background is well understood. It is rarely the case that we have a complete model for the background, so what is the next best approach? If we examine the spectrum again, we see that there is evidence that the background in the vicinity of the peak is unusually high. In fact, if we draw a background line through the data *in the vicinity of the peak*, it is quite plausible that the background level is approximately 2 units. (Remember, according to the exponential rule, about $1/e$ of the data should be above the line, on average.) Had the background level been estimated in this manner, then the mistaken "high confidence level" prematurely estimated for the data would not have been derived.

The only way to be safe is to be conservative. For a peak truly to exist, it must be well above estimates of local background. Of course, if you are conservative, then you might miss some discovery. The result is that the rule has been frequently broken in the literature by scientists whose papers got accepted only because they were reporting a discovery. Another result is that the literature is full of reported frequencies that have turned out not to exist, once better data were available.

3.12 PLOTTING POWER, LOG(POWER) AND AMPLITUDE

When plotting spectral power, the choice of axes becomes important, and controversial. The plot that shows spectral power on a linear scale is very popular, but many statisticians object strongly. They insist that only log plots are acceptable, and anyone who makes a linear plot is simply showing their ignorance of statistics. Many paleoclimate papers contain all their spectra using an ordinate proportional to the square-root of the spectral power. They label the axis "amplitude".

The choice of axis is not just a matter of taste. Each of the plots has its advantages and disadvantages. Each of the plots shows something to its best advantage, and shows something else poorly. To illustrate this, we will use a particular data set known as the SPECMAP stack; we will describe these data more fully in Section 4.2. We show the same spectrum (of SPECMAP data for the period 0–600 kyr) in three plots. Fig. 3.32 shows the spectral power; Fig. 3.33 shows the logarithm (to base *e*) of the spectral power, and Fig. 3.34 shows the square-root of the spectral power.

The value of the first plot, linear in spectral power, is that it gives an immediate sense of how much of the variance comes from each cycle and how much comes from noise. This is useful since variance has the nice property that amplitude doesn't have: it is additive. By that we mean that the variance of the data is equal to the sum of the variances of the individual components. So, for example, you will often read in the literature statements such as "the 41 kyr peak accounts for 50% of the variance". To the extent that such statements are useful in evaluating signal strength, using an ordinate that is proportional to the spectral power is best. If the data are normalised to unit variance and zero mean, by subtracting the mean and dividing every point by the sample standard deviation, then the area under the peak shows the fraction of the variance contained in any frequency interval. Subtraction of the mean is also useful

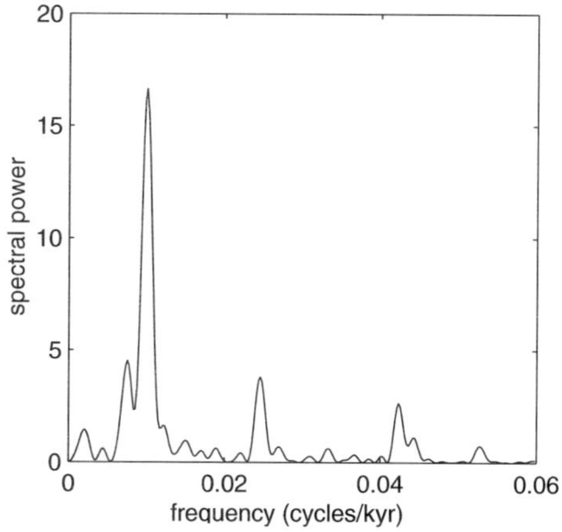

Fig. 3.32. Spectral power for SPECMAP.

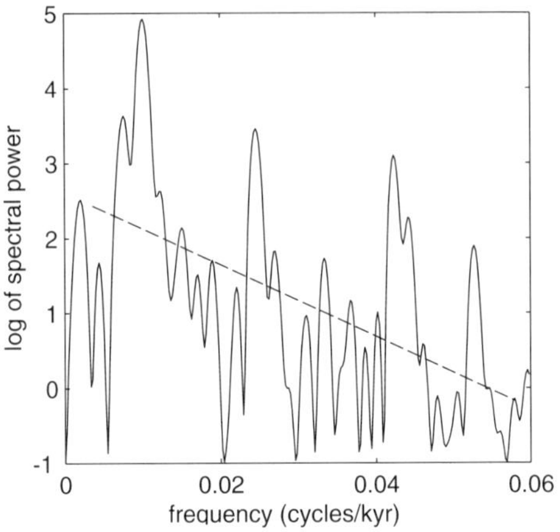

Fig. 3.33. $\log_e$ of spectral power, SPECMAP. The dashed line is an estimate of the background. Roughly one-third of the data points should rise above this line.

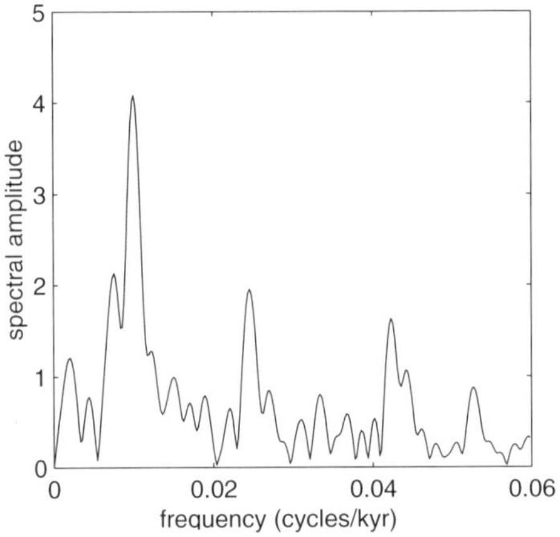

Fig. 3.34. Square-root of spectral power, SPECMAP.

since any line at zero frequency contaminates the estimate of low frequency power. Normalising also aids in comparison of spectra from different sites.

The linear plot also gives the most direct sense of the width of the peaks, since the full-width at half maximum can be seen without even reading the y-axis. The linear plot is the best plot if you are concerned about detecting evidence of an incorrect time scale (due to chatter or drift, which we discuss in Section 5.1). This sensitivity to time scale error is sufficient to justify its use in many cases.

The log plot puts the logarithm of the power on the y-axis. This plot is sometimes called a "semilog" plot, since only the ordinate is logarithmic, and not the abscissa. This kind of plot is the best one for making the statistical significance of peaks evident, particularly for small peaks. As we showed in Section 3.11, the statistical significance of a peak of power P over a background B is $\exp(-P/B)$. In paleoclimate data the background B is often different at different frequencies. In the log plot, peaks of equal statistical significance will be the same linear distance above the local background level. That's why the log plot is favoured by statisticians whose primary interest is evaluating the statistical significance of small peaks. However, a log plot makes it very difficult to see time-scale error, since the multiple peaks characteristic of that kind of error are hidden among a welter of background whose presence is emphasised in the log plot.

In addition, the log plot can easily give a misimpression of the *climate* significance of each peak. A peak can be statistically significant simply because the background level in its region is small. But that doesn't mean that the peak plays an important role in climate. (This distinction is often confused in the public eye. A release of radiation that is statistically significant can be reported to the public as a "significant release", and they will misinterpret it.)

Some people believe that the best ordinate to give a sense of climate significance is the square-root plot. The square-root of the spectral power is the magnitude of the spectral amplitude, and the square-root plot is often labelled "amplitude". The amplitude tells you the peak-to-peak variation in the climate proxy. Unlike the spectral power, however, the sum of the spectral amplitudes does not equal the total amplitude variation in the signal.

In the amplitude plot, Fig. 3.34, it is relatively easy to see that the amplitude of the 41 kyr peak is two to three times less than the amplitude of the 100 kyr peak. (The uncertainty of 2 to 3 arises because of the uncertainty in the placement of the background level.) One disadvantage of the amplitude plot is that the statistical significance of the peaks is hidden. To calculate it you must find R the ratio of the peak height to the background level, square it, and then calculate $\exp(-R^2)$. This is not hard to do—but it is not as evident on the plot itself. All three of these plots have been popular in the paleoclimate literature; often two kinds are given in the same paper.

3.13 MONTE CARLO SIMULATION OF BACKGROUNDS

Monte Carlo simulations have proved to be indispensable in physics research. Typically, a simulation consists of the computer generation of signal buried in computer-generated noise. The combination is then studied as if it were real, to see whether the numerical methods applied to the data will pull true signal out of noise. Such tests are frequently done before an experiment is even built, to see if it can achieve the sensitivity that is needed.

A typical application is as follows. A student has performed a spectral analysis of $\delta^{18}O$ data, and notices what appears to be structure near the base of the peak, a "shoulder". Does this indicate that the data are not properly tuned, i.e. that the time scale is not correct, so some spectral power is being leaked to adjacent frequencies? Or does it indicate that the peak has real structure? Or is it just a statistical fluctuation? One way to address this question is to run 100 or more Monte Carlo simulations. For each simulation, the computer generates a real signal that is a pure sine wave, with no shoulder, and a "random background signal" that is added to the sine wave. The spectral power is calculated for each Monte Carlo simulation, and the student looks for a shoulder. If they see one in, say, 3 out of the 100 simulations, then they know that random events can simulate the effect, but not very often. They might properly conclude that the shoulder is not a random event but real, with a "97% confidence level" since it did not occur in 97% of the simulations. But could it be due to a poor time scale? To do this test, another 100 simulations are run, but for each one the time axis is now changed in a small way that simulates what might happen with real $\delta^{18}O$ data. Again, the student studies the simulated events to see how often a shoulder appears.

The key technique in Monte Carlo simulation is the creation of background that behaves in a way similar to the background you have in the real data. If the background in the simulation is not an accurate mirror of the background in the data, then the conclusions drawn may be wrong. In this section we will discuss several ways of generating background, and we give Matlab programs for these methods in Appendix 2.

We have already presented one example of Monte Carlo simulation for Fig. 3.31. We will repeat the procedure here. We create our background by generating random numbers every 10 kyr, and doing a linear interpolation in between. This yields data and spectra shown in Fig. 3.35. The spectrum plot shows the spectrum generated from the data shown. Superimposed on the spectrum plot is the average spectral power found by repeating the entire process 100 times, and averaging the results.

As you can see in the spectral power plot, there appears to be a strong peak near frequency 0.017 cycle/kyr. Yet we know that these data were generated from random numbers, so we know this peak is not real. It is an result of the fact that spectral power shows statistical fluctuations which are exponential in their nature, and so large excursions are more common than in other types of analysis.

We now estimate the statistical significance of this peak. The mean value of the background in the vicinity of the peak is at spectral power = 2. The peak rises to a value of 10. Therefore the probability of having it occur by chance, at this frequency, is approximately Prob $= \exp(-10/2) = 0.0067$. There are about 45 frequency bins in

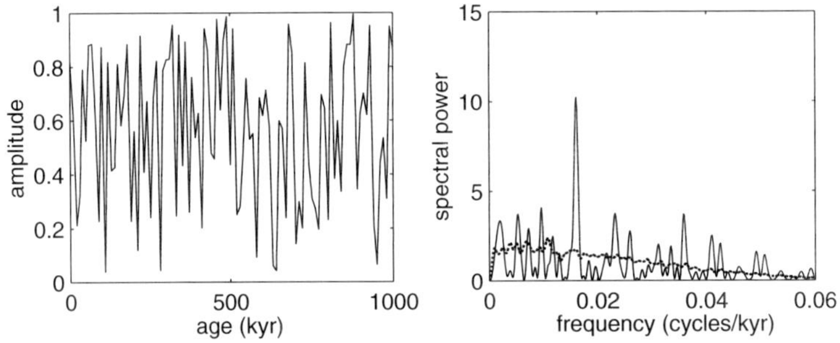

Fig. 3.35. Monte Carlo background and spectrum.

the plot, and we would have been equally interested in any other peak that was similarly high above the background. Thus, the probability that we would find something of similar height is $0.0067 \times 45 = 0.30 = 30\%$. So in 30% of our Monte Carlo simulations, we expect to see peaks with comparable significance—even in data that we *know* contain no real peaks.

Fluctuations such as seen in Fig. 3.35 are common, yet even experienced scientists can be fooled. We strongly urge every student of spectral analysis to spend a considerable amount of time generating Monte Carlo plots, before publishing anything from real data. There is no technique that is equally effective in showing how easily we can fool ourselves. And the primary difference between a scientist and a layperson is that the scientist is aware of how easily he can be fooled.

3.13.1 Red and white noise

In both of the examples given so far, Figs 3.31 and 3.35, the average spectrum tended to peak near the low frequencies. This is an example of "red noise". The name derives from optical spectra, where the red signal is the low frequency component. Many phenomena in physics show red noise. In many cases, the power spectrum of the background shows a strong enhancement at the low frequency end proportional to $1/f$. Such noise is called (without showing much originality) "one over f noise", or sometimes "flicker noise". Noise in paleoclimate work is frequently red. This may be because processes in the climate and in the ocean records tend to suppress rapid variations, rather than because low frequencies are enhanced.

In Fig. 3.36 we have generated another simulation of background, using a slightly different technique. It was created by generating a new random number for every thousand years, and then taking a 15 kyr running average of the results. This is the kind of background we would have, for example, if climate were random but the sea floor sediment was continually mixed to a depth of a few centimetres.

The plot on the left is an example of the data generated in this method. The plot on the right is the spectral power for this data set, and superimposed on it (as before) is a plot of the average spectral power from 100 such simulations. The choice of smoothing period allows us to adjust the redness of the spectrum. For example, if we smooth the same random numbers we used above over 10 kyr, rather than 15, we get

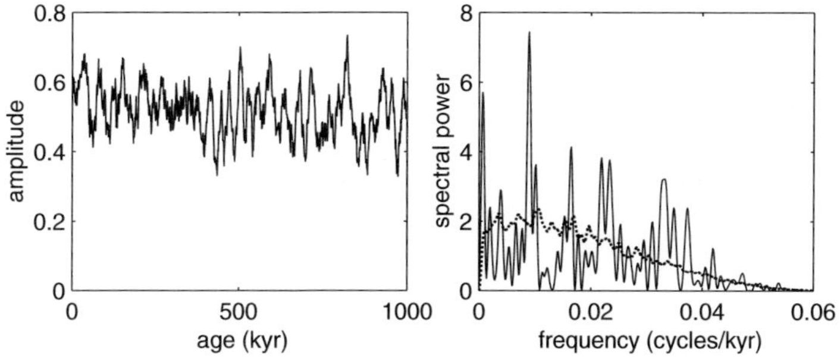

Fig. 3.36. Background from 15 kyr smoothing, and spectrum.

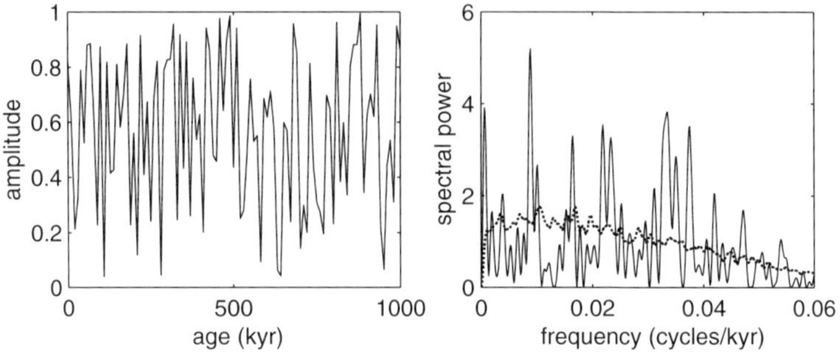

Fig. 3.37. Background from 10 kyr smoothing, and spectrum.

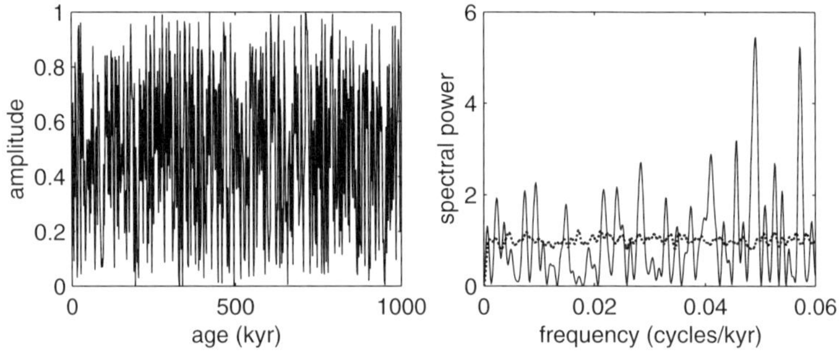

Fig. 3.38. White background, and spectrum.

the data and spectrum shown in Fig. 3.37. Note that the spectrum is less red, i.e. its average value does not fall off as quickly at high frequencies.

For contrast, we generate data with no smoothing. Let every data point, every 1 kyr, be random. The data and spectrum are shown in Fig. 3.38. Note the flatness of the average spectrum. This is the kind of signal known as white noise, since all frequencies are equally present. (This is the terminology, even though for the colour "white" in optics, all that is demanded is equal contributions of red, blue and green, the three colours of the eye sensitivity.)

A small spectral peak, appearing on top of a red background, will have the frequency at which it achieves maximum power shifted slightly towards lower frequencies. This is an example of "bias". It is particularly troublesome if the goal of your analysis is the precise determination of frequencies. To minimise the bias, you can use a method called "prewhitening" that tends to flatten the background. This method is discussed at length in the standard books on spectral analysis. Once the spectrum is computed, you must then compensate for the prewhitening by adjusting the measured spectral powers.

3.13.2 Prewhitening

Prewhitening can be used to compensate for red noise, or for the presence of an adjacent large spectral peak. The techniques are quite different. For red noise, prewhitening is sometimes done by doing the spectral analysis not on the data, but with a new set of points equal to the differences of adjacent data points. This is equivalent to the process of taking a derivative of the data. We will show below that taking the derivative is equivalent to multiplying the Fourier transform by a factor of frequency f. If the Fourier transform is defined as

$$H(f) \equiv \int_{-\infty}^{\infty} h(t) e^{2\pi i f t}\, dt$$

the Fourier transform of the derivative of the data is

$$\begin{aligned} FT\left(\frac{\partial h(t)}{\partial t}\right) &= \int_{-\infty}^{\infty} \left(\frac{\partial h(t)}{\partial t}\right) e^{2\pi i f t}\, dt \\ &= -\int_{-\infty}^{\infty} h(t) \left(\frac{\partial}{\partial t} e^{2\pi i f t}\right) dt \\ &= -2\pi i f \int_{-\infty}^{\infty} h(t) e^{2\pi i f t}\, dt \\ &= -2\pi i f H(f) \end{aligned}$$

where we integrated by parts to shift the derivative to the exponential term. Thus taking differences multiplies the Fourier transform by a term proportional to the frequency f. The spectral power is multiplied by f^2. Taking differences often over-compensates for the observed red noise, which is typically proportional to $1/f$. Blackman and Tukey (1958) recommend that you try milder prewhitening, a partial differencing: replace the data h not by Δh (equivalent to taking the derivative) but instead by ($h - 0.6\Delta h$). But they emphasise that you must look at the effect that such prewhitening has on your data, and then change it to do a better job at flattening the spectrum. There is no single prescription that works for all data.

Given the high signal to background levels seen in paleoclimate data, and the uncertainty in the time scale, there is little to be gained by the prewhitening methods discussed above, and most scientists no longer use them for such data. It is usually sufficient to subtract the mean from the data (thus assuring that the zero frequency bin, which represents the constant offset, will be zero), and sometimes to remove the "trend", i.e. the best fit to a constant slope. Removing the mean is common, but many experts prefer not to remove the trend since doing so may bias both the power and the frequency estimates.

Another form of prewhitening that is valuable is the systematic removal of strong peaks, as we discussed in Section 3.7. It consists of systematically finding the peaks, and removing them one at a time, before proceeding with further analysis. When the peaks have been removed, the bias that they might have caused is eliminated. It is an important technique, not so much for reducing bias, as it is for discovery of weak peaks on the tails of stronger peaks, and for the estimate of the statistical significance of such weak peaks.

3.14 FREQUENCY ESTIMATION: THE HWHM RULE

The best estimate for the frequency of a spectral peak can be obtained by using the maximum likelihood method: make a preliminary estimate of the frequency of the strongest peak. Using the parameters of this peak as a starting point, perform a maximum-likelihood fit (or a minimum chi-square fit) to the data. In this fit, the frequency as the starting point should be a variable. Unfortunately, this method is cumbersome, and rarely worth the effort. It is much simpler, and usually accurate enough, to pad the data with lots of zeros, and then find the frequency from the maximum of the periodogram.

But how accurate is this estimate? The primary error often comes from the inherent bias from noise and, indeed, the elimination of such bias forms the bulk of many texts on spectral analysis. Many methods have been developed to reduce such bias, but most of these methods depend on the presence of well-characterised noise. And lack of bias does ensure accuracy. The noise itself is a true climate variation often reflecting local or even global conditions. So, for example, the benthic $\delta^{18}O$ signal may reflect global ice, but it also includes a component reflecting local climate variations such as systematic temperature and salinity variations (as you would get, for example, if the deep thermohaline circulation changed).

The net result of this complexity is that you are often reduced to estimating the effects of such bias, rather than attempting to compensate for them. The uncertainty should be reflected in the error estimate for the frequency. For much paleoclimate data the result is relatively simple: the uncertainty in the frequency estimate will be approximately equal to $\pm$w, where w is the half-width at half-maximum of the peak in the spectrum. Thus the frequency is determined only over a range of the full-width at half-max of the peak. We call this the "HWHM rule". Obviously, this is not a precise result that works in all cases. But we will show below why it is a reasonable approximation for much paleoclimate data. More importantly, it is rare that one can determine the frequency to an accuracy better than this. You will see papers in which the period of the peak is stated to be 100 kyr, and the authors will claim this is incompatible with a theoretical prediction that the period should be 95 kyr. But although the peak of the spectrum may be at 100 kyr, that does not mean that it is an accurate determination of the true frequency, since background variations can shift the peak.

The uncertainty in frequency arises from the effect of background on the peak. Let us assume that the amplitude at maximum, in absence of background, is A_M. There is, of course, a phase associated with the amplitude. Typically the phase will vary as rapidly as the amplitude, so that as we go away from the maximum value to the half power point the phase will change by 45 degrees. The background also varies in amplitude and phase over a similar change in frequencies.

To get an idea of the relative values of the signal and background, we show in Fig. 3.39 the $\delta^{18}O$ spectrum for Site 552 (Shackleton and Hall, 1984). The peak near $f = 0.01$ cycles/kyr stands well above the background. Let us try to estimate the background. It appears to be higher at low frequencies than at high ones, and the high frequencies may include some non-background signals (e.g. the peak at 41 kyr

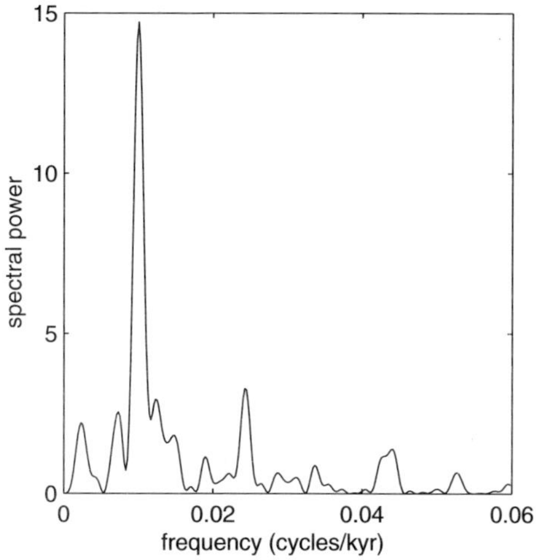

Fig. 3.39. Site 552 spectrum.

period). There does appear to be structure on either side of the 100 kyr peak, at a level of about 2 units. (The plot is normalised so that the average power is one unit.) For the sake of discussion, let us accept this level as the background estimate. The power of the peak is at 14 units. So the ratio of amplitudes of S/B is approximately $\sqrt{14/2} = 2.6$.

The question at hand is: how much does the presence of B affect our estimate of S, in particular, our estimate of the frequency at which S reaches its maximum? We will estimate this by looking at the effect of changing the background, while keeping the signal unchanged. (This can easily be done with Monte Carlo simulations, and the reader is encouraged to try it.) Suppose, for example, that the amplitude S_M of $\sqrt{14} = 3.7$ were partly due to constructive interference. To get its true value, subtract the amplitude of the background $B = \sqrt{2} = 1.4$, to get $S_{\text{TRUE}} = 2.3$. If, instead of constructive interference, the background B happened to give destructive interference, then the amplitude at this point would have been $2.3 - 1.4 = 0.9$. The peak would not have been visible at all at this frequency. Back at the present half-power points, the relationship between the background and the signal could differ by 90 degrees, so the amplitude at this frequency would be $\sqrt{(2.32^2 + 1.42^2)} = 2.7$, and the power at this frequency would be $2.7^2 = 7.3$, well above the noise. The peak would be seen, but far removed from the correct frequency.

Anyone who wants to put uncertainty limits on a frequency estimate will learn a great deal by doing Monte Carlo simulations. However it is clear that a peak cannot be moved much further than its half maximum. In addition, it is clear that unless the signal to background level is unusually high, that a movement of that much in frequency has a reasonably high probability of happening. A more sophisticated analysis requires a better characterisation of the background.

To illustrate the usefulness, and limitations, of the HWHM rule, we generated 40,000 Monte Carlo data sets, each containing a true 100 kyr peak, but with varying

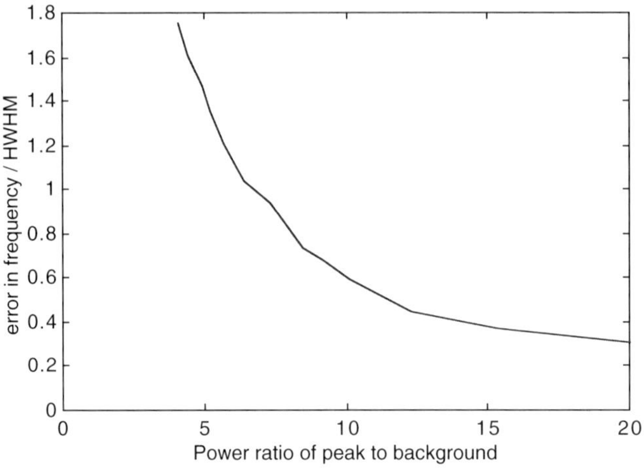

Fig. 3.40. Frequency determination vs. signal-to-background.

levels of random background. (The background was generated by choosing random numbers every 10 kyr, and interpolating in between.) For each data set, the periodogram was calculated, and the frequency of the maximum power (with period between 80 and 120 kyr) was recorded. The average error was the frequency difference between this value and the true value (0.01). In Fig. 3.40 we show the error, normalised to the half width at half maximum, plotted as a function of background level.

The plot demonstrates that at very high signal/background levels (20), the frequency can be determined to within about 30% of the half width of the peak. For more typical data (signal/background = 7), the accuracy is equal to the half width. When the signal/background drops to 5 or below, the frequency determination becomes very inaccurate.

3.15 SPECTRAL WIDTH AND SHAPE DETERMINATION

Suppose we have a climate proxy $h(t)$ which consists of a pure sine wave present for a time T. Moreover, assume that $h(t)$ is zero outside of this time. Then the untapered periodogram will show a peak proportional to the square of a sinc function:

$$\left(\frac{\sin(2\pi(f-f_0)T)}{2\pi(f-f_0)T}\right)^2$$

If the observed spectral peak has this shape, then nothing intrinsic to the shape is observable, and we say the peak is *unresolved*. However, if the signal is the sum of two or more sine waves, or has an amplitude that varies with time, then the shape can become complicated. The case of two sine waves is particularly interesting, since such *doublets* are common in orbital parameters. The other important case for paleoclimate work is the effect of background on our ability to detect and measure such a doublet.

If two spectral lines are sufficiently close in frequency, then they will appear to be a single peak with a somewhat broader shape. How close can they be? This issue has received a great deal of study and analysis; see, for example, a lengthy section in Priestley (1981). The limit is set, in part, by the shape of the spectral peak, and that, in turn, depends on the method of analysis. The most widely used equation is the "Rayleigh Criterion", originally applied to images in a telescope. Rayleigh decided that two stars of equal brightness could be "resolved" if the maximum intensity of one star fell on the first zero (minimum intensity) of the other. However, in optics the telescope image of a star is equal to the square of the Fourier transform of the aperture—i.e. it is the spectral power! So the Rayleigh criterion can easily be used in spectral analysis. For a simple sine wave, the first zero in the spectral power occurs at

$$\Delta f = \frac{1}{T}$$

This looks very similar to the uncertainty principle criterion—with good reason, since the width of the spectral lines determines the resolution. We will illustrate this equation with an example. The eccentricity of the Earth's orbit contains oscillations with period 95 and 125 kyr. Let us consider a simple function consisting of two sine waves with these periods:

$$h(t) = \sin(2\pi t/95) + \sin(2\pi t/125)$$

They should be "Rayleigh resolved" when

$$T = \frac{1}{\Delta f} = \frac{1}{\dfrac{1}{95} - \dfrac{1}{125}} = 400$$

To illustrate this, we show in Fig. 3.41 the spectral power for $h(t)$ for the time interval $t = 0$ to 400 kyr.

Several features of this plot are worth noting. The first is that the two peaks appear to be "super resolved"—the spectral power in between the two peaks goes to zero. This is much better than Rayleigh required. The reason is that the two signals, at frequencies 0.008 and 0.0105 cycles/kyr, interfere with each other to

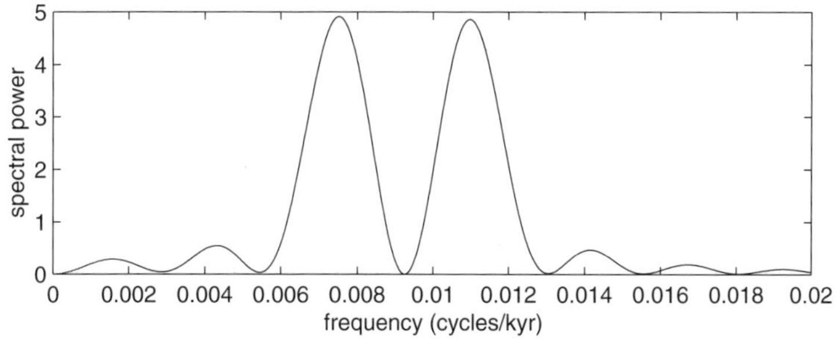

Fig. 3.41. Rayleigh Criterion example: 95/125 kyr peaks.

give the cancellation. A similar feature does not occur with stars, because stars are *incoherent*, and do not interfere with each other. But the super resolution is not always achieved. Since it requires a cancellation, it depends on the relative phases of the two signals. If we add a 90 degree phase shift to one of the signals, the dip between the two peaks is not nearly as deep.

A second important feature is that the maxima of the individual peaks do not occur at the expected frequencies. The peak on the left actually reaches its maximum value at $f = 0.0075 = 1/133$, i.e. it corresponds to period 133 kyr, not at 125 kyr. This disagreement is a consequence of the fact that this peak is sitting on the side-lobes of the other peak. This tilts the peak and, in this case, moves the maximum to a lower frequency (higher period). The degradation for close peaks is known as the Cramèr-Rao bound (Rife and Boorstyn, 1974, 1976; van Trees, 1968). In general, the estimate of the frequency degrades when the frequencies become closer than $2/T$, and that is the effect we are seeing here. However, if precise frequency estimation is important (and only if low levels of background suggest it is possible; see Section 3.14) it is easy to avoid the Cramèr-Rao restriction by doing the frequency estimation using the pre-whitening approach advocated by MacDonald (see Section 3.7). In paleoclimate data, the Cramèr-Rao limit is rarely important, since the uncertainty (the half-width at half-maximum) will be much larger than the bias.

In Fig. 3.42–Fig. 3.44, we plot the separation of the two peaks for various time intervals T. For $T = 250$ kyr (Fig. 3.42), the peaks are unresolved; however it is clear

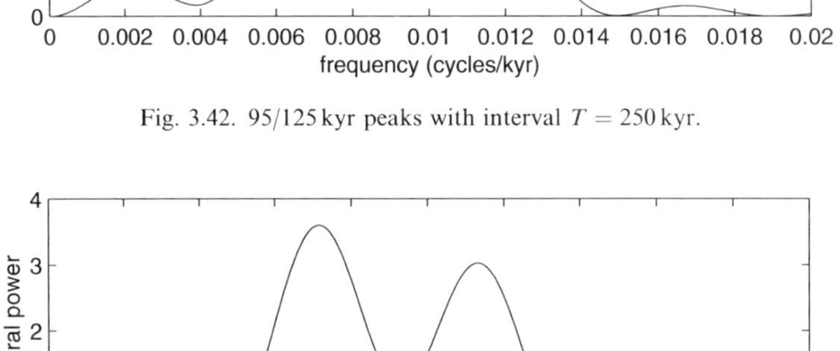

Fig. 3.42. 95/125 kyr peaks with interval $T = 250$ kyr.

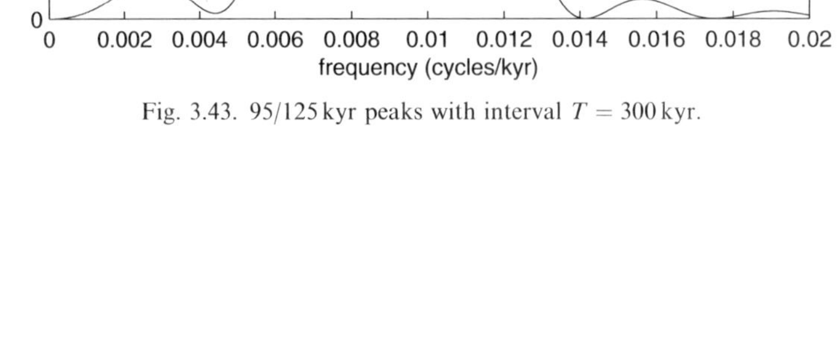

Fig. 3.43. 95/125 kyr peaks with interval $T = 300$ kyr.

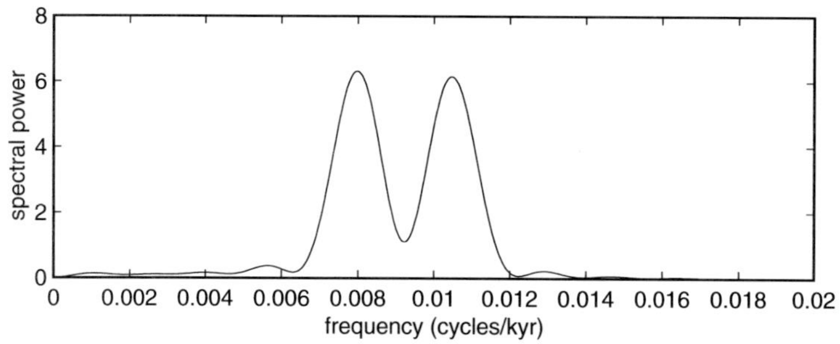

Fig. 3.44. 95/125 kyr peaks with interval $T = 600$ kyr.

that it is broad and is more than a single frequency. For $T = 300$ kyr (Fig. 3.43), the peaks are not separated, yet are resolved.

The spectral power for $T = 600$ kyr is shown in Fig. 3.44. Note that the peaks are not as well separated as they were for $T = 400$ kyr. With this time interval, the bias in the frequency maxima is much smaller, but still present. The low frequency peak reaches its maximum at $0.00796 = 1/125.5$, rather than at $0.08 = 1/125$. This Cramèr-Rao bias could be serious in the communications industry, but it is far too small to be important in paleoclimate work.

Next we consider the effects of background on shape determination. Such effects can be substantial, since background fluctuations can be large. If the background level is B, then the probability that it reaches a level C for any given data point is $\exp(-C/B)$. (This is an immediate consequence of the fact that the background has an exponential distribution, as discussed in Section 3.11.) This effect of background is best illustrated by a numerical example, similar to the one we gave when discussing frequency estimation. Let the spectral power of a peak, in the absence of background, be P. Let the average spectral power of the background, in the vicinity of the peak, be A. Let us assume that $P = 3B$. That seems like a reasonable ratio; it means that the probability that a random background peak could reach as high as P is $\exp(-3) = 0.05$, corresponding to a 95% confidence level. But although it is not terribly likely that the background will fluctuate that high, it is much more likely that the background will cancel the peak into insignificance. How can that be? Let us assume that the actual background in the vicinity of the peak has value B, but that it also happens to have the opposite phase from P. (This will happen about 25% of the time.) The Fourier amplitude of the peak is $\sqrt{P}$. The Fourier amplitude of the background is $\sqrt{B}$. The net amplitude is $\sqrt{P} - \sqrt{B} = (\sqrt{3} - 1)\sqrt{P} = 0.732\sqrt{P}$, and the observed power will be $(0.732\sqrt{P})^2 = 0.54P$. Recall that by itself, P was three times background. This partially cancelled peak will only be 1.6 times background. The experimenter will determine that its statistical significance is only $\exp(-1.6) = 0.2$. That places it at the 80% confidence level.

The lesson here is that background can be more bothersome in its tendency to *cancel* real peaks, than in its ability to fluctuate up and cause false ones. This is particularly important in the case of theories that attribute the 100 kyr peak to eccentricity. The eccentricity model attributes the 100 kyr peak to the 95 kyr cycle

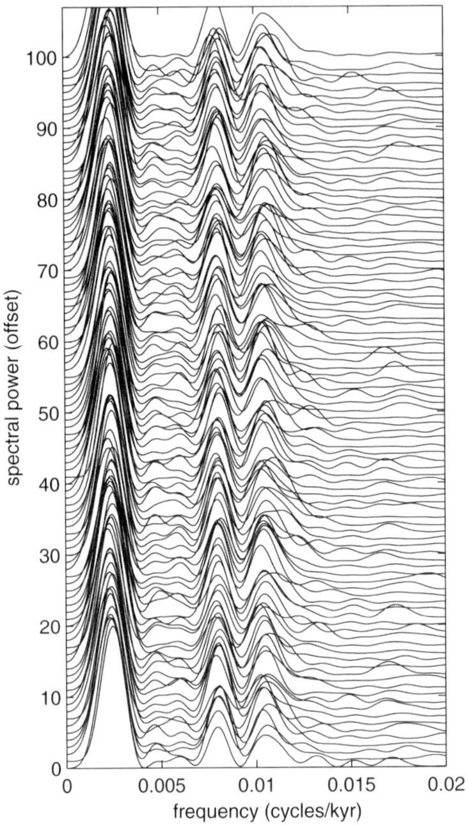

Fig. 3.45. Monte Carlo simulations: eccentricity + background.

of eccentricity, and the statistics are usually compatible with this identification (since you can rarely measure the frequency of a peak in paleoclimate data to better than a full-width at half-maximum). But the 125 kyr peak is missing. Could it have been accidentally cancelled by background? The answer is yes—if the background level is as high as one-third of the peak level.

However, if the background is significantly lower than one-third of the peak level, such cancellation becomes increasingly improbable. In Fig. 3.45 we show 100 Monte Carlo simulations of eccentricity added to background, in which the *average* level of background was ten times lower than the height of the expected 125 kyr peak. In the 100 spectra shown (each one with a different random background), the doublet nature of the peak is not significantly cancelled. This is not surprising, since to cancel would require both that the phase of the background be opposite to that of the peak (a 25% chance) and that the background spectral power fluctuate upwards to a significant fraction of the peak power.

4

Climate proxies

"There is something fascinating about science. One gets such wholesale returns of conjecture out of such a trifling investment of fact."

Mark Twain, *Life on the Mississippi*, 1874

There are many aspects of ancient climate that we would like to know. What was the temperature during the ice ages? How much did it vary? What were the patterns of ocean currents and winds? Was the sky clear, or cloudy, or mixed as it is today? Did greenhouse gases, including water vapour, methane and carbon dioxide, have a different concentration than now? How intense were the storms? If we had all this information, then there is some hope that we would be able to understand the behaviour of climate—and perhaps be in a better position to understand potential human-induced effects. But that is optimistic. We have such information for the present day, and yet we still can't predict present-day climate.

Some parameters of the past can be directly measured. Entrapped air in the Greenland and Antarctic glaciers give us samples of the atmosphere back to about 420 kyr, so we can directly measure the concentrations of greenhouse gases. The temperature of the ice ages has been partially preserved in the temperature of the glaciers (Cuffey and Clow, 1997). The extent of glaciers in North America and on mountains in the tropical Andes can be estimated from scour marks, moraines and erratic boulders.

Most of the most detailed information we need, particularly for older periods, comes from indirect records. Forams (short for foraminifera) are microscopic organisms whose life cycles depend on local temperature and whose fossils preserve samples of ancient material. These can be picked, species by species, from sediment, and analysed. The abundance of a certain species of planktic (surface dwelling) forams depended on the local sea-surface temperature. In the jargon of paleoclimate, we say that such forams represent a "proxy" for sea surface temperature. One of the most remarkable proxies is the ratio of oxygen isotopes in benthic (bottom dwelling) forams. This ratio, in ancient sediment, is believed to reflect the total amount of ice that existed on the Earth at the time the sea beds were formed. We'll explain why in the next section. This ratio is called a proxy for global ice.

Unfortunately, most proxies are dependent on more than one aspect of climate. The oxygen isotopic ratio depends not only on global ice, but is somewhat sensitive to the patterns of ocean currents and temperature. We can try to separate out the various contributions of each variable to each proxy; in principle, we can do this if we have more proxies than variables, and they respond linearly. The separation method is called principle component analysis. Deuterium abundance in ice samples, for example, is believed to respond strongly to local temperature; however, there is some contribution from global ice. The global ice component can be subtracted (it is argued) by choosing the correct linear combination of deuterium and oxygen isotopes. In practice, this is difficult to accomplish, and it is even more difficult to know how well you are doing.

The art of the proxy has become highly developed in the paleoclimate literature. That does not mean it is accurate, as most paleoclimatologists will readily admit. Do not be surprised if, in reading the literature, you find a proxy that makes little physical sense to you. You may be right. Many proxies are used in a manner that would suggest, to the uninformed, that they mean more than they do. You might, for example, find a plot of hydrogen isotope ratios labelled "temperature", despite the fact (as we will discuss below) that there is no simple or reliable conversion. But, at the same time, proxies have proven to be extremely useful. To the extent that the isotopes of carbon have made it possible to discover changes in the history of the deep ocean currents, it doesn't particularly matter what the precise meaning of this proxy is. To the extent that oxygen isotopes indicate periods of glaciation, maybe it doesn't really matter if they respond to global ice, temperature, or just to some unknown combination of the two.

In this chapter we will discuss the primary proxies that are important for understanding the role of astronomy in paleoclimate. These proxies are the isotopes of oxygen and hydrogen. We will also discuss some of the other proxies that are important for unravelling the details of ancient climate (such as amplification mechanisms and nonlinear processes), but we will not describe them in great detail unless they are directly related to issues of the astronomical driving force. We refer the reader who wants to know more about these other proxies to detailed discussions in the texts by Faure (1986) and Bradley (1999).

Many of the samples that we will discuss came from seafloor cores; others came from cores taken in Greenland or Antarctic ice. In the geologic community, the cores are named things like V22-174, RC13-110, and DSDP-806. If you are in an oceanography department at a university, then you probably already know the origin of these names. If you are in the physics department, then you almost certainly have never heard of them. The cores are usually named after the ship or the programme that collected them. So, for example, core V22-174 was collected by the research vessel Vema on its 22nd voyage, at its 174th station. Following is a short list of the prefixes you will encounter, with a little information about each.

V	Vema, a converted yacht (61 metres long, with sails) operated by the Lamont-Doherty Earth Observatory of Columbia University. It began collecting samples in 1951.
RC	Research vessel Robert Conrad, a successor to Vema, operated by Lamont-Doherty.
DSDP	The Deep Sea Drilling Project. Operated from 1968 to 1983 by the Scripps Institution of Oceanography. The ship Glomar Challenger drilled 1092 holes in the sea floor, at 624 sites. Don't confuse it with the Glomar Explorer, which was run by the CIA to recover a Russian submarine. Mnemonic: the Challenger was exploring, and the Explorer was challenging.
ODP	Ocean Drilling Program. Continued the DSDP programme, as an international collaboration. The larger ship Joides Resolution replaced the Glomar Challenger. Managed by Texas A&M.
GRIP	European-based GReenland Ice-core Project. Recovered a core 3028.8 metres long, to the base rock at the Greenland Summit, 72°34'N, 37°37'W (Johnsen et al., 1992; Dansgaard et al., 1993).
GISP2	US-based Greenland Ice Sheet Project #2. Recovered 3050 metres of ice core from the Greenland summit, 28 km from the GRIP location (Grootes et al., 1993).
Vostok	Russian station on the East Antarctic ice plateau (78°S, 106°E). The core samples at this site reach a depth 3350 metres, and date back about 426,000 years.
MD	The research vessel Marion Dufresne, operated by the French since early 1990s.

This list is not meant to be complete; there are many more sea floor cores, and many more ice samples available. For example, there has been a very productive sampling of ice in equatorial regions. Some of these glaciers may have samples going back 500 kyr; see, for example, Thompson et al. (1997, 1998). The cores have been preserved in core "libraries" that are more like museums. If you have a hypothesis that can be tested by doing an experiment on a core, then you can make an application to the library to obtain a sample. Certain cores (e.g. at the last termination) have been heavily sampled; others are virtually untouched. A fraction of each core has often been set aside as an archive, in anticipation of the invention of a new measurement method that will yield new information.

4.1 OXYGEN ISOTOPES

One of the most useful climate proxies is the ratio of oxygen isotopes found in ice, trapped air, or in fossils such as benthic or planktic forams. Oxygen consists of three stable isotopes: 99.759% is ^{16}O, 0.037% is ^{17}O, and 0.204% is ^{18}O. Even though it is only one part in 500, variations in the fraction of ^{18}O can be measured with high accuracy and are extremely useful. The change in ^{18}O is usually designated by the quantity $\delta^{18}O$, which is defined as the fractional change in the $^{18}O/^{16}O$ ratio, compared to a standard. This simple concept looks more daunting, and takes more space, when described by an equation:

$$\delta^{18}O \equiv \left(\frac{\left(\frac{^{18}O}{^{16}O}\right)_{\text{Sample}}}{\left(\frac{^{18}O}{^{16}O}\right)_{\text{Reference}}} - 1 \right) \times 1000$$

The factor of 1000 is there since, by convention, the results are expressed in parts per thousand (‰) rather than in percent. (Some people believe that this makes it more *metric*.) The reference sample is presently taken from standard mean ocean water (SMOW) defined by Craig (1961) from samples collected by the US National Bureau of Standards.

4.1.1 Physical mechanisms

Oxygen isotope separation occurs because of the isotopic differences in vapour pressure and chemical reaction rates, which, in turn, depend on temperature. Some of the most important geophysical processes that lead to changes in $\delta^{18}O$ are:

1. Evaporated water is lighter than the remaining liquid. Water containing ^{16}O has a higher vapour pressure than if it contains ^{18}O, so it evaporates more readily (Craig and Gordon, 1965).
2. Precipitated water molecules are heavier than those in the residual vapour. $H_2{}^{18}O$ condenses more readily than $H_2{}^{16}O$, so as water vapour is carried across a continent (e.g. to Greenland or to central Antarctica), the residual becomes lighter. But there are exceptions: under certain conditions, the formation of snow crystals can enrich the light isotope, because its higher diffusivity enhances its ability to find its way into a crystal (Jouzel and Merlivat, 1984).
3. Oceanic $\delta^{18}O$ is nonuniformly distributed. This is a consequence of the first two phenomena, but it means that changes in the pattern of winds that carry vapour and change the source will also change $\delta^{18}O$. At present, the difference in surface water is 1.5‰ from pole to equator.
4. Biological activity enriches heavy isotopes. The ^{18}O in the calcium carbonate of shells is 40‰ greater, on average, than in the water in which the organism lives. To make matters more complicated, this enrichment is temperature dependent.

The net result of these effects is that glacial ice is light, with $\delta^{18}O$ typically 40‰ (i.e. 4%) lower than seawater, so in glacial ice $\delta^{18}O$ is negative. If the precipitation returns to the oceans as runoff, there is no enduring change in the oceans. However, when large volumes of ice are stored in ice-age glaciers, then there can be considerable depletion of the light isotopes in the oceans. This effect is illustrated in Fig. 4.1.

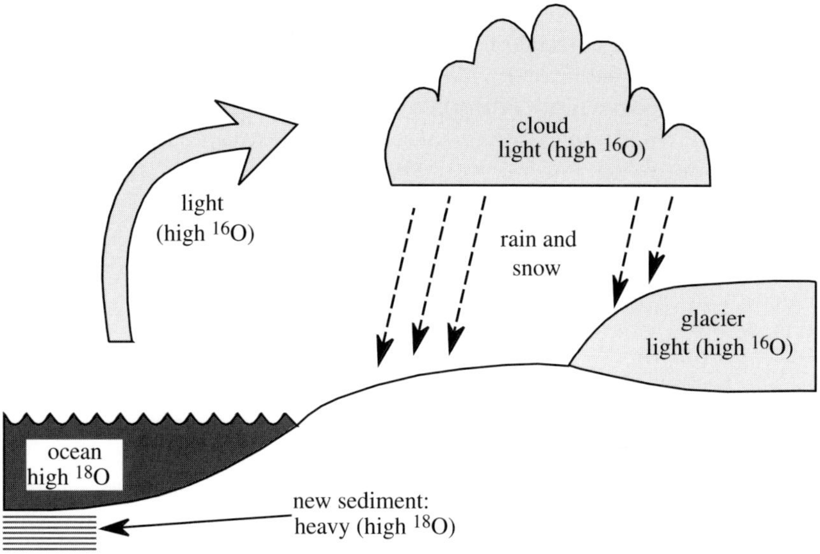

Fig. 4.1. ^{18}O in sediment reflects global ice.

To get an estimate of the resulting depletion of ^{16}O in ocean water, consider the last ice age. Based on measurements of coral terraces (Bard et al., 1990a, b; Broecker et al., 1968), sea level during the glacials was estimated to be more than 100 metres lower than it is today. (The coral method uses radiometric methods to date corals that show the presence of *Acropora palmata* which is known to grow only within two metres of sea level.) Since the average depth of the oceans is 3800 metres, this means that $100/3800 = 2.6\%$ of the oceans was stored as ice on the continents. Since the missing water was enhanced in light isotopes by about 4%, the remaining oceanic water was heavy by about $(4\%) \times (2.6\%) = 1‰$. This agrees with typical depletion levels observed in sediments. To the extent to which this effect dominates, $\delta^{18}O$ is a valid proxy for global ice. Since sea water around the world mixes in about a thousand years, the $\delta^{18}O$ records in the northern Atlantic are similar to those in the equatorial Pacific. (Section 4.1.2 discusses this "universality".)

Remember, however, that the isotopic separation can also reflect local temperatures and other climate conditions, even if glaciation has not changed appreciably. A discussion of the effect of temperature on isotope enrichment will be found in Section 4.3, in the discussion on deuterium. For now, let it suffice to say that you will find a good deal of theory and empirical equations in the literature that address this subject, but most of these equations are valid only in very limited circumstances. For example, in 1964 Dansgaard and colleagues (Dansgaard, 1964) derived an equation relating isotope enrichment to temperature, based on the assumption that the precipitated water does not re-evaporate. They also showed that measurements of isotopic enrichment in ocean water as a function of latitude yield the following approximate relationship between temperature T and $\delta^{18}O$:

$$\delta^{18}O \approx 0.7T - 13.6$$

But although this empirical equation has theoretical justification, it may be misleading, since it represents a correlation that may not be causal. The factors that lead to changes in $\delta^{18}O$ are complicated, as outlined in the list above. Thus the empirical equation could just be the result of the fact that both temperature and $\delta^{18}O$ vary with latitude on the Earth for unrelated reasons, and that if we go back to an earlier time when the temperature was lower, it would not necessarily imply that $\delta^{18}O$ would also be lower. In fact, when several measurements are made at the same latitude, the effect is argued to depend primarily on the amount of precipitation and not on temperature (Rozanski et al., 1993). There is an extensive discussion of this point in the literature; for more detail see Bradley (1999).

There are other issues to consider. In planktic fossils, we might expect $\delta^{18}O$ to reflect surface conditions, and therefore be most sensitive to temperature and/or salinity changes. We might expect $\delta^{18}O$ in benthic forams to be most sensitive to global ice, since there is little temperature variation on the sea floor. To the extent that temperature and global ice track each other, the measurements of $\delta^{18}O$ from planktic and benthic forams may be equivalent. But we must be careful in examining data, since some records may reflect extreme local conditions. In general, the $\delta^{18}O$ signal in benthic forams is probably the most dependable measure of total ice volume. In planktic forams there is a much larger contribution from temperature. In other samples (e.g. ice, trapped air or calcite), a substantial fraction of the $\delta^{18}O$ record, sometimes the dominant component, may represent temperature, not ice volume.

4.1.2 Universality

The pattern of oxygen isotopes is remarkably similar in sea floor records around the world. In Fig. 4.2 we show ten measured $\delta^{18}O$ records, from regions that include the Pacific, Atlantic and Indian Oceans. All were made from foraminifera extracted from sea floor cores, except for the samples taken from Devils Hole, which was from a calcite deposit in a cave (see Section 4.10.1). The ordinate for each is in parts per thousand. A word of caution: the ages of the samples are not known, so for the purposes of the plot, we assumed that the sedimentation rate for each location was constant; we chose the scale for each to make the ice age terminations (dashed lines) occur at roughly the same times. The fact that the terminations do not line up precisely with each other suggests that the sedimentation was not perfectly constant for each record.

There is at least one important oxygen isotope record that is notably different from the others: in the Vostok ice core, the measured $\delta^{18}O$ is from entrapped atmospheric air, rather than from ice or from fossils. And this proxy, in Vostok, shows a behaviour that is strikingly different from the behaviour of the $\delta^{18}O$ measured elsewhere. We elaborate on this special case in the discussion of the "Dole effect" in Section 4.7.

Given the complexity of the isotopic separation phenomena that we discussed in the last section, it is remarkable that the pattern of oxygen isotope variations appears to be so similar around the world. Most of the records show the same pattern of

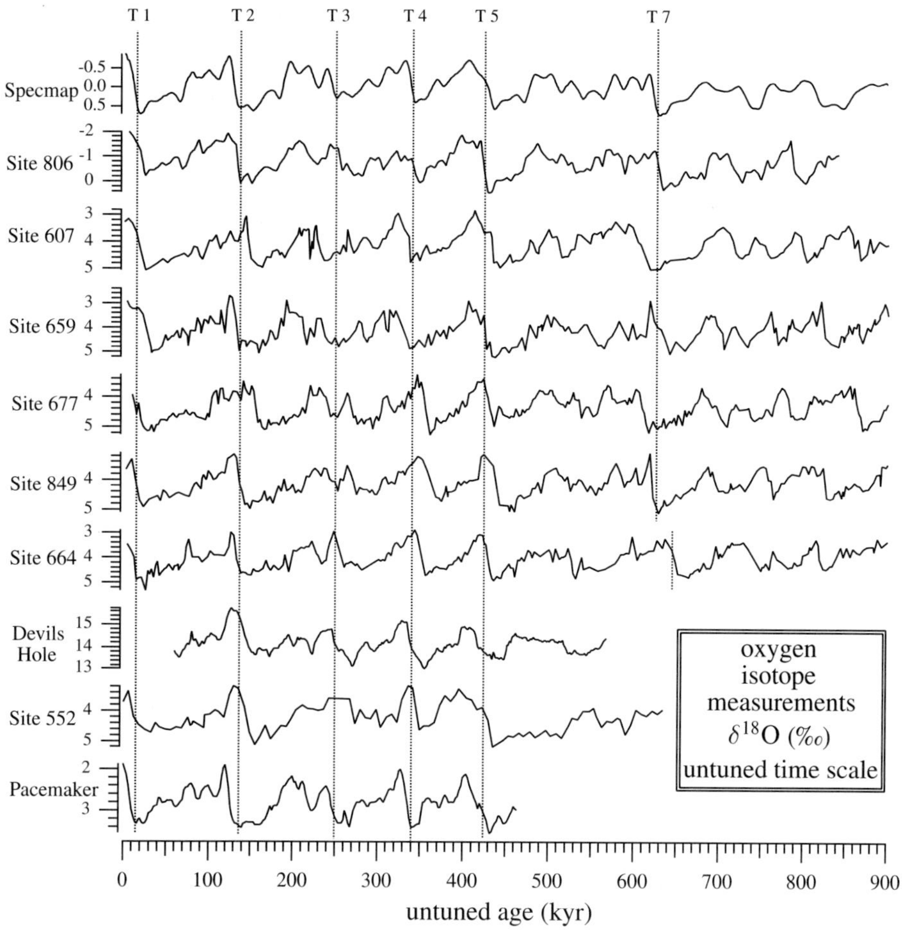

Fig. 4.2. Universality of oxygen isotope behaviour.

100 kyr cycles. The ends of the ice ages, the terminations, often appear to be abrupt; these have been marked with dotted vertical lines, and each labelled with a termination number. (We did not mark termination 6 in the plot, since it is not as clear as the others. Termination 3 is also weak in these records, yet we included it since it is very strong in others, e.g. the Vostok deuterium record.) The location of the sites and the fundamental references for the data appear in Table 4.1.

Although there are differences in the details, the most striking aspect of these data is the similarity of the patterns. We now recognise that the primary reason the data are so similar is that the biggest effect comes from the presence of global ice, and this has a 100 kyr oscillation that is clearly visible even in the presence of variations caused by local climate (e.g. temperature) effects. Although there are systematic offsets in the absolute level of $\delta^{18}O$ (which are latitude dependent), the basic character of the oscillations is that they are remarkably similar.

Table 4.1. Locations of sea floor cores.

Site 659	18N 21W	Coast of Africa	Tiedemann et al. (1994)
SPECMAP			Imbrie et al. (1984)
V30-40	0S 23W	Equatorial	Imbrie et al. (1984)
RC11-120	44S 80E	South Indian Ocean	Hays et al. (1976)
V28-238	1N 160E	Western Pacific	Shackleton and Opdyke (1973)
V22-174	105 13W	Southern Atlantic	Thierstein et al. (1977)
DSDP 502b	11N 79W	Caribbean north of Panama	Prell (1982)
DSDP 607	41N 33W	North Atlantic	Ruddiman et al. (1989)
ODP 806	0N 159W	West equatorial Pacific	Berger et al. (1993); Berger et al. (1994); Janecek (1993); Party (1991)
ODP 664	0N 23W	Equatorial Atlantic	deMenocal (1995); Raymo et al. (1997)
ODP 677	1N 84W	East equatorial Pacific	Shackleton et al. (1990); Shackleton and Hall (1989)
ODP 849	0N 111W	East equatorial Pacific	Mix et al. (1991)
DSDP 552	56N 23W	Northern Atlantic	Shackleton and Hall (1984)

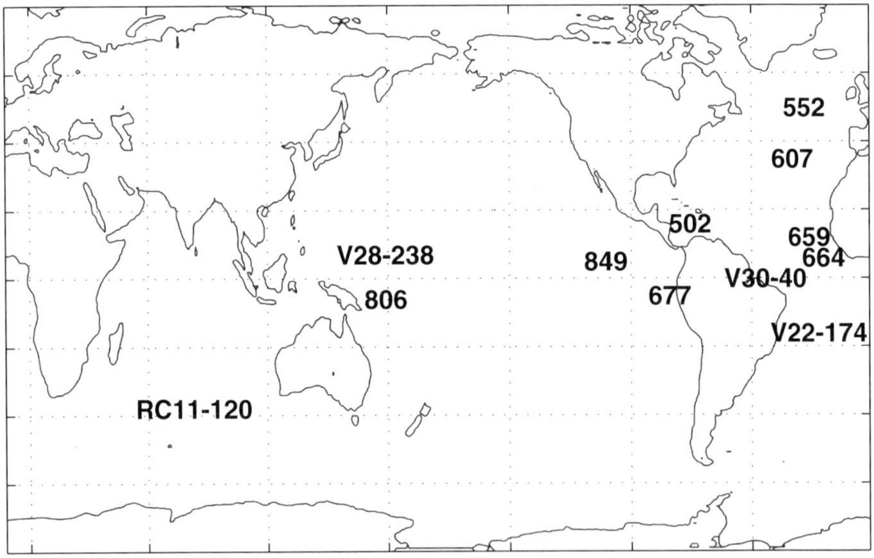

Fig. 4.3. Map of site locations.

4.2 THE SPECMAP STACK AND MARINE ISOTOPE STAGE NUMBERS

Several attempts have been made to extract the underlying $\delta^{18}O$ signal that is common in the records. One of the most respected, and certainly the one that is most referred to in other publications, is known as the SPECMAP stack (Imbrie et al., 1984). The stack was a combination of five $\delta^{18}O$ records from five cores: V30-40, RC11-120, V28-238, V22-174 and DSDP502b, which came from the Indian, Atlantic and Pacific Oceans.

Although the time scale for this stack is now known to contain serious errors in the period older than 600 kyr (see Section 5.4), it is still a widely-used template for more recent times. Many stacks have been devised subsequently to correct the error in SPECMAP; one of the most widely used is the "low latitude stack" of Bassinot et al. (1994); this stack will be discussed in greater detail in Section 5.9.3.

The original SPECMAP stack is shown in Fig. 4.4. Indicated on this plot are the marine isotope stage numbers. These follow a convention originally invented by Cesare Emiliani, and which is now universally used. Each interglacial is indicated with an odd number, beginning with the present as Stage-1, and counting backwards. The glacials are given even numbers. The only exception is a small warming that was given the name Stage-3 by Emiliani (Emiliani, 1966); a possible reason for this anomaly is described in Section 8.1. The spectrum for this stack is shown in Fig. 4.5, in both a linear and a log plot. The prominent spectral peaks are labelled by their periods (1/frequency). In addition to the narrow 100 kyr peak, there is a strong peak at the obliquity period of 41 kyr, and smaller but statistically-significant peaks at the precession periods of 24, 22 and 19 kyr. The statistical significance of the smaller peaks is particularly evident in the log plot, where they can be seen to be substantially higher than the background power on either side. (For methods of determining the statistical significance of peaks, see Section 3.11.)

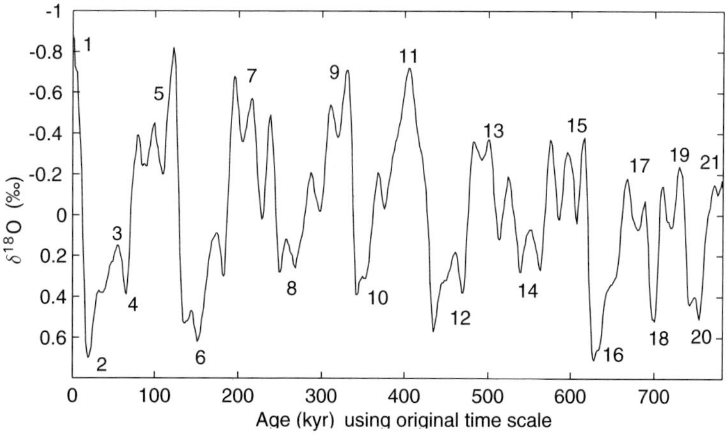

Fig. 4.4. SPECMAP $\delta^{18}O$ stack with marine isotope stage numbers.

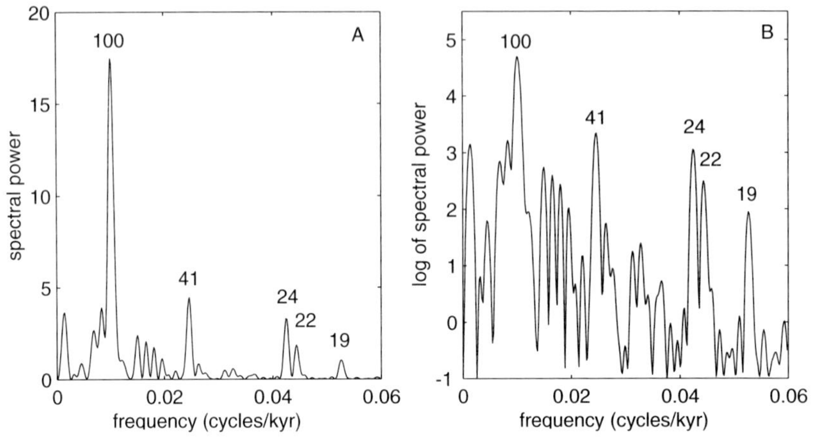

Fig. 4.5. Spectrum of original SPECMAP stack.

4.3 DEUTERIUM—A TEMPERATURE PROXY?

A water molecule that contains one deuterium atom weighs only 5.5% more than ordinary water; in contrast, water with ^{18}O weighs 11% more. So, naively, one might expect the isotopic effects of deuterium to be less than those of ^{18}O. However, exactly the opposite is true. The reason has to do with the vibrational energy levels of the hydrogen-oxygen bond, which are sensitive to the fractional change in mass of the atom, which is a factor of two for D, and only a factor of 1.06 for ^{18}O. The consequences are dramatic. Bonds formed with deuterium tend to be much more stable than those with light (ordinary) hydrogen. For humans, this makes heavy water into a poison: we die if more than one third of our body fluids have their light hydrogen replaced by deuterium.

The implications of deuterium's unusual chemical properties on its use as a climate proxy are subtle. A particularly important effect for paleoclimate is that the physical chemistry of deuterated water is more sensitive to temperature than that of ^{18}O water. This is clearly seen in the "fractionation factor" which describes the equilibrium between liquid and vapour. The fractionation fraction for deuterium is defined as the ratio of D/H in a liquid to D/H in a vapour that is in equilibrium with that liquid. The fractionation factor for HDO is approximately 1.08, and it varies with temperature as shown in Fig. 4.6. So when water evaporates, the vapour has about 8% less deuterium than the liquid. In contrast, the fractionation factor for ^{18}O is approximately 1.008. In Fig. 4.6 we also show the fractionation factor vs. temperature for ^{18}O. (The figure is adapted from Faure (1986), which is in turn based on measurements of Dansgaard (1964).)

The fractionation that takes place depends on the quantity $\alpha - 1$. If the fraction of an isotope that has condensed from the vapour is f, then the rate of remaining

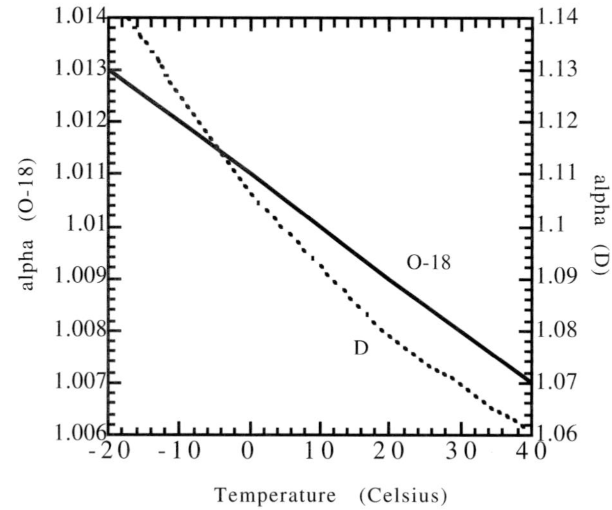

Fig. 4.6. Fractionisation coefficient vs. temperature.

fractionation is given by the Rayleigh distillation equation (Broecker and Oversby, 1971):

$$\frac{R}{R_0} = f^{(\alpha-1)}$$

where R is the isotopic ratio (either $^{18}O/^{16}O$ or D/H) of the remaining vapour. To illustrate the temperature sensitivity of water that is heavy with deuterium vs. oxygen, let's consider the case where half of the vapour remains uncondensed, i.e. $f = 1/2$. In Fig. 4.7 we plot the isotopic enrichment as a function of temperature for these two isotopes. From this plot, you can see that condensation in the deu-

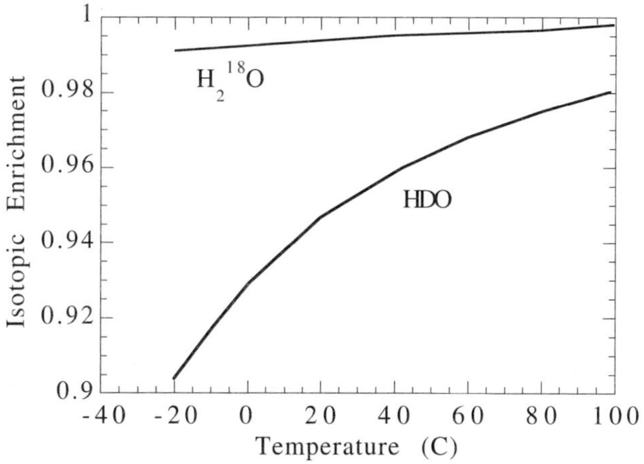

Fig. 4.7. Isotopic enrichment vs. temperature.

terised form of heavy water is significantly more sensitive to temperature variation than is the ^{18}O form.

Because of its temperature sensitivity, deuterium is often considered a temperature proxy rather than a global ice proxy. It is reasonable, however, to try to untangle the two effects, i.e. to use a combination of oxygen and deuterium measurements to deduce the global ice and local temperature. For example, a temperature scale was devised for the Vostok ice core (Petit et al., 1999) by assuming the equation

$$\Delta T = \frac{\Delta \delta D_{ICE} - \Delta \delta^{18} O_{SW}}{9}$$

In this equation, the $\delta^{18}O_{SW}$ refers to the sea floor isotope record from the "low latitude stack" of Bassinot et al. (see Section 5.9.3). There is a danger, of course, when combining new data with old that the combination may contain biases; we argue in Section 5.9.3 that the Bassinot stack does not use an accurate time scale.

4.4 INTEGRATIVE PROXIES AND PHASE DELAY

If a proxy, such as deuterium, measures the instantaneous temperature, then it gives a record of the climate at the moment it is trapped. Some proxies, however, represent the accumulation of a climate variable. For example, if we measure the $\delta^{18}O$ in bottom dwelling (benthic) foraminifera, then the proxy represents not the instantaneous climate, but the accumulated global ice. The $\delta^{18}O$ doesn't reflect the climate at the time it is laid down; rather, it represents the isotopic content of the oceans, which in turn reflects the amount of global ice that has accumulated over an extended period of time.

We illustrate this phenomena with a simple triangular model. Let us assume that the rate of ice accumulation follows the triangular function shown in Fig. 4.8 as the dashed line. The accumulated ice is shown as the solid smooth curve. It is the integral of the triangular curve, taken from age = 350 kyr to the time plotted. (Note that we integrate from right to left.) The total ice volume is called an "integrative proxy" since it measures the accumulation, i.e. the integral, of the snowfall. Although the

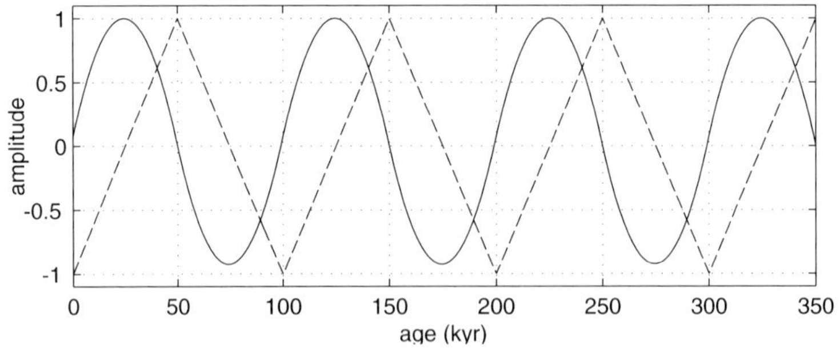

Fig. 4.8. Lag between driving force and response.

smooth curve looks like a sine wave, it is really a collection of side-by-side parabolas; each half is the integral of the linear function that makes up the side of the triangular function.

At age 350 kyr, the accumulation rate is at a maximum value (= 1, in this plot). We define the total accumulated ice to be zero at this time. While the accumulation rate is dropping (from 350 kyr to 325 kyr), the amount of ice is still growing, as shown by the smooth curve. That is simply because although the rate of ice accumulation is dropping, it is still positive. However, note that at 325 kyr, the accumulation rate goes negative. That represents melting ice, and the accumulated ice begins to drop from its peak.

Note the lag between the triangular instantaneous driving force, and the smooth accumulated amount. The driving force reaches peaks at 250, 150 and 50 kyr, but the peaks in the accumulation rate are at 225, 125 and 12 kyr—a lag of 25 kyr, or one-quarter of the 100 kyr cycle. A one-quarter cycle lag is typical, and results from many models.

For another example, assume that the accumulation rate is sinusoidal:

$$\frac{dh(t)}{dt} = \sin(2\pi f t)$$

The accumulated ice is the integral of this equation:

$$h(t) = -\frac{1}{2\pi f}\cos(2\pi f t)$$

The driving force is proportional to the derivative; the response, $\delta^{18}O$, is proportional to $h(t)$, not to the derivative. Again, the phase difference between the rate and the accumulation is the phase difference between a sine function and a cosine function: $\pi/4$ radians, or 1/4 cycle.

Phase lags of this magnitude are very common in climate. For example, the accumulation of snow in the northern mountains during the winter usually does not peak on December 21—the day of minimum insolation. In California, the snow typically peaks in March, which is one-quarter of a year later. On the other hand, in New York City the maximum snow is often on the same day as the maximum storm—since the accumulation does not last from one storm to the next. Thus, if there is a rapid mechanism for the removal of accumulated snow, then we expect that the phase delay can be appreciably less than 1/4 cycle.

This very simple point has been missed in much of the paleoclimate literature (see, for example, Berger (1999)). It is often assumed that a more reasonable phase lag between insolation and $\delta^{18}O$ should be 10 kyr, which is only one-tenth of a 100 kyr cycle, or less. Such short phase lags are *possible*, but to achieve a realistic theory we must postulate a relatively short time constant for the natural removal of ice, so that the accumulated ice represents the instantaneous rate. (Such is obviously not the case in Greenland or Antarctica, where the accumulated ice dates back over a hundred thousand years.) It takes time for ice to accumulate, and it is difficult to create a model in which it is the accumulated ice, rather than the ice formation rate, which instantaneously reflects the driving force.

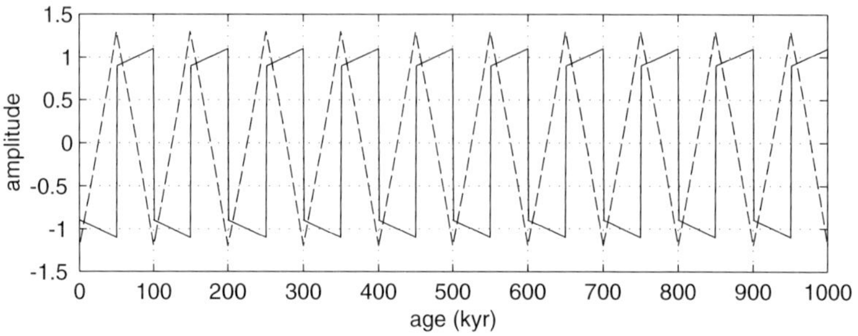

Fig. 4.9. Response with a 180 degree phase shift in the peak.

Large phase delays are also created in models which postulate a resonant response in the Earth. Note that this is a different kind of delay than the integrative delay we have been discussing. As we discussed when we gave an equation for the phase shift in Section 3.1.3, at resonance, the delay between the response function and the driving function is 1/4 cycle. Can the phase delay ever be more than 1/4 cycle? For a resonance, if the driving frequency is much higher than the resonant frequency, then the expected phase delay is π radians, i.e. an entire half cycle, as we discussed in Section 2.1.1. Phase shifts *less* than 1/4 cycle are characteristic of responses that are far below resonance.

There can be additional delays that depend on the way in which the delay is measured. If we measure delay by the time between the peak of the driving force and the peak of the response, then the delay can be much longer—up to a full cycle, even in absence of resonance. We illustrate this in Fig. 4.9, in which the driving force is shown as the solid curve. It is a semi-square wave with a tilted top and bottom. The integral of this driving force is shown as the dashed curve and is triangular in shape. The triangular function is the integral of the driving force. Note that the maxima of the response function is a full half cycle after the maximum of the driving force. If we measure the "centres" of each cycle, instead of the peaks, we get a delay that is more than 1/4 cycle, but less than 1/2 cycle. If we take a sawtooth driving function, then the response can be lagged by 360 degrees! We leave it as an exercise for the reader to show this.

4.5 CARBON-13

Carbon on the Earth contains two stable isotopes, ^{12}C with an abundance of 98.9%, and ^{13}C with an abundance of 1.1%. The ratio of these two isotopes is described by the quantity δ^{13}C, defined by an equation analogous to that for the oxygen isotopes:

$$\delta^{13}\mathrm{C} \equiv \left\{ \frac{\left(\frac{^{13}\mathrm{C}}{^{12}\mathrm{C}}\right)_{\mathrm{Sample}}}{\left(\frac{^{13}\mathrm{C}}{^{12}\mathrm{C}}\right)_{\mathrm{Reference}}} - 1 \right\} \times 1000$$

The reference value is often taken to be a sample known as the "Peedee belemnite" (PDB), but its $\delta^{13}C$ value is very close to that of mean sea water. The current value for the atmosphere is $\delta^{13}C = -7.2‰$. If seawater were in equilibrium with the atmosphere, it would have (at 2 C) $\delta^{13}C = +3.5‰$. The fact that it doesn't reflects biological activity in the oceans.

The lighter isotope, ^{12}C, is more readily incorporated into the organic tissue of plants, leading to negative values for $\delta^{13}C = -20‰$ to $-25‰$. In regions in which photosynthesis is active, this removes typically 10–20% of the dissolved inorganic carbon in seawater, leading to a ^{13}C enrichment in the surrounding water of $\delta^{13}C \approx 3‰$. (A similar biological effect occurs with oxygen, but it has a much smaller effect on the isotopic ratio since the reservoir of oxygen is much larger than that of carbon.) Because different regions of the world have different activity, there is a geographic variation. Warm surface water has the highest $\delta^{13}C$; deep Pacific water has the lowest $\delta^{13}C$. This makes $\delta^{13}C$ a valuable tracer for oceanic currents.

In contrast, there is only small separation of carbon isotopes that takes place in the formation of calcium carbonate shells. Thus the measurement of $\delta^{13}C$ reflects the composition of the ocean water at the time and location in which the shell grew. This isn't completely true, since respiration of the forams has a slight effect on the compensation, but there are methods to take this into account, for example, by taking advantage of a correlation between the size of this effect and the shell size. Thus ^{13}C plays an important role in understanding the role played by biology in climate, and in tracking the flow of deep ocean water. When it reflects biological phenomena, it can yield spectra that are more complicated than those of oxygen and deuterium. Much of its value is found in the time domain rather than the frequency domain; ^{13}C has been used as a primary tool in tracking the thermohaline circulation of the oceans (see, for example, Raymo et al., 1989).

$\delta^{13}C$ records climate change. During glacial periods, biological activity was reduced by advancing glaciers and colder temperature, and light carbon was released into the atmosphere and eventually mixed into the oceans (Shackleton, 1977). $\delta^{13}C$ from benthic (bottom dwelling) forams is typically 0.35‰ lower during glacials than during the interglacials. In contrast, no such change has been seen in planktic (surface) forams. The use of this effect is complicated by the fact that not all benthic foram species record the same values of $\delta^{13}C$.

A complication arises from the exchange of CO_2 between the oceans and the atmosphere. Atmospheric ^{13}C reflects this, as well as the activity of terrestrial plant life. Global distribution of ^{13}C reflects the nutrient content and organic productivity of the water mass. It is more sensitive to biological effects than ^{18}O.

^{13}C is an extremely important isotope for paleoclimate studies, largely because it responds to the presence of life. Yet this is the very reason why we will make only limited use of it in this book. Our focus will be on the links between astronomy and climate, and ^{13}C will prove too complex to offer us much help. In Fig. 4.10 we show the $\delta^{13}C$ data for Site 607 for the last 600 kyr, and the corresponding $\delta^{18}O$ data for the same site. The glacial terminations are evident in the $\delta^{13}C$ plot, once you know where they are, but the pattern is not as simple nor as universal as the pattern for $\delta^{18}O$.

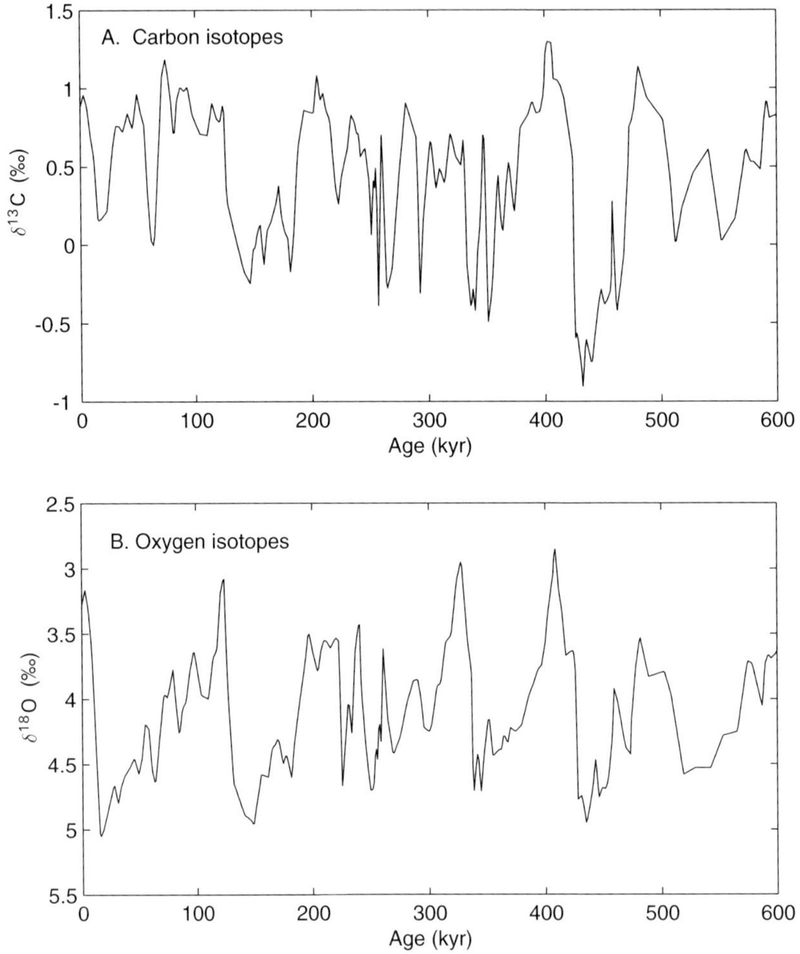

Fig. 4.10. Carbon and oxygen isotopes for Site 607.

4.6 VOSTOK

The ice core recovered from the Vostok site in Antarctica (Petit et al., 1999), located at 78°S and 107°E, covers the longest period of time of any ice record. It reached a depth of 3623 metres. A time scale was derived based on ice accumulation and glacial flow models, fixed to two tie points (at ages 110 and 392 kyr), but not tuned to any orbital model; the accuracy of the time scale is estimated to be better than ± 15 kyr. The untuned time scale makes the Vostok data unbiased with regard to many of the features we are interested in. The record extends back to 423 kyr, which is adequate for many tests of climate theories. For example, the variations of eccentricity contain a component at 400 kyr, and although the Vostok record is long enough to show only one cycle of this variation, one cycle is enough that it

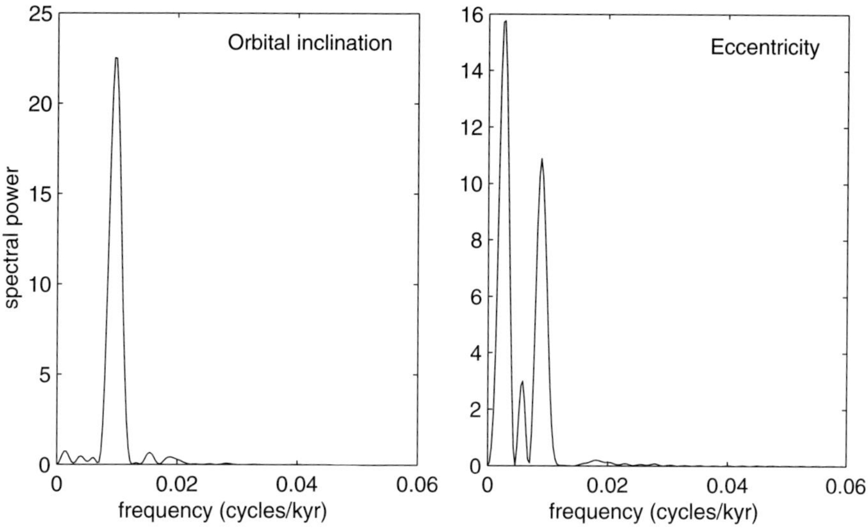

Fig. 4.11. Spectra of inclination and eccentricity, Vostok time interval.

should appear clearly in the spectrum. To illustrate this, we give the spectra of orbital inclination and of eccentricity derived from orbital mechanics for the Vostok time interval in Fig. 4.11.

Many proxies of climate interest were measured in the Vostok core, including atmospheric methane, atmospheric oxygen, deuterium in the ice, dust content and sea salt. Atmospheric methane (Fig. 4.12) is produced by the biological activity of anaerobic bacteria, and it has a very short lifetime in the atmosphere, at present about ten years. Its existence in paleoclimate data is presumed to reflect the area of the Earth covered by swamps and wetlands. Many of these areas are either drained by the dryness accompanying glaciation, or actually covered by the ice itself; the result is that the glacials show low methane in the atmosphere. The spectrum (Fig. 4.13) shows a strong 100 kyr period, but no power near period 400 kyr; it is a much closer match to orbital inclination than to eccentricity.

The Vostok dust record is shown in Fig. 4.14. The observed dust increase during the glacials is believed to reflect reduction in vegetation during those periods and the accompanying increase in wind-blown erosion. The spectrum is shown in Fig. 4.15. It has a strong 100 kyr cycle, and also a strong low frequency variation that is within a full-width at half-max of the 400 kyr eccentricity period.

The sodium concentration (Fig. 4.16) reflects the presence of sea spray aerosols blowing over the Vostok region. The spectrum from these data is shown in Fig. 4.17. There is a strong 100 kyr period. There is also suggestive power near the frequency corresponding to both the 23 kyr precession period and the 41 kyr obliquity period. A minor adjustment in the time scale ("minimal tuning"; see Section 5.2.1) makes both of these peaks strong and statistically significant. The spectra with the tuned time scale will be discussed in Section 8.10.

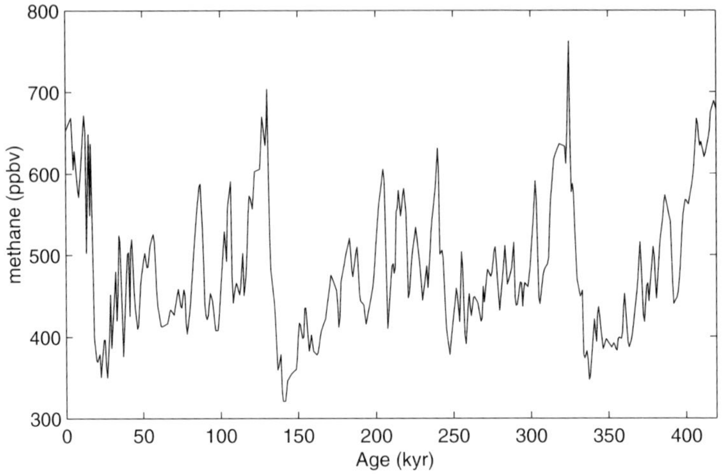

Fig. 4.12. Atmospheric methane record from Vostok.

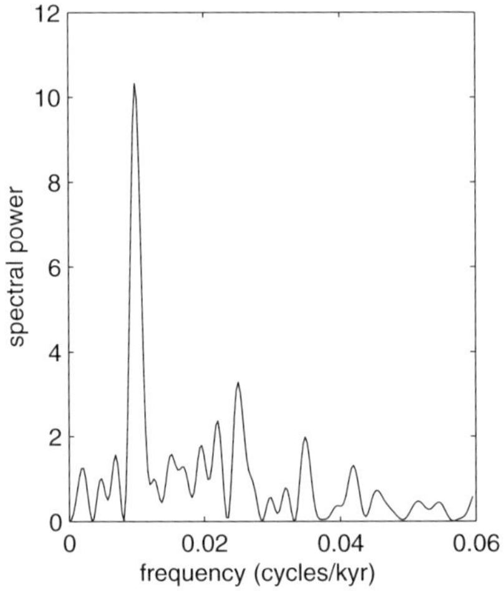

Fig. 4.13. Spectra of Vostok methane.

4.7 ATMOSPHERIC δ^{18}O AND THE DOLE EFFECT

The oxygen in the atmosphere has a δ^{18}O of +23.5‰ compared to that of mean ocean sea water. The difference, called the "Dole effect" (Dole et al., 1954), is attributed to the preferential removal of the lighter isotope ^{16}O from the atmosphere by biological activity.

The Dole effect is often assumed to be time-independent. This implies that the oxygen isotope ratio in the atmosphere is an accurate measure of the oxygen that will be found in the ocean water. We regard this assumption to be a serious mistake. An

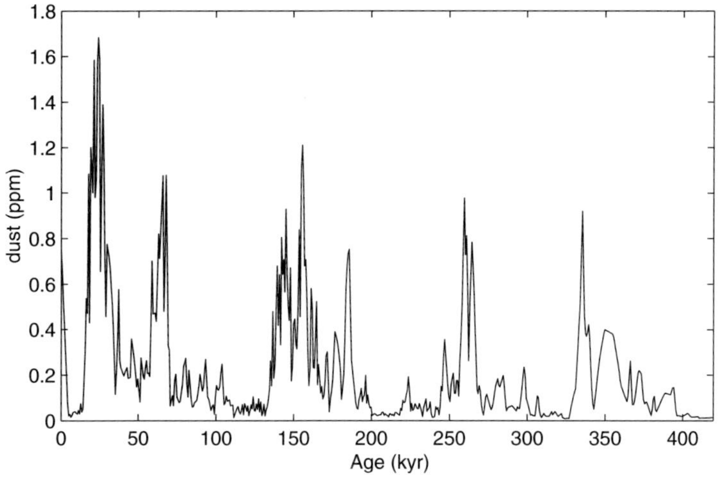

Fig. 4.14. Dust record from Vostok.

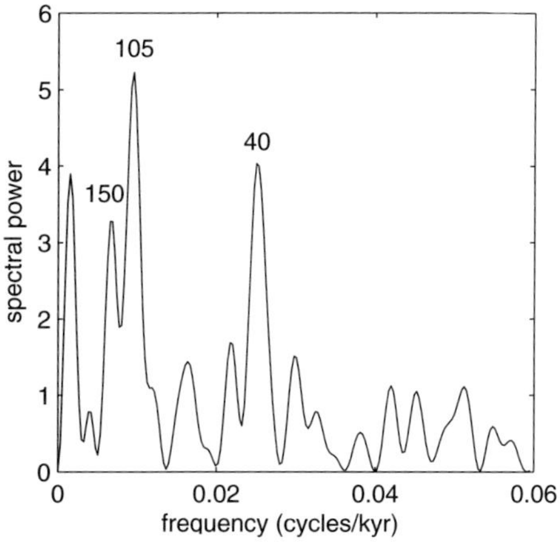

Fig. 4.15. Spectrum of Vostok dust.

example is in the $\delta^{18}O$ records of the SPECMAP stack, compared to those of the atmospheric $\delta^{18}O$ from Vostok, shown in Fig. 4.18.

There are clear similarities. Both signals show the 23 kyr precession cycle. The spectra are shown in Fig. 4.19. Although the same frequencies are present in both data sets, the ratios of the amplitudes are very different. These differences may be called the Dole effect, but it is not correct to assume that, given one signal, there is a reliable way to calculate the other one. Records from Greenland ice that have both atmospheric and ice $\delta^{18}O$ show significant differences that change with time; see, for example, Sowers and Bender (1995) and Bender (1994).

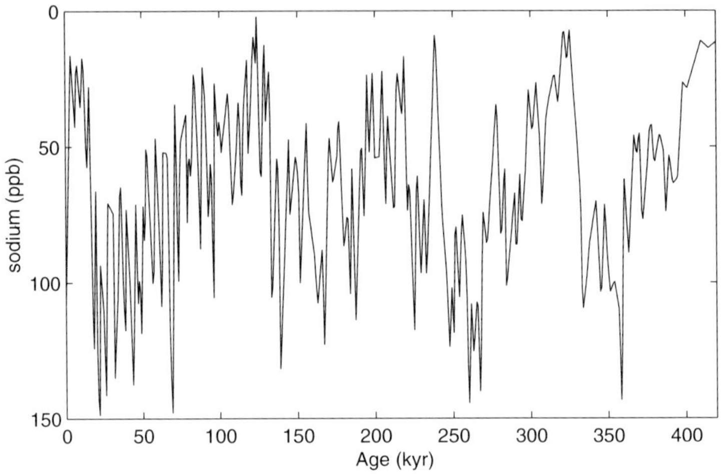

Fig. 4.16. Vostok sodium (sea salt).

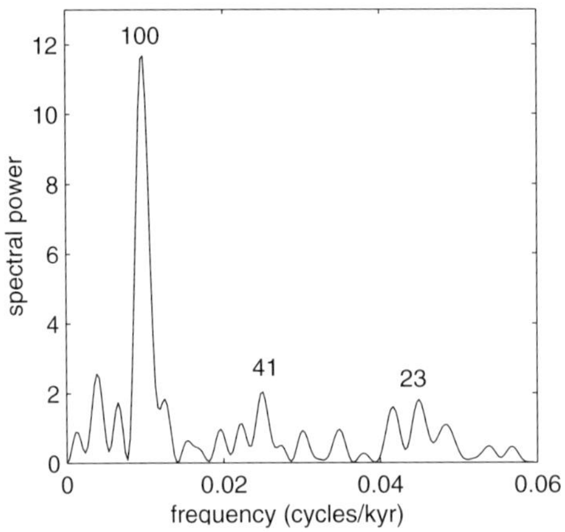

Fig. 4.17. Spectrum of Vostok sodium.

4.8 $\delta^{18}O/CO_2$ MYSTERY

The difference between ocean and atmospheric $\delta^{18}O$ is almost all attributed to biological activity. Since carbon dioxide is also driven by biological processes, one might expect similar behaviour in this record. Atmospheric CO_2 going back 420 kyr has been measured in the Vostok ice core. The spectrum is shown in Fig. 4.20. (For this analysis, we are using the tuned time scale published with the Vostok data, discussed in Section 5.2.1.

The strong peaks at 0.042 and 0.053 seen in the oxygen signal spectrum (Fig. 4.19) are indicative of the contribution of the precession parameter. But these peaks are virtually absent in the carbon dioxide record. Yet both are supposed to be caused by

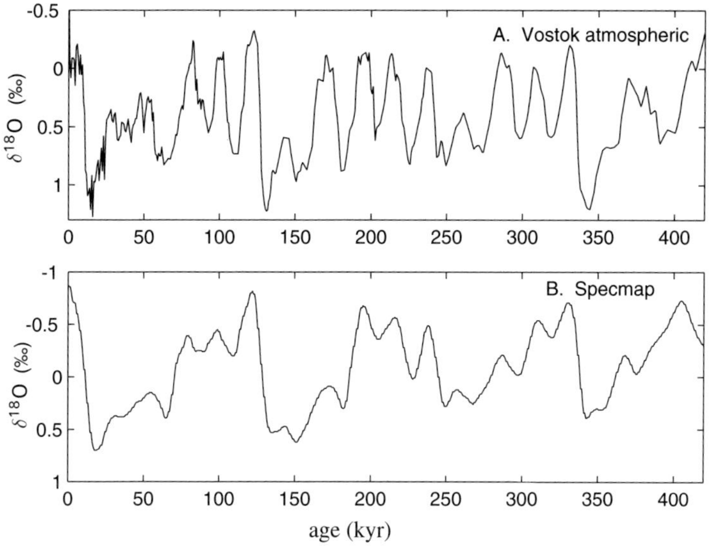

Fig. 4.18. Vostok and SPECMAP $\delta^{18}O$.

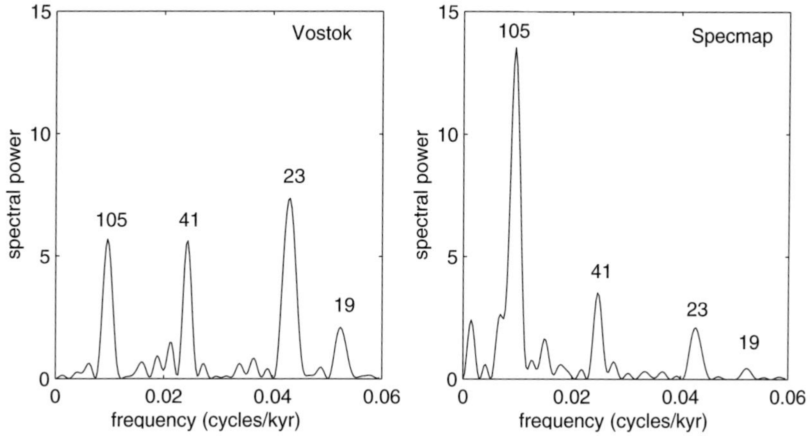

Fig. 4.19. Spectra of Vostok and SPECMAP $\delta^{18}O$.

changes in the amount of biological activity with climate shifts. This is yet another mystery to unravel, and we offer no suggestion. There are still a lot of potential Ph.D. theses left in these data.

4.9 OTHER SEA FLOOR RECORDS

4.9.1 Terrigenous component

The terrigenous component of sea floor sediment is that fraction which is believed to come from land, usually in the form of wind-blown dust. Off the coast of Africa,

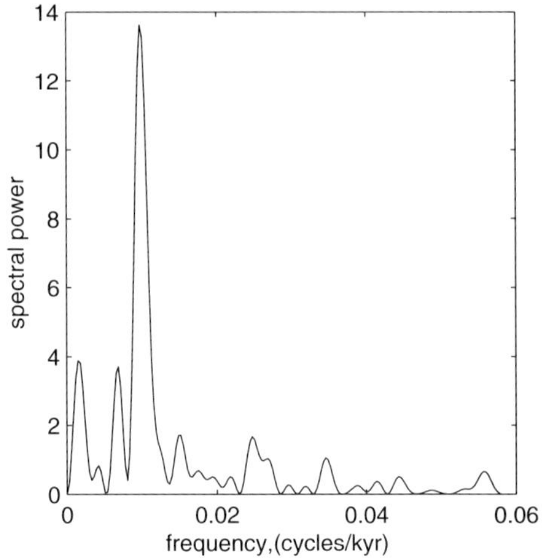

Fig. 4.20. Spectrum of Vostok CO_2.

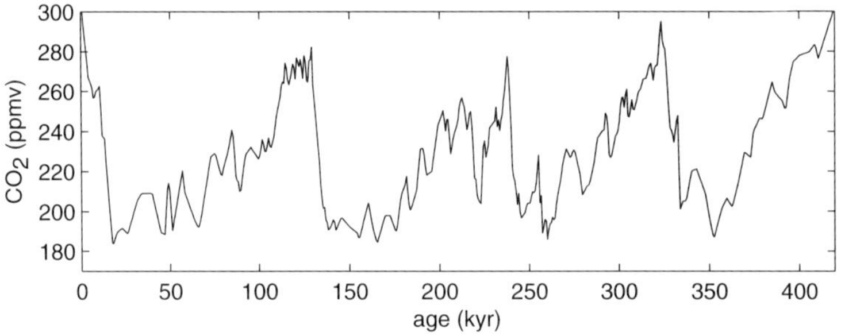

Fig. 4.21. CO_2 measured in Vostok ice.

such dust gives a record of the aridity of the surrounding land. Interesting measurements of the terrigenous detritus were published by Peter deMenocal (1995) for Sites 721 and 722 off the coast of Africa. The data are shown in Fig. 4.22. From these data, even without a spectrum, we can see that the data are dominated by a cycle of about 23 kyr. For example, note that there are 8.5 cycles that occur in the period 5000 to 5200 kyr. This kind of frequency determination in the time domain works well only if one cycle (or a clustered group of cycles of similar frequencies) dominates.

For Fig. 4.22, we assumed constant sedimentation rate; the ages of the beginning and ends of the records were taken from deMenocal's estimates. He also plotted the data using a time scale that was tuned to an insolation record. This forced any oscillations present to match the expected 23 kyr cycle, so it does not provide independent confirmation of the presence of this cycle. We discuss the consequences of such strong tuning in Chapter 5. To try to extract spectral information that does not represent a reflection of the tuning target, we "detuned" the data as follows. We took

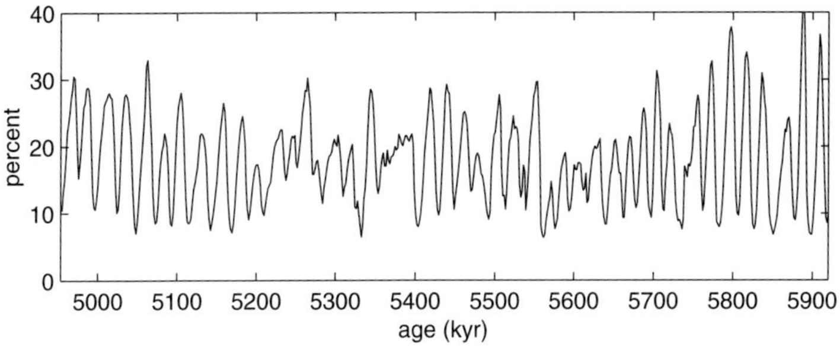

Fig. 4.22. The terrigenous component of Site 721.

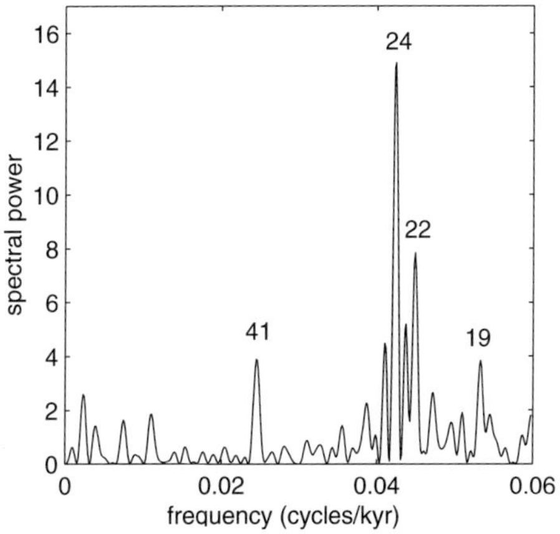

Fig. 4.23. Spectrum of detuned terrigenous component Site 721.

deMenocal's tuning points, and eliminated all of them except for one point every 100,000 years. That gives data that we call "minimally tuned" (see Section 5.8 on minimal tuning). This is because our Monte Carlo studies of tuning have shown that it is unlikely that one tuning point every four cycles (100 kyr for a 23 kyr cycle) will falsely produce a nonexistent cycle. When that was done with the Site 721 data for the age interval 4953 to 5920 kyr, we obtained the spectrum shown in Fig. 4.23.

The most significant frequencies are marked with their periods: 41, 24, 22 and 19 kyr. These are the periods expected if the signals were dominated by insolation. This record shows some of the clearest evidence we have seen that insolation played a major role in paleoclimate. We will discuss another example of the simultaneous analysis of dust and $\delta^{18}O$ for Site 659 in Section 8.7.

4.9.2 Foram size: the coarse, or "sand", component

Many quantities are measured in a sea floor core. One that has not received much attention is the "coarse component", sometimes called "sand". By these terms, oceanographers mean particles that are sand-like in size, specifically, between 1/16

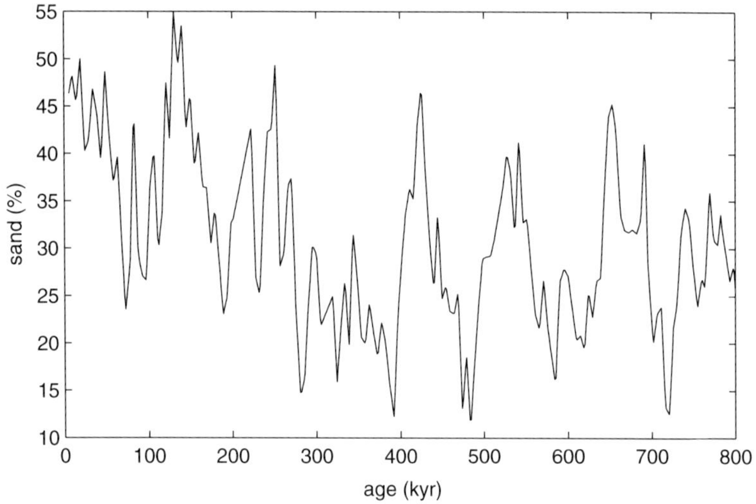

Fig. 4.24. Sand (large foram) component of Site 806.

and 2 mm in diameter.[1] The main component of the sand is frequently large forams; thus, the coarse component reflects an interesting change in the ecology of the oceans. The percentage of sand is measured by sieving and weighing.

In Fig. 4.24, we show the coarse component from Site 806 in the western Pacific. These data play an interesting role in our understanding of paleoclimate, because it was in this component that a clear eccentricity signal was detected in a core that already showed a clear absence of eccentricity in the $\delta^{18}O$ component. We will discuss these points later in Section 8.4.1.

4.9.3 Lysocline: carbonate isopleths

The solubility of calcium carbonate depends on pressure, and therefore on the ocean depth. At a certain depth, the shells of fossil plankton begin to dissolve, and are no longer well preserved. This boundary is called the lysocline. It can be quantified by the percentage of calcium carbonate in the sediment, as a function of depth. One can plot, for example, the depth at which the 60% lysocline is found, as a function of age, and this depends on the depth of the oceans at that age. Figure 4.25 shows the 60% lysocline depth for the past 800 kyr, based on 15 cores measured in the equatorial Pacific by Farrell and Prell (1991).

There are many other phenomena that could affect the depth of the lysocline (for a summary, see Broecker, 1995). Nevertheless, the behaviour of the data is remarkably simple. It appears to be dominated by a 100 kyr cycle, as would be expected if the primary driving force were the depth of the oceans, determined

[1] We marvel at the use of a fraction such as 1/16 in metric units! But it gets even better. Silt is defined as particles in sediment with diameter between 1/16 mm and 1/256 mm.

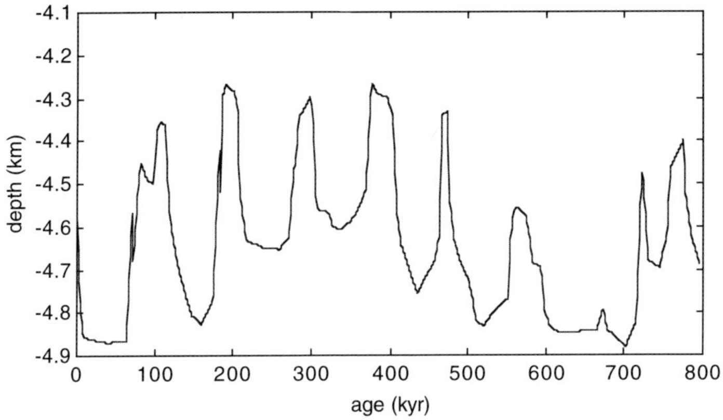

Fig. 4.25. Lysocline depth estimates. The data shown are the 60% carbonate isopleths, in km (Farrell and Prell, 1991).

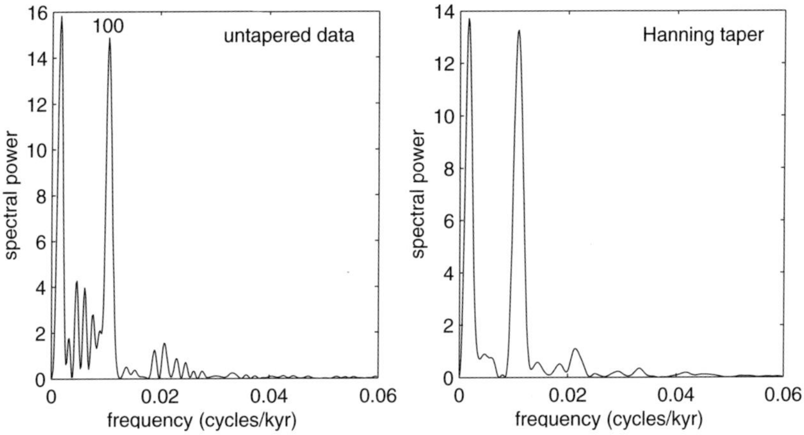

Fig. 4.26. Spectrum of lysocline depth measurements.

(in turn) by the amount of ice accumulated on land. A spectral analysis of the signal confirms this, and is shown in Fig. 4.26. Because of the noise that occurs near the peak, we have included a second spectrum, this one done with a Hanning taper. Such tapers or windows can reduce the background if it is caused by an inaccurate time scale. The taper also causes some loss of resolution (see Section 3.4).

It is clear that the 100 kyr cycle consists of a narrow, single peak. This feature will have interesting implications in the elucidation of the driving mechanism. A second, slightly stronger peak, occurs at $f = 0.00157$ cycle/kyr, corresponding to a period of 650 kyr. This peak derives from the slow modulation of the signal seen in the data.

4.10 TERRESTRIAL RECORDS

4.10.1 Devils Hole

There are several data sets for which the time scale is directly known, and this allows direct testing in the time domain. One of the most important of these is Devils Hole. Devils Hole is the memorable name given to a cave in Nevada that has provided a fascinating record of paleoclimate, and has been the source of a raging controversy. In this cave, $\delta^{18}O$ in calcite on the walls is believed to reflect the isotopic composition of the groundwater that feeds the cave. What makes Devils Hole so important is that the calcite has been extremely well-dated by measuring the ratios of radioactive isotopes of uranium and thorium. (^{238}U decays to form ^{230}Th which decays with a half-life of 77 kyr; the ratio gives a measure of the age.) The interpretation of the data is somewhat uncertain, since it is unclear how to interpret the meaning of $\delta^{18}O$ in groundwater, although higher values of $\delta^{18}O$ are believed to reflect warmer conditions. Note that the plot does not have an inverted *y*-axis. $\delta^{18}O$ was plotted increasing in the upward direction, in order to make the plot look similar to those in the sea floor cores. Of course, the spectrum is completely insensitive to this inversion.

The similarity of the Devils Hole $\delta^{18}O$ variations to those from ocean and ice records (see Fig. 4.2) makes it tempting to accept it as a well-dated, perhaps the best, record of global climate change over the period 60–568 kyr.

But this conclusion has not been readily accepted. The reason is that one of the glacial terminations evident in the Devils Hole data, the rapid rise that takes place between 150 and 135 kyr (Fig. 4.27), turns out to be incompatible with the usual Milankovitch theory. It occurs at a time when the northern hemisphere summer insolation has not yet risen, and yet it is that insolation that is supposed to be the cause of this termination. This has given rise to one of the most debated controversies in the last decade. We discuss this controversy in substantial detail in Section 8.3.

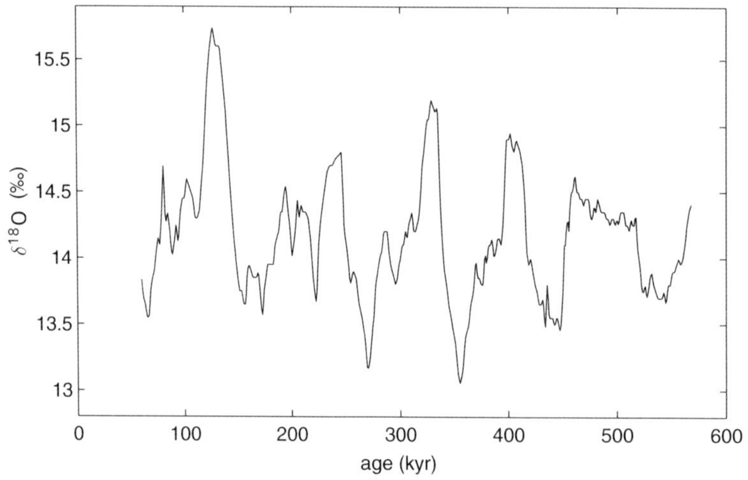

Fig. 4.27. Devils Hole $\delta^{18}O$ measurements.

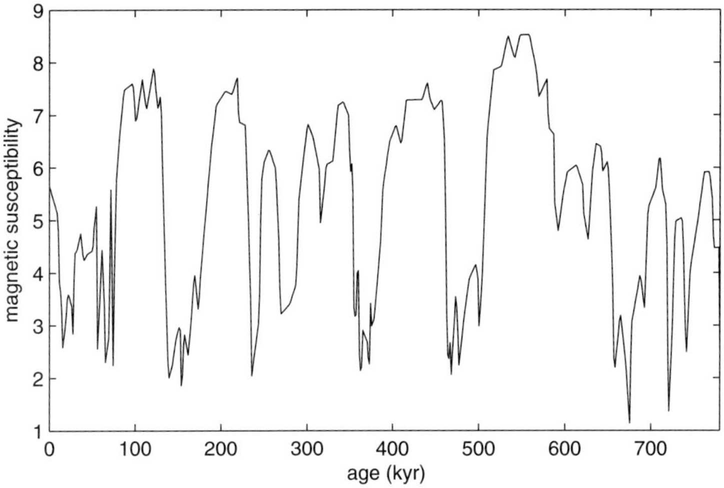

Fig. 4.28. Chinese loess susceptibility.

4.10.2 Loess

Over a million square miles of land in China, Europe and the Americas are covered with a fine yellowish sediment known as loess. Its origin was first correctly identified by the German geologist Ferdinand von Richthofen, who argued it was dust carried by strong down slope winds coming off the mountainous glaciers. George Kukla played a central role in the analysis of these layers in terms of the glacial cycles when he showed that cycles evident in the loess matched those seen in sea floor sediments (Kukla, 1970).

The loess is not deposited at a constant rate, but it is believed that the variation can be taken out by studying the density of entrained magnetic minerals, whose rate of production is assumed to be constant. These can be measured by the magnetic susceptibility of the loess. In Fig. 4.28 we show the derived magnetic susceptibility vs. age for the Chinese loess, based on this compensated time scale (Kukla et al., 1988). The spectrum for the loess is shown in Fig. 4.29. Because of evidence of high background (the small fluctuations in power seen just to the left and to the right of the high peak), we plot the spectrum first with no taper, and then with a Hanning taper (see Section 3.4). The spectra of the loess show a single narrow peak with a period of about 100 kyr.

4.11 FEATURES OF THE DATA

In the data that we have shown, there have been several features which are critical to our understanding that have shown up repeatedly. In this chapter we will discuss the nature of these features in greater depth.

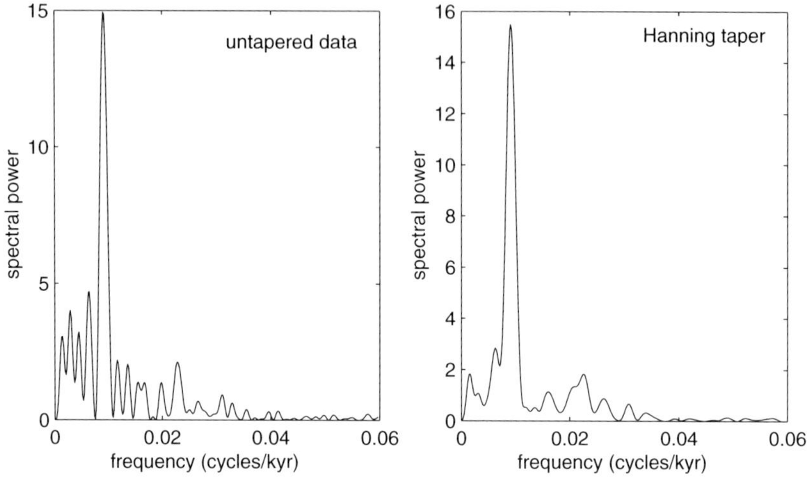

Fig. 4.29. Spectrum of magnetic susceptibility of Chinese loess.

4.11.1 The 100,000 year cycle

The existence of the 100 kyr cycle was discovered by Broecker and van Donk (1970) in $\delta^{18}O$ data from site V12-122, shown in Fig. 3.9. The strength of this cycle has been confirmed in essentially every $\delta^{18}O$ record that covers this age interval, both from benthic and planktic forams. It shows as the dominant oscillation in virtually all the ice records, including deuterium, sodium (sea spray), dust, methane and carbon dioxide. In fact, the only Vostok record that is not dominated by the 100 kyr cycle is the atmospheric $\delta^{18}O$ signal! (See Section 4.7.) This cycle also dominates many other climate proxies, including the lysocline level (Section 4.9.3), and the magnetic susceptibility of the loess formations (Section 4.10.2). The exceptions are interesting, and helpful in elucidating ancient climate. The absence of this cycle in older records is also important. A clear example of such absence is found in the terrigenous component of sediment found in cores off the coast of Africa (Section 4.9.1).

4.11.2 Sudden terminations and sawtooth shape

Broecker and van Donk (1970) also drew attention to the sudden terminations and to the "sawtooth" shape of the cycles. Abruptness of the terminations was also noted by Kukla (1970) in his studies of loess. Nowhere is the abruptness of the terminations more evident than in the Vostok deuterium data, which we plot in Fig. 4.30.

Notice that the "sawtooth" nature is much less evident in these data, although the deepest part of glaciation often occurs just prior to a termination. The interglacials appear as short spikes less than 20 kyr long. We also see that it is not only the terminations of the glacials that are sudden, but that the terminations of the interglacials can also be abrupt.

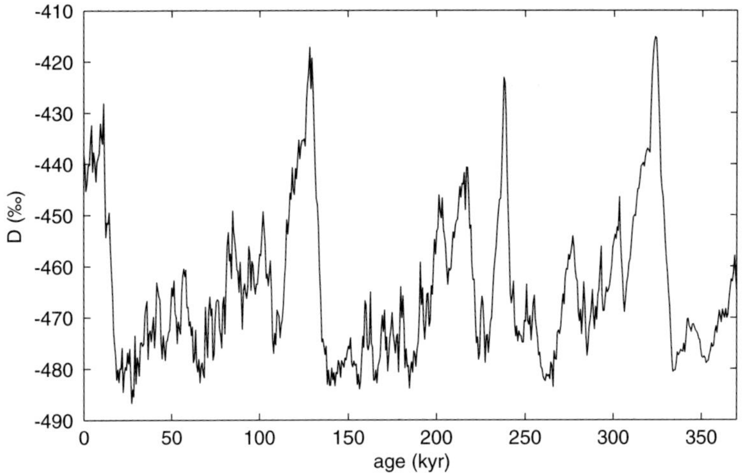

Fig. 4.30. Deuterium record from Vostok ice.

The presence of the sawtooth signal is not readily detected in the frequency domain. In Fig. 4.31 we plot the spectrum from the Vostok deuterium, and next to it we plot the spectrum that we would get from a pure sawtooth at the same frequency. The maximum power occurs for a frequency of 0.0095 ± 0.0021, corresponding to a period of 105 ± 21 kyr. (The frequency uncertainty was estimated using the HWHM rule; see Section 3.14.) The sawtooth signal contains multiples of this fundamental frequency—but note how small they are. One cannot conclude that they are absent in the deuterium spectrum, since the background level is

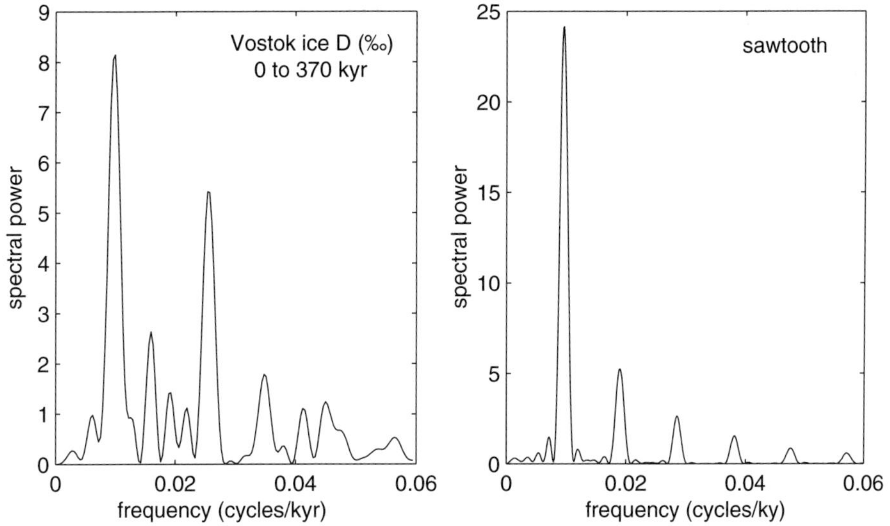

Fig. 4.31. Sawtooth in Vostok? Spectra compared.

comparable to the expected signal size for these peaks; background can just as readily cancel a signal as enhance it. So arguments about the sawtooth shape have to be made in the time domain.

The cause of the sudden terminations is not understood. They do not occur exactly every 100 kyr, even though the spectra show that the bulk of the ice obeys the 100 kyr pattern. All the known astronomical driving forces behave as sinusoids or slowly modulated sinusoids, not as sawtooths. One possible explanation is that the driving signal is sinusoidal, but the response is sudden because of nonlinearities. Catastrophic terminations are common in ordinary climate change. For example, the ice covering rivers in the north tends to disappear very rapidly; as soon as the ice "breaks" at the end of winter, the restoration of surface flow tends to remove the remaining ice very rapidly. Likewise, when the snow cover in the Sierra Nevada is reduced to the point that earth shows, the increase in solar absorption warms the mountains and hastens melting. These processes would suggest that the terminations will be somewhat haphazard, and that their timings may be difficult to account for in detail. On the other hand, it is possible that the terminations are triggered by a combination of astronomical causes. M. Raymo has proposed that the sudden terminations are driven by the relationship between precession amplitudes and linearly changing carbon dioxide levels on the Earth. We discuss this theory in more detail in Section 6.4.4. W. Berger also accounts for the sudden terminations in terms of the precession cycles; see Section 6.4.8.

For some records, the sawtooth shape (or part of it) can be an artifact of variations in the sedimentation rate. For example, assume that the global ice is driven by a sinusoidal signal of frequency f, and that the resulting $\delta^{18}O$ is given by

$$\delta^{18}O(t) = \sin(2\pi f t)$$

and that the sedimentation rate is given by

$$r = \frac{dD}{dt} = 1 + A\cos(2\pi f t)$$

where D is the depth in the sediment and A is a constant. Integrating, we get $D(t)$:

$$D(t) = t + \frac{A}{2\pi f}\sin(2\pi f t)$$

Under these assumptions, the amplitude $\delta^{18}O$ vs. depth is as shown in Fig. 4.32. (We took the constant $A = 15$ for this plot.)

Note that the function of $\delta^{18}O$ vs. depth is a sawtooth, with "sudden terminations" similar to those observed in many records. We do not suggest that this is the sole origin of the sudden terminations. However, it is possible that in some records this effect tends to exaggerate them.

4.11.3 The 41,000 year cycle

The presence of the 41 kyr peak was first seen in the "Pacemaker" stack of Hays et al. (1976) and it was immediately identified as the expected line due to obliquity

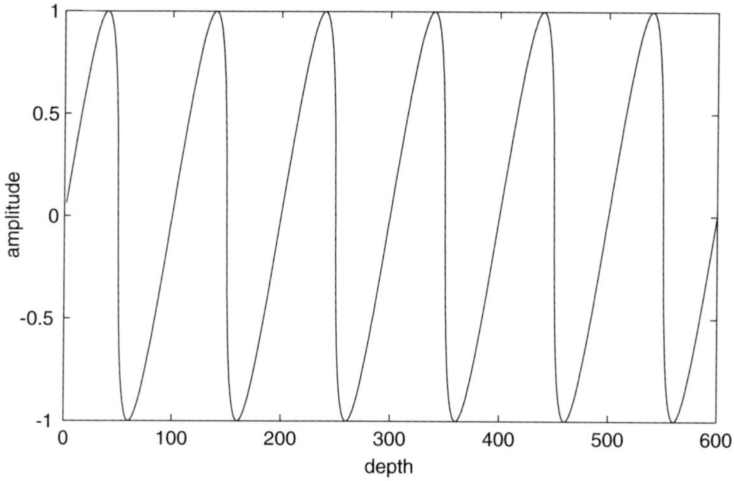

Fig. 4.32. Artificial sawtooth.

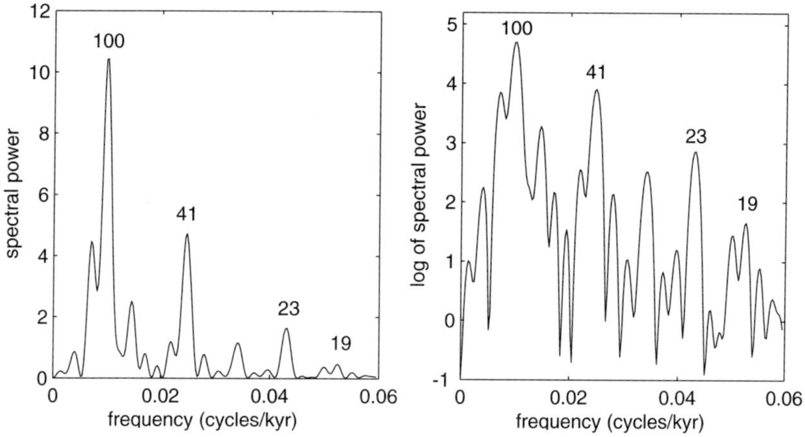

Fig. 4.33. Spectrum of Pacemaker stack, 0 to 464 kyr.

variations. Figure 4.33 shows the spectrum of these data, which cover the period from the present back to 464 kyr. Although the peak shows clearly in the spectrum, it is difficult to see in the data directly, since it is overwhelmed by the 100 kyr oscillation. In Fig. 3.19 we showed the δ^{18}O data from the Pacemaker stack. In Fig. 3.20 we showed the same data with the 100 kyr cycle removed, and in this plot, the 41 kyr cycle can be seen. The cycle is much easier to see in the spectrum.

For δ^{18}O data between 1 and 3 Ma, the 100 kyr signal becomes very weak and intermittent, and the 41 kyr oscillation dominates. This can be seen in Fig. 1.6. You should be able to look at that figure, count cycles, divided by the time interval, and get the approximately correct 41 kyr period. The spectrum of data in this region is

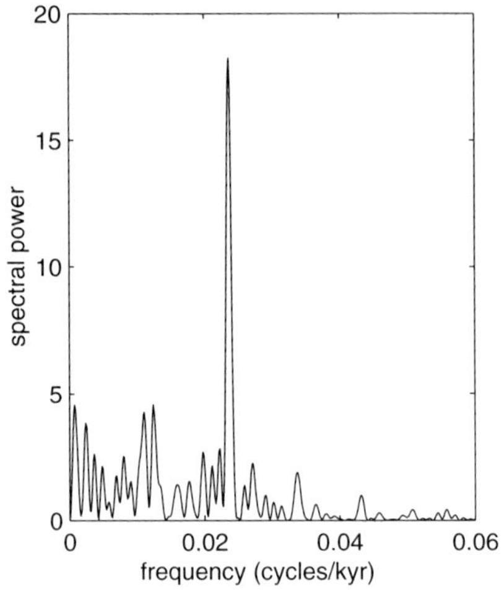

Fig. 4.34. Spectrum $\delta^{18}O$ from Site 607 1.5 to 2.5 Ma.

shown in Fig. 4.34. Although the peak in this plot appears stronger than the corresponding peak in the Pacemaker stack, that is an artifact of the normalisation. As before, the plots are normalised such that the average value is unity; thus the absence of any other strong peaks in the older data set means that presence of the 100 kyr peak in the Pacemaker plot reaches a higher level, by normalisation alone.

Because variations in sedimentation rate can hide the 41 kyr peak (see Section 5.1), it is not easy to measure its strength versus time. The 41 kyr peak is much more susceptible to being hidden by sedimentation variations than is the 100 kyr peak. This can be understood as follows: suppose that the sedimentation rate varies enough to create a 20,000 year error in the age scale. Such an error has little effect on the 100 kyr cycle, but since it is half of a period for the 41 kyr cycle, it would bring a negative excursion to a location in the sediment where one would expect a positive excursion. That would tend to cancel the Fourier amplitude, which assumes (in the calculation) exact phase coherence. A signal that changes its phase with time is said to be "incoherent", and its power in the spectrum can be reduced to zero. It was to pull these signals back that Hays et al. used the process of tuning.

One possible cause for the disappearance of the 41 kyr cycle in some records could be variation of the sedimentation rate caused by the 100 kyr cycle. If climate is changing with a 100 kyr period, then it is certainly plausible that the sedimentation rate would change with that same period. This does not appear to be the case, however. Rau et al. (1999) searched to see if an assumed 100 kyr modulation in sedimentation rate could significantly enhance the 41 kyr peak (or the 23 kyr peak), and concluded that in the data that they looked at, it could not.

It is also true that the short period oscillations such as the 41 kyr peak can easily be artificially enhanced. Minor adjustments to the sedimentation rate of ± 20 kyr can bring *any* random peak into phase with an expected 41 kyr target oscillation. We call this process overtuning, and it is discussed further in Section 5.6. This will create an

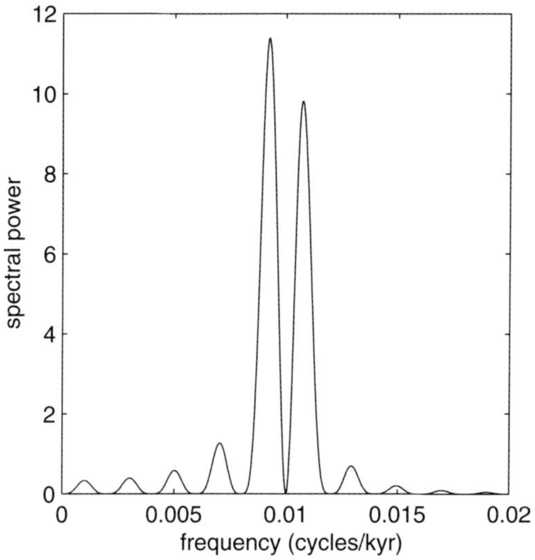

Fig. 4.35. Spectrum of an incoherent sine wave.

artificially high value for the spectral power. It is very difficult to achieve the right balance and to avoid overtuning.

4.11.4 Narrow spectral peaks

When a signal stays in phase with itself for the duration of the record, its spectrum shows a single narrow peak. It is interesting to look at what happens when that condition is not satisfied. Suppose, for example, we generate an artificial wave which is a sine wave for the first 500 kyr, and then changes phase by 180 degrees for the next 500 kyr. (This is indistinguishable from it becoming a negative sine wave for that period.) The power spectrum of such a signal is shown in Fig. 4.35. (Compare with Fig. 2.4.)

The individual peaks are still narrow, but the spectral power is now spread out over a frequency range that is more than twice as broad. If we use the full width at half maximum as the measure, the width is three times as large as the width of an individual peak. The width of the individual peaks is determined by the length of the record; the individual components are sometimes called "speckles", because the speckles seen in a laser beam arise from an identical process in optics. (When you look at a laser beam, the size of the individual speckles is equal to the resolving power of the pupil of your eye.)

If you are new to spectral analysis, then we strongly urge you to use a computer program to duplicate the curve shown in Fig. 4.35. It is also worthwhile staring at it while remembering where the split peak came from. When you work with paleoclimate data, you will see many peaks structures that look similar to this one, and it is important to recognise the simple behaviour that can give rise to it.

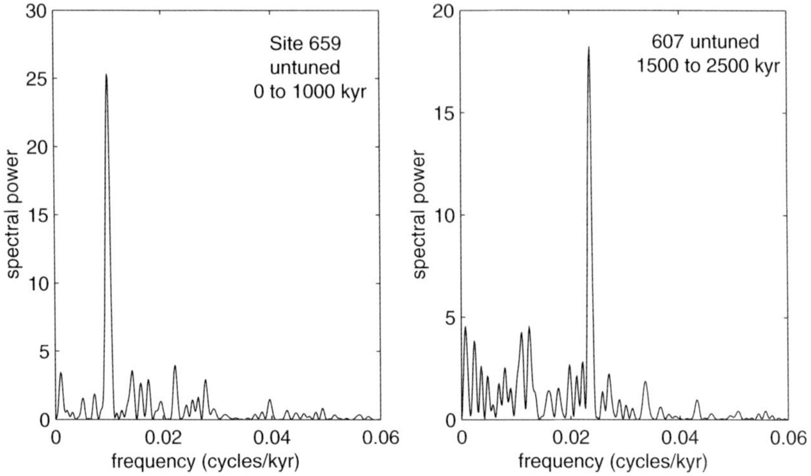

Fig. 4.36. Narrow peaks at 100 and 41 kyr periods.

Given the fact that a sine wave can be so easily broadened, it is truly remarkable that much paleoclimate data show a single narrow peak. Moreover, the fact that they do so in untuned data, i.e. data for which the time scale was derived by assuming a constant sedimentation rate, implies that the sedimentation rate must have been close to constant. In Fig. 4.36 we show two of the most interesting spectra. On the left is the spectrum from $\delta^{18}O$ from Site 659, located off the coast of Africa (see Fig. 4.3) covering the last million years, when the 100 kyr cycle dominated. On the right is a plot from Site 607 in the North Atlantic, covering the period 1.5 to 2.5 Ma, when the 41 kyr cycle dominated. Each spectrum is dominated by a single narrow peak, despite the fact that the sedimentation rates were assumed to be constant.

In Section 2.1 we made the argument that these narrow spectral lines imply that the oscillations are driven by astronomical forces. Models that give rise to such narrow peaks will be discussed in Chapter 6.

4.11.5 The 23,000 year cycle

According to the Croll/Milankovitch theory, the primary driving force for climate is insolation. Spectral analysis of insolation shows that it is dominated by frequencies $f = 0.024, 0.042, 0.045$, and 0.053 cycles/kyr, corresponding to periods $1/f = 41, 24$, 22 and 19 kyr, although other cycles (e.g. period 29 kyr) sometimes appear. The 41 kyr peak is the primary oscillation of obliquity (the tilt of the north pole towards the Sun); the other three periods are those of the precession parameter (the delay between perihelion and summer). The relative amplitudes of these cycles change with time, so for comparison with data, the spectrum of insolation should be calculated for the identical time interval as the data. In addition, the relative strengths depend on the latitude and month that is chosen for insolation.

Example: Vostok atmospheric $\delta^{18}O$

The atmospheric oxygen signal captured in the Vostok ice core shows one of the most beautiful insolation signals seen in paleoclimate data. Figure 4.37 shows this $\delta^{18}O$ record from Vostok.

For comparison, Fig. 4.38 shows the insolation calculated for July, 65N. Although the signal is not identical to the Vostok oxygen, the similarity is striking. (We'll show an even more striking similarity soon.)

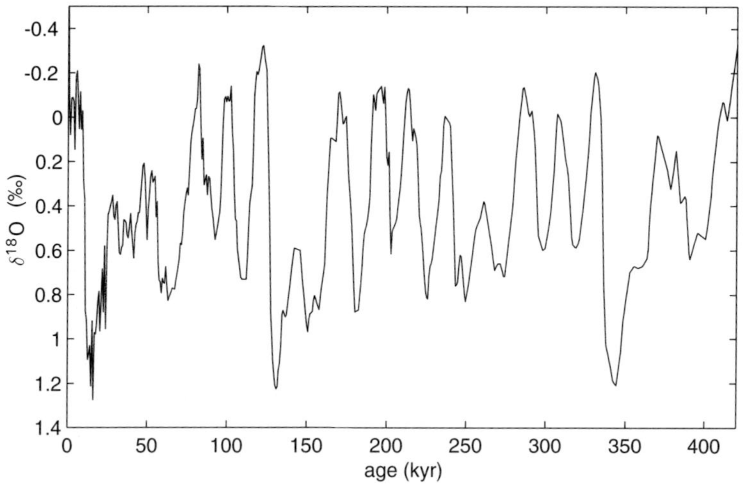

Fig. 4.37. Atmospheric $\delta^{18}O$ from the Vostok core.

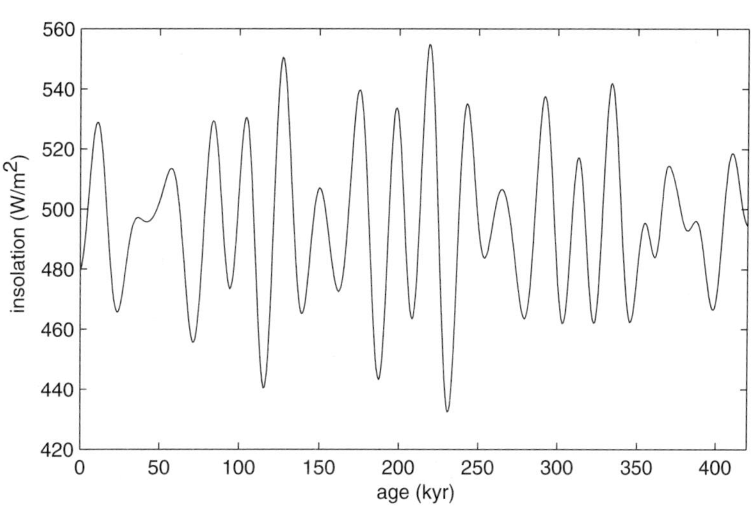

Fig. 4.38. Insolation, July 65N. Note similarity to Fig. 4.37.

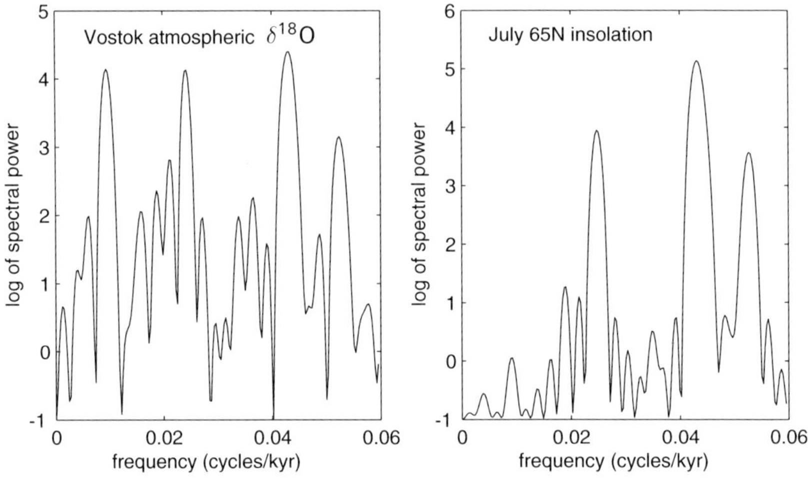

Fig. 4.39. Vostok $\delta^{18}O$ spectrum compared to July 65N insolation.

In Fig. 4.39 we show the spectra of the $\delta^{18}O$ and the insolation on a semilog scale;[1] for the Vostok data on a linear scale, see Fig. 4.20. The peak at 0.043, present in both spectra, is the primary 23 kyr peak of precession. In longer data sets (which thereby have better resolution) the 23 kyr peak is split into a doublet with periods 22 and 24 kyr. There is another precession peak at 0.053/kyr (see in the right plot); although the Vostok data are noisy, the power seen near this frequency region in the $\delta^{18}O$ data is compatible with this peak.

It is worthwhile to expand on the role of background in affecting data. In the Vostok data, the background is not random—it is true climate variation. We call it background only because it has no clear identification as being of astronomical origin. If we ignore the clearly astronomical peaks in the Vostok data, we estimate the background to be at the level of approximately 1. When the background level is comparable to the expected signal, as it is here for the 19 kyr peak, the noise could enhance the signal, doubling its strength and quadrupling its power; on the other hand, it is equally likely that the background could completely cancel the signal.

From the spectrum we immediately see an aspect of the data that is not evident in Fig. 4.37: the presence of a strong 100 kyr (frequency 0.01) signal. The Vostok signal shows a strong peak at 23 kyr, another at 41, and possibly a third at 19, similar to those seen in the insolation spectrum. The 23 and 19 come from the role of precession in determining insolation, and the 41 comes from obliquity.

June or July are usually the months chosen for insolation calculations, but we found that the closest match between insolation and the Vostok $\delta^{18}O$ signal occurs

[2] For this analysis, we are using the tuned time scale published with the Vostok data, discussed in Section 5.2.1.

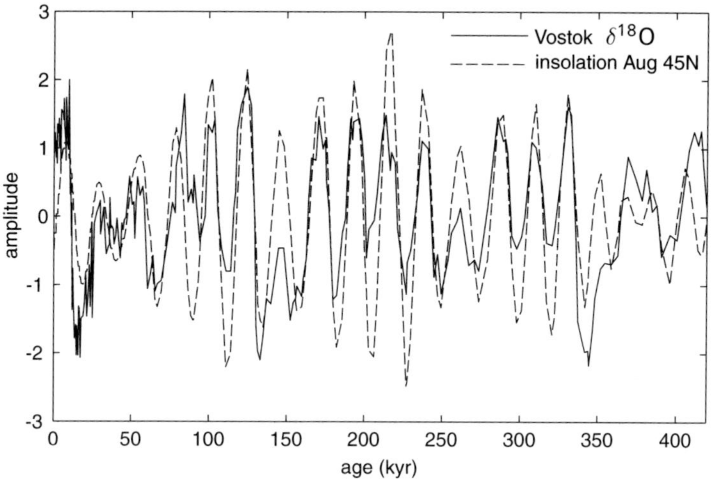

Fig. 4.40. Vostok oxygen and August 45N insolation. No phase lag used.

for August insolation. In Fig. 4.40 we show the August 45N insolation superimposed on the Vostok δ^{18}O.

Example: Terrigenous component

Another example in which the insolation signal is clearly seen is in data showing the terrigenous component of sediment collected from Site 721 (deMenocal, 1995); see Section 4.9.1. As published, the data were tuned to insolation, so to avoid the possibility of circular reasoning in our demonstration, we have retuned the data by removing all tie points except for one every 100,000 years. From our experience with Monte Carlo simulations, we have found that this will qualify as "minimal tuning", and is unlikely to cause important artifacts in the data (see Section 5.8). These minimally-tuned data are shown in Fig. 4.41.

The spectrum of the Site 721 terrigenous component is shown in Fig. 4.23 on page 125. Because it represents a longer time period (1 million years) it has a higher

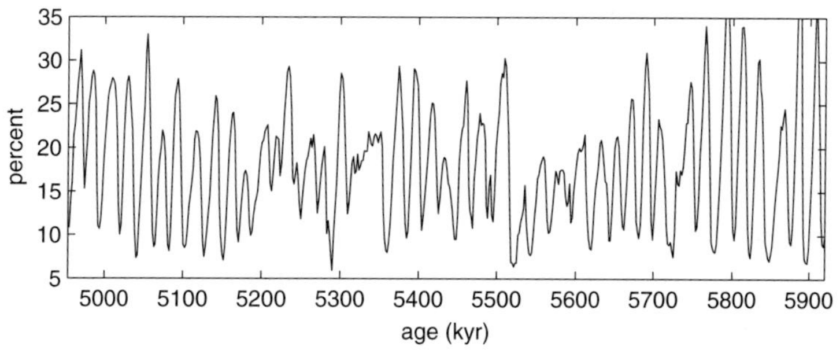

Fig. 4.41. Terrigenous component of sediment, Site 721.

resolution (narrower peaks) than the Vostok spectrum. In this spectrum there appears to be good evidence for the split nature of the 23 kyr peak, as expected if it is due to precession. Likewise the 19 kyr peak appears to stand above the noise, as does the 41 kyr peak. (Methods for determining the statistical significance of such peaks were discussed in Section 3.11. For a peak to be significant it must stand well above the local background level.) The total absence of the 100 kyr peak is worthy of note. Also of interest is the relatively low power in the 41 kyr obliquity peak. This relationship is similar to that seen in high latitude insolation spectra (see Fig. 2.14).

5

Time scale and tuning

"They had mastered the ultimate technology, the ability to alter physical reality by will alone. They called this ability tuning."

Dr Schreber, in the movie 'Dark City' by Alex Proyas

As we discussed in the preceding chapters, we have a truly remarkable record of past climate, recorded in the "tape recorders" of the sea floor sediment and glacial ice. Some of the signals have a reasonably clear interpretation, such as the global ice and temperature proxy, $\delta^{18}O$. But even for these records, the tape recorder ran at a variable speed. The sea floor accumulation rate has not been constant, so it is difficult to compare the record to the predictions of astronomy, which generally give their predictions vs. time rather than vs. depth. Variation in sedimentation rate has sometimes been referred to as "chatter", in analogy to old tape recorders, for which low chatter in the tape speed was one indicator of quality.

When we say we have a "time scale", we mean that for any sample we have a means of estimating the age. In tree rings and for the upper layers of glaciers, the age can be obtained by counting layers. For the GISP2 ice core, layer counting can be used back 40,000 years although, because of ambiguities, the age scale is assumed to be accurate only to 1%. For some samples (e.g. Devils Hole), the age is directly measured using radioisotope dating techniques. In paleoclimate, the most useful radioactive isotopes have been C-14, K-40 and U-238. It is important to recognise, however, that very few sediment records contain sufficient radioactive material to allow them to be dated. If you want to make a contribution that will revolutionise our understanding of the past, and perhaps win the first Nobel Prize for geology, invent a general method for determining the age of sediment or sedimentary rock!

The best time markers in sea floor cores have proven to be geomagnetic reversals. When these can be identified in sediment, they provide a horizon that allows the oceanographer to relate the event to one that has been dated elsewhere. Such reversals provide "tie points" consisting of a (depth, age) pair to which other estimates of age can be related. Unfortunately, good tie points are hard to find. The last geomagnetic reversal, called the Brunhes/Matuyama reversal, was 790,000 years ago. If sedimentation rate were constant since then, there would be no problem, but this was a time full of great cycles of glaciation. We certainly cannot depend on the fact that sedimentation was constant when the climate underwent such huge changes. But even though we can't depend on it, it sometimes turns out to be true. We will show that there are several records which show remarkable evidence for a virtually constant sedimentation rate for periods up to a million years.

5.1 EFFECTS OF CHATTER AND DRIFT

Chatter refers to variations in the sedimentation rate that occur on a time scale short compared to the changes of climate. If the sedimentation rate is not constant, but we incorrectly assume that it is, then we will assign incorrect ages to the samples. Some glacial cycles will appear to occur before their true time, others after, even if the sedimentation rate is correct on average (since it might, for example, be tied to a radiometric point).

To make a simple chatter model we can study, let us assume that the sedimentation rate is varying in such a way as to distort the age scale in a sinusoidal manner. (All variations in sedimentation rate can be written as a sum of sine waves, so this example is quite general.) Assume that the true variation of the measured δ^{18}O, is given by

$$\delta^{18}\mathrm{O} = A \sin(\Omega t)$$

but because of the sediment variation, the signal appears to have a phase modulation:

$$\delta^{18}\mathrm{O} = \sin[\Omega t + \varepsilon \sin(\omega t)]$$

We will call the variation "chatter" only if $\omega \gg \Omega$, i.e. if the sediment rate varies more quickly than does the climate. This would be true, for example, if a 23 kyr insolation cycle varied sedimentation, and distorted the signal from a 100 kyr glacial cycle.

The spectrum, with a correct time scale, would have a single narrow spectral line. But because of the chatter, the spectrum of the observed data is broken into several lines. This can be seen by using a mathematical expansion for the phase-modulated sine wave:

$$\sin[\Omega t + \varepsilon \sin(\omega t)] = J_0(\varepsilon)\sin(\Omega t) + \sum_{n=1}^{\infty} J_n(\varepsilon)[\sin(\Omega + n\omega)\,t + (-1)^n \sin(\Omega - n\omega)t]$$

Don't worry if you are not fluent in Bessel functions (the J_n symbols in the equation). The important thing is that this expansion shows the power that was originally confined to one peak at angular frequency Ω, is now distributed among a series of frequencies, with frequencies $\Omega \pm n\omega$, where n is an integer. The key observation is that a small variation at a high frequency (i.e. with $\omega \gg \Omega$) will move spectral power to high frequencies, not to nearby frequencies. This reduces the statistical significance of the peak, and causes an underestimate of the power. The amplitude left in the original peak is given by the Bessel function J_0. If $\varepsilon \ll 1$, then $J_0 \approx 1 - \varepsilon^2/4$, so not much power is lost. (That's all you really have to know about Bessel functions for this purpose!)

In addition to chatter, sedimentation rate can also show *drift*, by which we mean variations that take place slowly compared to climate variations. Drift corresponds to $\omega \ll \Omega$. Once again, the power is taken from the main peak and dispersed to frequencies $\Omega \pm n\omega$, but since ω is small, these frequencies are now close to the main peak. Drift thus tends to broaden peaks. The broadening can also be understood by recognising that if the sedimentation rate changes slightly, from S to $S + \delta S$, then the apparent frequency f will be changed to $f(1 + \delta S/S)$. It is important to realise that a broadened peak is not just one that is wider, but may appear as a split peak, or as a cluster of peaks.

Note that the fractional change in frequency is independent of frequency:

$$\frac{\delta f}{f} = 1 + \frac{\delta S}{S}$$

This means that the amount of broadening is proportional to frequency. The highest frequencies are the most strongly affected. In fact, a useful method for estimating the effect of drift on weak, high frequency peaks, is to look at the broadening of the larger, lower frequency peaks. So, for example, if a 100 kyr peak is broadened over its minimum width by 30%, then we expect the higher frequency peaks to be similarly broadened. A 30% broadening may have a small effect on a 100 kyr cycle, but it can spread a 23 kyr cycle over a wide range of frequencies, and make it statistically unobservable. For this reason we must be cautious when we see weak signals at high frequencies. The signals may be strong, but hidden by drift.

5.2 TUNING

Tuning is the process of refining a time scale by taking advantage of the known periodicities of the proxy signals. As the term was used by Hays et al. (1976), the assumption was that an approximate time scale already existed, based on linear interpolation between radiometric tie points. This time scale could be "tuned up" by taking advantage of the fact that the cycles present in the data should match known astronomical cycles. Hays et al. took care to point out (p. 1130) that their tuning procedure adjusted the time scale only a small amount, "within the absolute error of the radiometric dates". The original goal was to fine-tune a time scale that had been determined by other means.

Since that seminal paper, tuning has continued to become ever more complex and diverse. It has been expanded to become the primary mechanism for determining a time scale, using a large number of adjustable parameters. There are now about as many ways to tune a time scale as there are ways to do spectral analysis. Virtually all tuning procedures, however, have several steps in common.

1. Choose radiometric tie points. If the data extend back to the Brunhes/Matuyama geomagnetic reversal, or to other identifiable geomagnetic events (e.g. the Jaramillo), then use the best available radiometric dates of these events to constrain your tuning curve.
2. Choose a target curve or "template". Pick one that the data should match. Many different target curves have been used in the literature. These include:
   - insolation, e.g. Petit et al. (1999)
   - obliquity, e.g. Berger et al. (1993)
   - nonlinear ice model, e.g. Berger et al. (1995)
   - eccentricity, obliquity, and precession (all treated equally), e.g. Imbrie et al. (1984)
   - precession, e.g. Shackleton et al. (1990)
   - obliquity + precession, e.g. Ruddiman et al. (1989)
   - previously tuned records, e.g. Tiedemann et al. (1994).
3. Filter the data and/or the target. If tuning to a weak signal, e.g. precession, in the presence of a stronger signal, then (optionally) remove the unwanted signal or isolate the desired signal by applying a band-pass filter (see Section 5.3).
4. Pick the control points. These are the set of depths which will be assigned ages. The set of points (depth, age) will constitute the "tuning curve".
5. Adjust the ages. Do this to optimise the match between the data and the target. This can be done in several ways:
   - matching the spectrum
   - matching events, e.g. terminations.
   - maximizing the cross-correlation or the coherence
   - matching the phase (especially if filtering was used).

5.2.1 Tuning example: Vostok

The initial time scale for the Vostok ice core was derived from an ice-flow model with the assumption (based on similarity of the record to sea floor cores) that two depths had known ages of 110 kyr and 390 kyr.

As we noted earlier, the atmospheric $\delta^{18}O$ from Vostok bears a strong similarity to insolation. This was chosen as the template for a tuned time scale published with the original Vostok data by Petit et al. (1999). They chose to align the peaks of insolation with the mid-points of the glacial terminations. The depths at which these midpoints occur are listed in the first column of Table 5.1, and the non-tuned aged from the ice flow model (the "GT4 age") is in column 2. The third column lists the ages of the insolation maxima. When we match the depths to the ages of the insolation maxima, and interpolate for intermediate depths, then we are using the tuned time scale.

The spectrum of the atmospheric $\delta^{18}O$ with the untuned time scale is shown in Fig. 5.1. On the left is the untapered periodogram. The width of the peaks suggests that the data suffer from drift (see Section 5.1), so in the right figure we show the Blackman–Tukey spectrum with lag of 1/3; this is the same as was published in the original paper (Petit et al., 1999). In the Blackman–Tukey spectrum there is evidence

Table 5.1. Vostok tuning tie points.

Depth (metres)	Age GT4 (kyr)	Insolation maximum (kyr)	Difference (kyr)	Precession maximum
305	10	11	1	12
900	57	58	1	60
1213	83	84	1	83
1528	105	105	0	106
1863	128	128	0	127
2110	150	151	1	151
2350	179	176	−3	176
2530	203	199	−4	198
2683	222	220	−2	220
2788	239	244	5	242
2863	255	265	10	267
2972	282	293	11	291
3042	301	314	13	313
3119	322	335	13	335

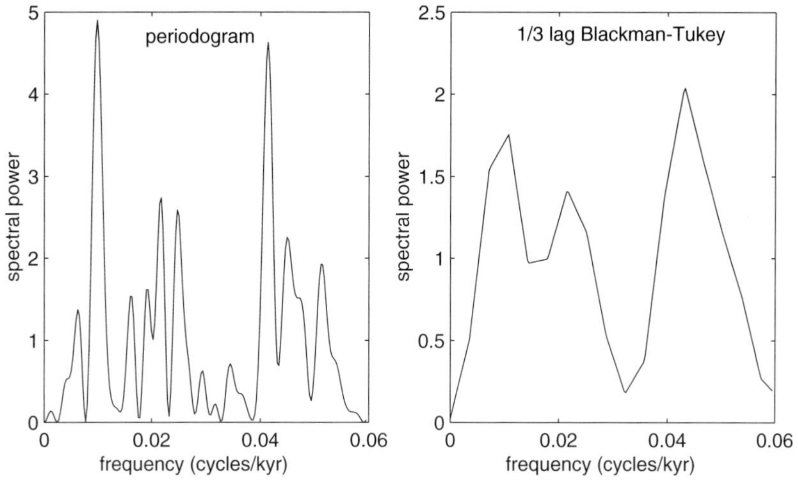

Fig. 5.1. Spectrum of Vostok $\delta^{18}O$.

for three peaks, one near period 100 kyr, one near the obliquity period of 41, and one near the precession period of 23 kyr. The spectrum of July insolation at 65N is shown, for comparison, in Fig. 5.2.

The spectrum for the tuned time scale is shown in the left side of Fig. 5.3. It shows sharp peaks at the obliquity and precession frequencies. The peak near 0.01 (100 kyr period) has become a little stronger, and the low-level background noise appears to

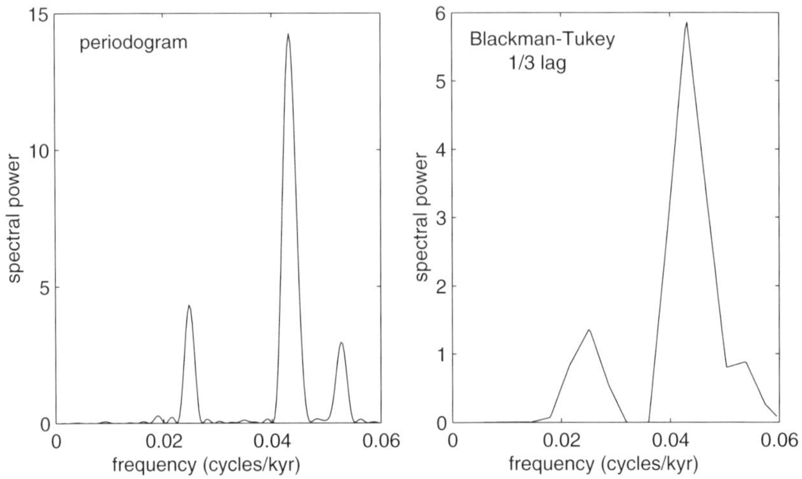

Fig. 5.2. Spectrum of July 65N insolation 0–420 kyr.

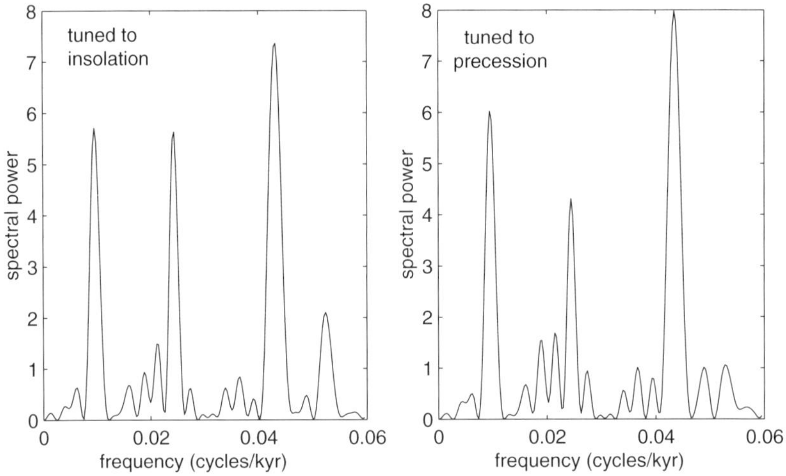

Fig. 5.3. Spectra of Vostok $\delta^{18}O$, tuned time scales.

have significantly decreased. Of course, the fact that the frequencies are such a good match to insolation may be an artifact of the fact that we tuned specifically to insolation.

It is possible that we overtuned, that is, that by adjusting the time scale we made the data match the target, even more than it would if the true time scale were accurately known. To avoid this possibility, we have redone the tuning, using precession alone as our target. We generated a precession curve, made a list of the precession maxima, and listed those in the last column of Table 5.1. These ages then provided a new time scale. The spectrum with this time scale is shown on the right side of Fig. 5.3.

Table 5.2. Vostok tuning points for minimal tuning.

Depth (metres)	2683	2788	2863	3119
New age	220	244	265	335

There are several important features to notice in this spectrum. It was tuned *only* to precession so, not surprisingly, the spectral power in the peak at 0.042 (23 kyr) is stronger than before. The peak near 0.024 (41 kyr) is still strong and prominent, even though the data were not tuned specifically to this peak. And the peak near 0.01 is even stronger than in the insolation-tuned version. This figure should really be compared to the spectrum of the untuned spectrum. The fact that two unrelated peaks were significantly increased and sharpened when tuned to precession is the best possible kind of evidence that we can adduce that the time scale is now more accurate.

You may have noticed in Table 5.1 that several of the tie points represented only a very small adjustment; six of them were adjustments of less than 2 kyr. It is reasonable to ask, are these points really necessary? With a little bit of effort, we obtained a smaller set of tie points listed in the Table 5.2.

If we count the end point (at 420 kyr) then we have effectively five tie points spanning 420 kyr, a significant reduction from the original 15 such points. A reduction of tie points makes it much more difficult to do the kind of minor adjustments that can match any data set to any target, whether it be a valid target or not. The spectrum that we obtain with this minimal tuning is shown in Fig. 5.4.

The process of reducing the number of tuning points as a test to see if your tuning is valid is one way to avoid the dangers of overtuning.

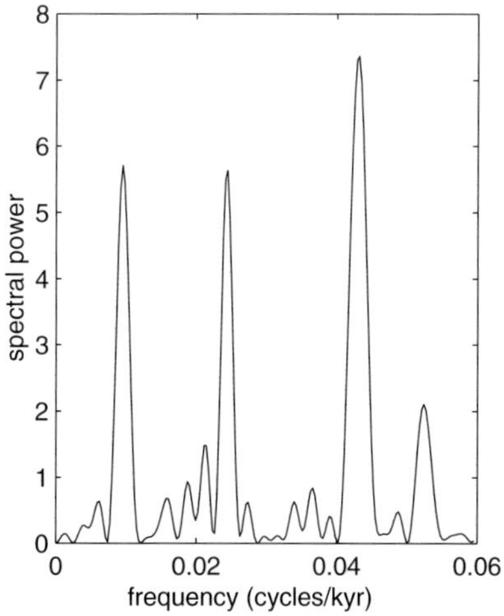

Fig. 5.4. Vostok $\delta^{18}O$, minimally tuned.

5.3 BAND-PASS FILTERING

If a signal is weak, but shows up in the spectrum, then it is sometimes useful to eliminate everything but that signal by the process called band-pass filtering, or "narrow banding". This is something that you do every time you tune a radio or television. Band pass filtering can be done either in the time domain (as it is for radio) or in the frequency domain. Since time domain filtering is useful for real-time signals (radio), a sophisticated art has developed to do this; however, we recommend that you do your band pass filtering in the frequency domain, where you can see more clearly what you are doing. The easiest way to do this is to perform a Fourier transform, cut out the unwanted frequencies, and then do the inverse transform. We give a routine for doing this in Appendix 2.

In Fig. 5.5 we show a random signal created by generating a random number every kyr, then averaging over a 10-kyr sliding window. The spectrum is shown in Fig. 5.6. We now apply a band pass filter with a minimum frequency of 1/18 cycles/kyr, and a maximum frequency of 1/26 cycles/kyr. This was done using the filter function given in Appendix 2. After the band pass filter has been applied, the data appear as in Fig. 5.7, and its spectrum is given by Fig. 5.8.

Note how the signal (both its time behaviour and spectral power) resembles a precession signal. This is not surprising, since we cut out all frequencies except those in the same band as precession signals. The important point is that a precession-like signal will appear out of completely random noise, if the noise is passed through a narrow band filter. Beware of this when you do narrow-banding, or you can fool yourself into seeing a signal that isn't really there.

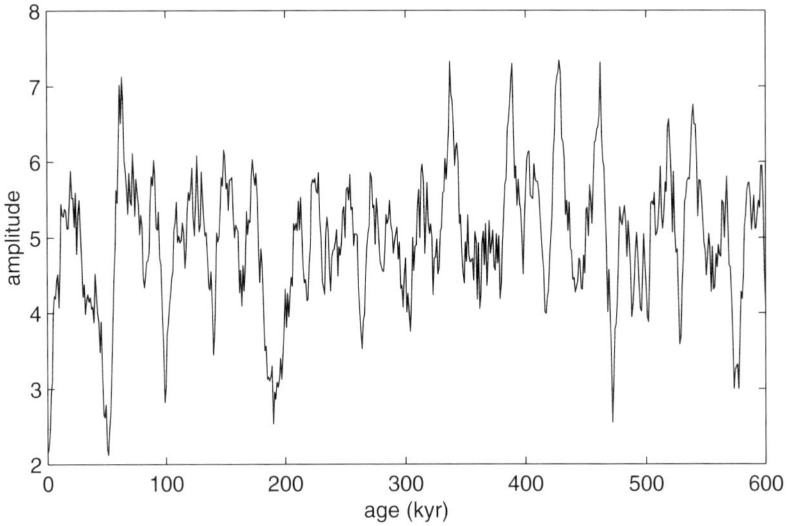

Fig. 5.5. Random signal, prior to narrow banding.

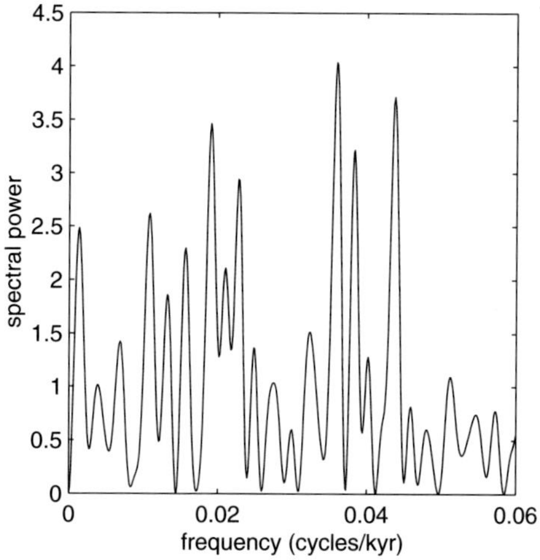

Fig. 5.6. Spectrum of random signal.

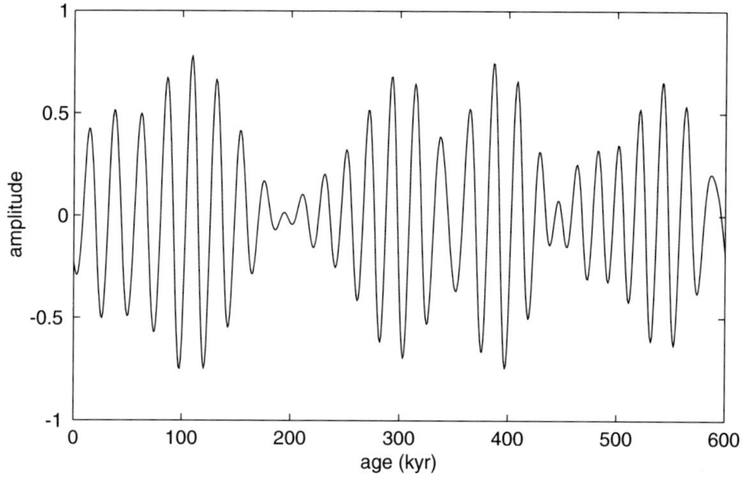

Fig. 5.7. Random signal, after narrow-banding.

5.4 SPECMAP: SUCCESS AND FAILURE ANALYSED

The most influential time scale published in the last two decades of the 20th century came from the "SPECMAP project", and was published by Imbrie et al. in 1984. This work not only became the prototype for future tuning work, but the tuned "stack" of five cores included in the work itself became a target curve for future tuning efforts. However, it is not only because of its importance that we devote a special section to SPECMAP. SPECMAP proved to be a fundamentally correct time scale for the period 0–600 kyr. We will argue that this came about because SPECMAP followed

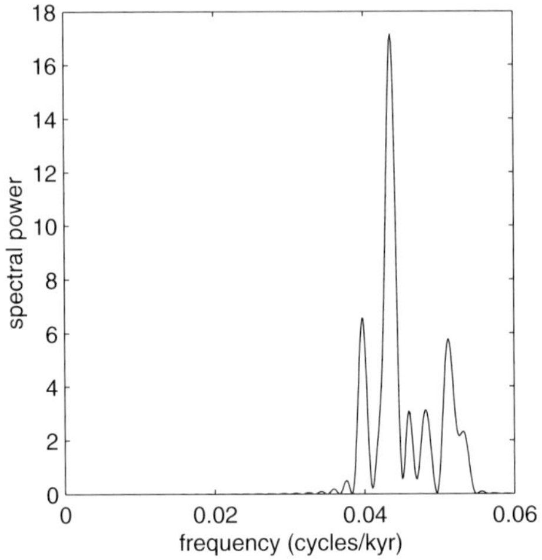

Fig. 5.8. Spectrum of random signal, after narrow-banding.

a procedure that we call "minimal tuning". But it is now recognised that SPECMAP, despite considerable effort to avoid bias, was seriously wrong for the period beyond 600 kyr. It is worthwhile, for anyone who plans to do tuning themselves, to see precisely where SPECMAP went wrong. (And, as far as we know, this point is not discussed in the published literature in any detail.)

SPECMAP was based on five sea floor cores: RC11-20 in the southern Indian Ocean, V28-238 in the western equatorial Pacific, DSDP-502 in the Caribbean just north of Panama, V30-40 in the equatorial Atlantic, and V22-174, also in the Atlantic but 10 degrees south of the equator. The data they used are plotted in Fig. 5.9, taken from their paper, using an untuned time scale.

The time scale for this plot was derived by linear interpolation between (and extrapolation beyond) radiometric control points at 127 kyr and 730 kyr.

Before describing their tuning in detail, it is very interesting to jump ahead to the results. In Fig. 5.10, we show their final $\delta^{18}O$ estimate, obtained by separately tuning the individual five data sets, and then adding them together.

It is worth taking some time to compare the SPECMAP stack with the data from which it is derived, Fig. 5.9. By looking at these curves, you can deduce quite a lot. For example, notice the briefness of the interglacial that occurs in DSDP502b, near 100 kyr. The tuning procedure stretched this peak to be in better agreement with the other peaks in the stack; to do this, it had to assume a rapidly varying sedimentation rate during the first 100 kyr. The individual tuned components are shown in Fig. 5.11, and you can see precisely how the peak in DSDP502b was stretched.

The SPECMAP tuning target curve consisted of obliquity and precession. They did not include any driving force for the 100 kyr cycle in their target, because "there is little agreement on the nature of the physical mechanisms responsible for the climatic oscillations around 100 kyr that dominate the isotopic record" (Imbrie et al., 1984, p. 274). This is an extremely important point. The fact that they did not include the

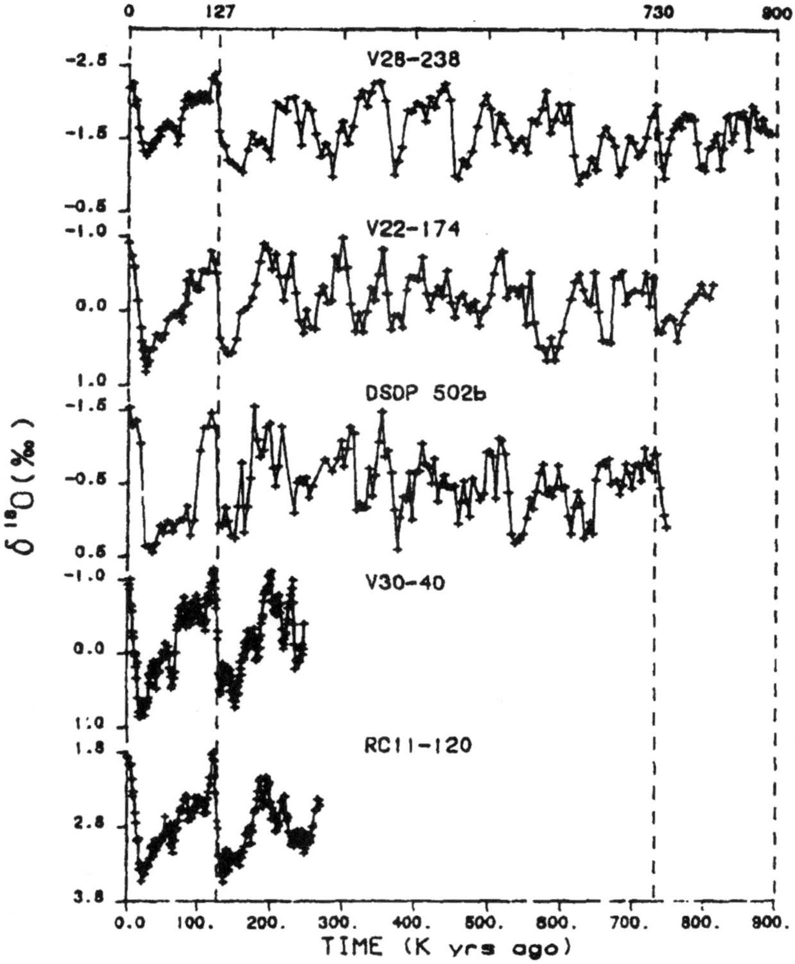

Fig. 5.9. SPECMAP data, untuned.

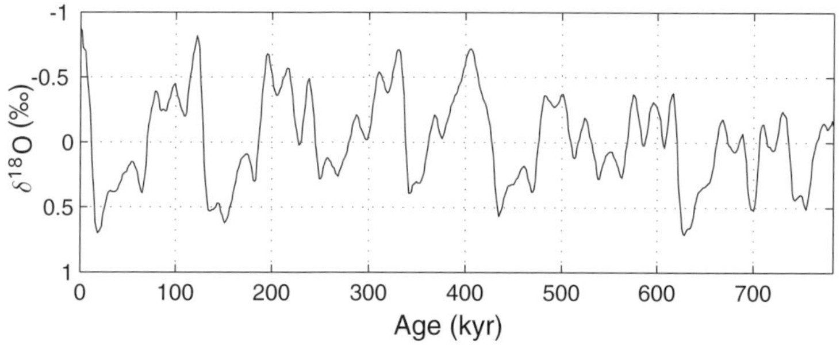

Fig. 5.10. SPECMAP δ^{18}O stack of five cores, mean set to 0.

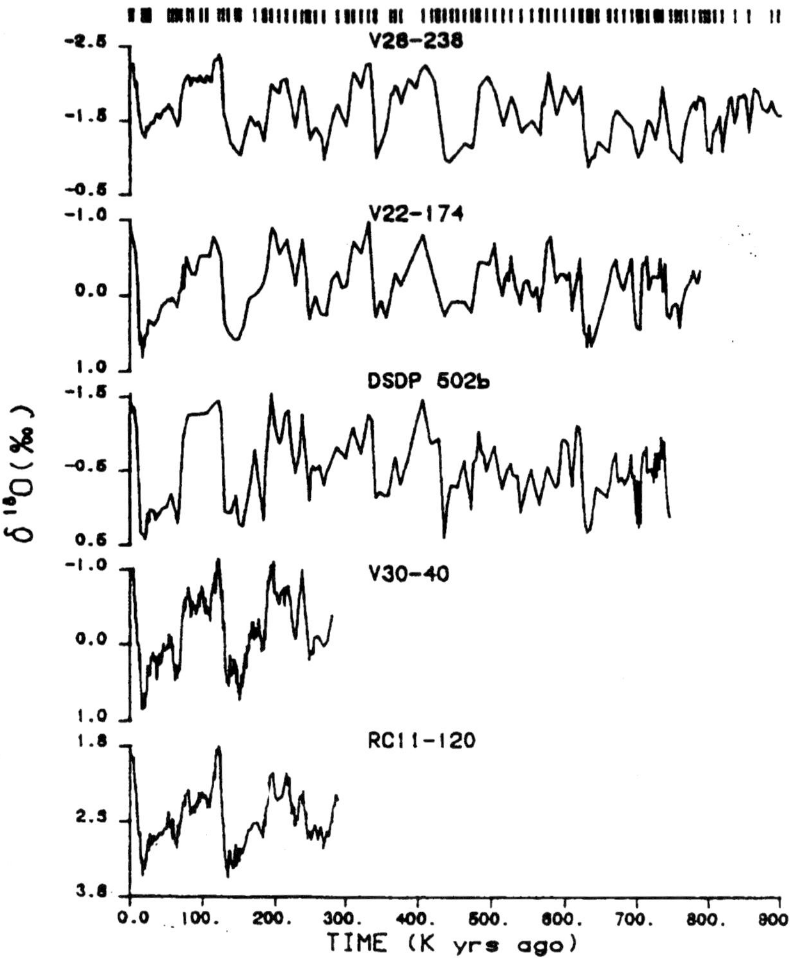

Fig. 5.11. SPECMAP tuned components.

100 kyr cycle in their target means that the 100 kyr cycle can be studied in these data without bias using the tuned time scale. We call the SPECMAP stack "minimally tuned" with regard to the 100 kyr cycle.

Tuning was accomplished by applying narrow-band filters (see Section 5.3) to each of the data sets, one centred around period 41 kyr, and the other centred around period 22 kyr. The filtered data were then compared to the target curves.

For each data set, control points were chosen for adjustment. The number of control points (excluding those determined radiometrically) is given in Table 5.3.

Orbital tuning of the stack started with the untuned set of Fig. 5.9. The narrow-band filters were applied to the data, and the phase compared to that of the targets. When differences were found, adjustments were made in the ages of the control

Table 5.3. SPECMAP control points.

Core	Number of control points	Duration of record (kyr)	Average spacing of control points (kyr)
V30-40	31	281	9.1
RC11-120	30	287	9.6
V28-238	71	892	12.6
V220174	54	784	14.5
DSDP502b	76	743	9.8
SPECMAP Total	262		

points, and then the filtering process repeated. The process continued to iterate 120 times. Imbrie et al. describe the work as "ungainly".

In the fit, Imbrie et al. initially constrained the data to match radiometric control points. One of these was the Brunhes/Matuyama boundary, for which K-Ar measurements had indicated an age of 730 ± 11 kyr (Mankinen and Dalrymple, 1979). Previous attempts to estimate the age of this reversal, based on correlation with astronomical frequencies, placed the age of the reversal at 728 kyr (Kominz et al., 1979; Kominz and Pisias, 1979), 738 kyr (Morely and Hays, 1981), and 790 kyr (Johnson, 1982). All of these were in agreement with the radiometric date except the one by Johnson, which was discarded. Ironically, we now know that the Johnson date was the only one that turned out to be right.

Half way through the iterations, the constraint on the radiometric control points was removed. Presumably this was done to see if a more accurate date would result. The final age for the Brunhes/Matuyama boundary that they obtained for the tuned SPECMAP stack was 734 kyr, 4 kyr older than the starting point. This attempt to obtain a better age is illustrative of the limitations of the tuning procedure. The best present estimate for the age of this boundary is 780 ± 10 kyr (Spell and McDougall, 1992), so the SPECMAP age was about 46 kyr in error. We will discuss more about this error later, including a retuning of the time scale by Shackleton et al. (1990) that helped lead the way to uncovering this error.

From the ages and depths of the control points, one can calculate the sedimentation rates versus time for each of the SPECMAP cores. These rates are plotted in Fig. 5.12.

Note that the variations in sedimentation rates on this plot are huge. For Site DSDP-502, the sedimentation varies by a factor of 20; for Site V30-40, it varies by a factor of 18. The sedimentation rates do not show any particular structure; there is no significant 100 kyr periodicity, for example. Nor do they correlate well with each other. The "stretching" of the peak near 100 kyr for DSDP502b was accomplished by the very low sedimentation rate for this peak between the years 80 and 105 ka; note that this is closest to the lowest sedimentation rate used for any data set for any period. These large sedimentation rate excursions are unphysical. Large spikes in the plot appear to be due to be the attempt by the tuner to bring a local maximum into

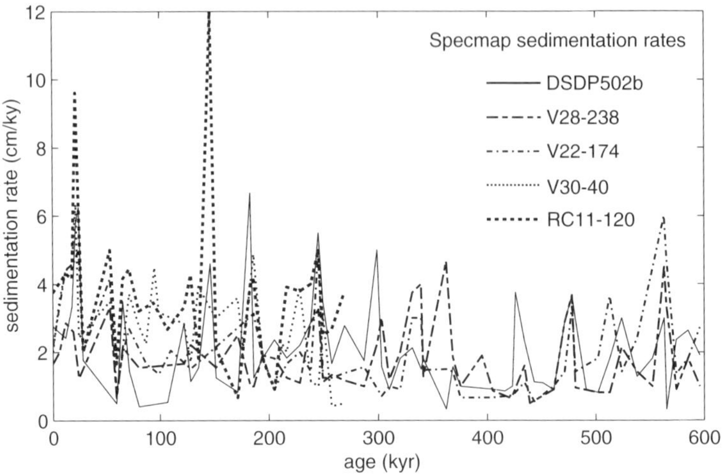

Fig. 5.12. Sedimentation rates derived from SPECMAP tuning.

agreement with a peak in the target precession curve. The spectral power of the SPECMAP stack is shown in Fig. 5.14.

5.4.1 Success

The SPECMAP stack included data from Site V22-174. The $\delta^{18}O$ spectra from this site, both before and after tuning, are plotted in Fig. 5.13.

It is important to note that the data were tuned only to frequencies of obliquity and precession, i.e. frequencies near 0.024, 0.043 and 0.053 cycles/kyr. Because these were part of the tuning target, the increase in power at these frequencies (compare the right figure to the left figure at these frequencies) could be artificial, or at least artificially enhanced. But note what happened to the peak near 100 kyr. Its level of significance rose dramatically, even though it was not part of the tuning target curve.

For an increase in significance of the 100 kyr peak, it is not important to get fine variations of the sedimentation rate correct, but it is important to reduce long-term drift. Thus the short pulses of sedimentation rate that we saw in the sedimentation rate plot will not affect the 100 kyr peak, as long as they are somewhat uniformly distributed. Note that the background in the vicinity of the 100 kyr peak has been reduced too. This is power that was moved into other peaks.

The spectrum of the entire stack is shown in Fig. 5.14. Peaks with high statistical significance appear at 0.01, 0.024, 0.042, 0.044 and 0.053 cycles/kyr, corresponding to 100, 41, 24, 22 and 19 cycles/kyr. Except for the first peak, these correspond precisely to the frequencies expected from insolation or insolation-based models such as the nonlinear ice model. Of course, since these frequencies were included in the tuning target, their amplitude may have been artificially enhanced. Not only could orbitally-driven oscillations be put into phase by the tuning process, but

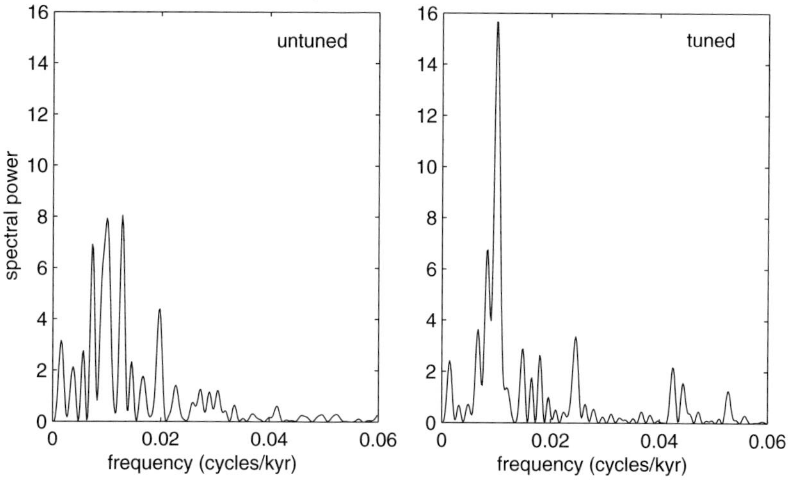

Fig. 5.13. V22-174 spectrum before and after tuning.

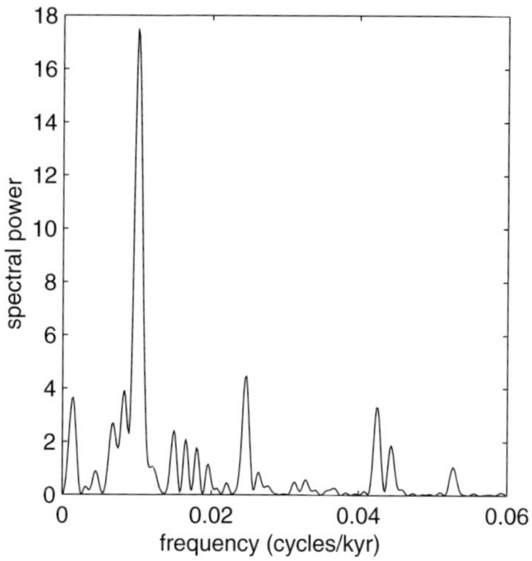

Fig. 5.14. Specmap stack spectrum.

signals due to local effects that are not coherent with orbital frequencies could have likewise been put into phase with the orbital oscillations. Therefore we conclude that the amplitude of the peaks near 41, 24, 22 and 19 represent upper limits of the true spectral power.

One important conclusion that can be drawn from the spectrum is the narrowness of the 100 kyr peak. Since this was not part of the tuning target, the fact that it is narrow confirms that the narrow width is significant, and that the time scale is basically correct—at least over enough of its time scale to make the 100 kyr cycles coherent. The presence of the shoulders on both sides of the 100 kyr peak might indicate the presence of some drift. Of course, we would expect similar shoulders around the other peaks, but they are too low to allow us to check this.

5.4.2 Failure of SPECMAP; triumph for tuning

The SPECMAP data was originally published to cover the period 0–730 kyr, but now the last part of that analysis is accepted (by the original authors) as being incorrect. Measurements of the Brunhes/Matuyama boundary, which appear at 730 kyr in the SPECMAP stack, suggested the age might be considerably older. Two years prior to SPECMAP, Johnson and colleagues (Johnson, 1982) had concluded that the correct age for the Brunhes/Matuyama reversal, based on astronomical tuning, was 790 kyr, but this was in disagreement with the radiometric age of Mankinen et al. (Mankinen and Dalrymple, 1979), and so had been discarded. But in 1990, based on the $\delta^{18}O$ signal from Site 677 in the eastern equatorial Pacific, Shackleton et al. (1990) concluded that the older age was correct. They proposed a time scale that agreed with SPECMAP for the last 620 kyr, but was significantly different for older ages. They concluded that if their work was correct, the age of the Brunhes/Matuyama boundary was "5 to 7% older" than had been assumed in SPECMAP. These older dates were subsequently confirmed radiometrically (Baksi et al., 1991, 1992; Spell and McDougall, 1992). The subsequent confirmation of this age by radiometric means was considered to be the greatest triumph of orbital tuning.

The reason that the Site 677 $\delta^{18}O$ data could make this revision possible was attributed by Shackleton et al. to the presence of a stronger precession component. A few years later, Bassinot et al. found an even stronger precession signal in $\delta^{18}O$ data from Site MD900963 (near the Maldives in the tropical Indian Ocean), and used this to construct a new time scale, with the Brunhes/Matuyama boundary placed at 775 ± 10 kyr, in good agreement with the radiometric dates (Bassinot et al., 1994).

It is interesting and educational to explore the SPECMAP time scale, and see precisely where the procedure went wrong. We will do such an analysis now. This is not done to criticise their methods; as we stated earlier, the SPECMAP time scale was remarkably accurate over most of its range, and it provides good evidence for the nature of the 100 kyr peak. But if we are to avoid making similar mistakes in the future, we must study what has happened in the past (to paraphrase Santayana again).

It is easiest to see what went wrong by dividing the SPECMAP stack into two segments, 0–650 kyr, and 650–730 kyr. The spectra for these segments are shown in Fig. 5.15.

The spectrum on the left, for the last 650 kyr, makes sense. There are peaks near the familiar periods: 100, 41, 23 and 19 kyr. But the spectrum for the older data in the right plot, 650 to 784 kyr, is peculiar. Due to the shortness of the time span, $T = 134$ kyr, the widths of the peaks are much broader than in the left figure, but this broadening has no physical significance. The widths are required by the uncertainty principle (Section 3.2.3) to be greater than $1/T = 0.008$ cycles/kyr. However, the observed structure of the spectrum is quite different from that in the left figure. There

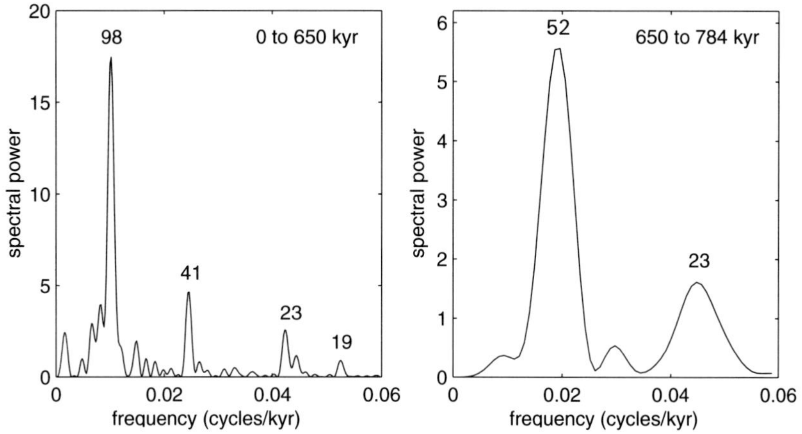

Fig. 5.15. SPECMAP spectrum 0 to 650; 650 to 784.

is a significant bump near 23 kyr, and a much larger bump near 52. We might be tempted to identify the 23 kyr feature with the precession parameter of the Milankovitch theory, but the large 52 kyr peak is a puzzle. The only cyclicity expected near 52 is a minor one attributed to the obliquity, and yet the major feature expected from obliquity (a line at frequency 0.0244 cycles/kyr, period 41 kyr) is totally absent.

However, if we ignore the unphysical broadening of the peaks, we note that the pattern in the left figure is similar to that seen in the right one, but with peak frequencies shifted to the right by a factor of 1.95. Such an artificial shift would come about if the SPECMAP time scale had consistently and uniformly overestimated the sedimentation rate for depths corresponding to times older than 650 kyr. In Fig. 5.16 we show the tuned sedimentation rate for one of the SPECMAP cores, DSDP-502. Note the rise in tuned sedimentation rate for ages older than 650 kyr. The average during this period is 2.6 times higher than it was in the period preceding 650 kyr. For all the cores in SPECMAP that cover this period, the average sedimentation rate, determined by tuning, increases by a factor of about 2.

The mistake made in the SPECMAP tuning was very simple: the 41 kyr obliquity cycle was misidentified as a 23 kyr precession cycle. How could this mistake be made? The answer is that tuning allows you to identify any cycle with any frequency you want, by adjusting the sedimentation rate. The fitting had been done originally to force the Brunhes-Matuyama geomagnetic reversal to be at 730 kyr. Although this constraint was "relaxed" half-way through the fit, the age of this reversal did not change significantly during the fit. The overestimate of the sedimentation rate turned a 41 kyr cycle into something that looked like a precession cycle.

In retrospect, having recognised this error, we can change the SPECMAP time scale in the post 650 kyr period by stretching it by a factor of 1.95, so that instead of covering the period 650 to 784 kyr, it now covers the period 650 to 910 kyr. The new spectrum in the region 650 to 910 kyr now looks as in Fig. 5.17. When this is done,

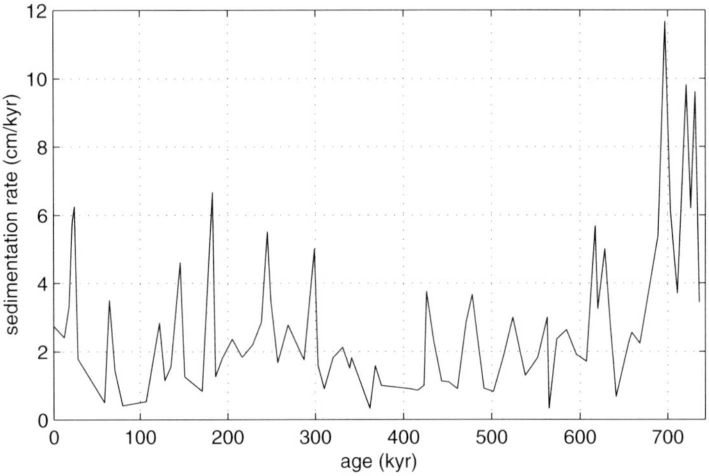

Fig. 5.16. Tuned sedimentation rate for DSDP-502.

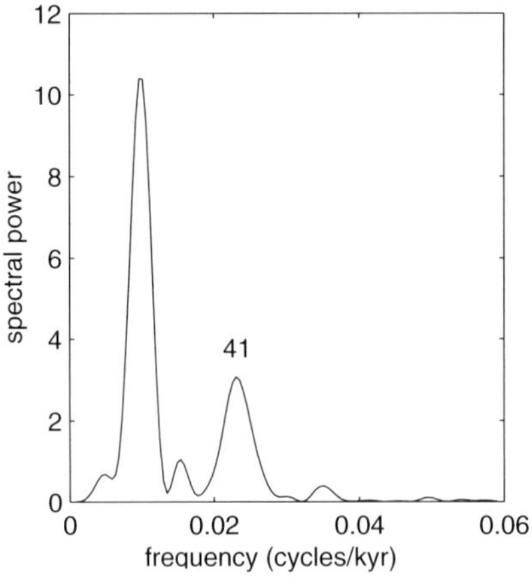

Fig. 5.17. Spectrum for adjusted SPECMAP, 650 to 910 ka.

the previously mysterious 52 kyr peak is transformed into a 100 kyr peak. The spectrum from 650 kyr to 918 kyr looks essentially identical to the spectrum prior to that, except for the relative weakness of the 23 kyr and 19 kyr peaks. This may be a consequence of the fact that these peaks are exaggerated in the 0–650 kyr region by the tuning process.

The SPECMAP stack, with both the original and adjusted time scales, is shown in Fig. 5.18.

The new age t_a is related to the SPECMAP age t_S by the following rules:

$$\text{for} \quad t \leq 650\,\text{kyr}, \qquad t_a = t_S$$

$$\text{for} \quad t \geq 650\,\text{kyr}, \qquad t_a = 650 + 1.95\,(t_S - 650)$$

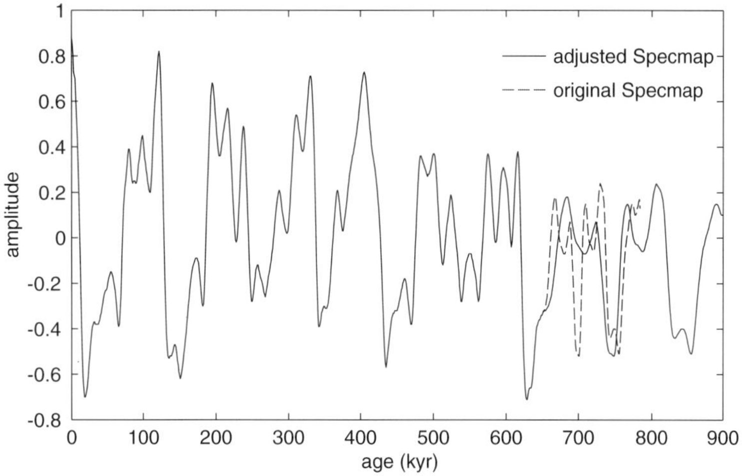

Fig. 5.18. Original, and adjusted SPECMAP.

Note that there is no significant bump near period 23 kyr ($f = 0.043$) and barely a weak hint near 19 kyr ($f = 0.053$). Although this appears to be a weakening of the "precession" line compared to the 0–650 kyr data, it may instead be a reflection of the fact that the strength of the peak in the region 0–650 kyr had been artificially enhanced by tuning. The region 650–910 kyr is the only region in the SPECMAP time scale that was not specifically tuned to enhance this peak, since the time scale for this period was, in effect, tuned on the obliquity line at 41. It is a useful exercise (left for the student) to redo the complete SPECMAP time scale, using the 100 kyr peak and the 41 kyr peak for tuning, and use this to search for the presence and strength of the precession frequencies. The possibility of doing the tuning in this manner was mentioned in the original SPECMAP paper (Imbrie et al., 1984), and then rejected because of the claim that "the length of a digital filter required to achieve the necessary resolution makes this strategy undesirable".

The SPECMAP data now indicate that the 100 kyr cycle dominated the glacial cycles back at least to 900 kyr. This conclusion agrees with the findings of Berger et al. (1993, 1994, 1995) for Site 806. With the adjusted SPECMAP time scale, we obtain a date for the Brunhes/Matuyama boundary of 806 ± 20 kyr. This is consistent with the age assigned by Johnson (1982), Shackleton et al. (1990), Berger et al. (792 ± 10 kyr) and with the radiometric dates of Izette and Obradovich (1991), Baksi et al. (1991), and Tauxe et al. (1992). However, our adjusted time scale differs significantly from that of Berger et al. in the period immediately preceding the Brunhes/Matuyama boundary; we assign 100 kyr to the next full cycle of $\delta^{18}O$, whereas Berger et al. assign approximately 80 kyr (i.e. two obliquity cycles).

As a test to see if the behaviour in the two parts is consistent, we performed least squares fits to sine waves separately to the segments of the revised SPECMAP: from 0 to 650 kyr and 650 to 900 kyr. The best fit frequencies for both segments matched 98 kyr within 3%. Perhaps more impressively, the phase at the 650 kyr boundary of

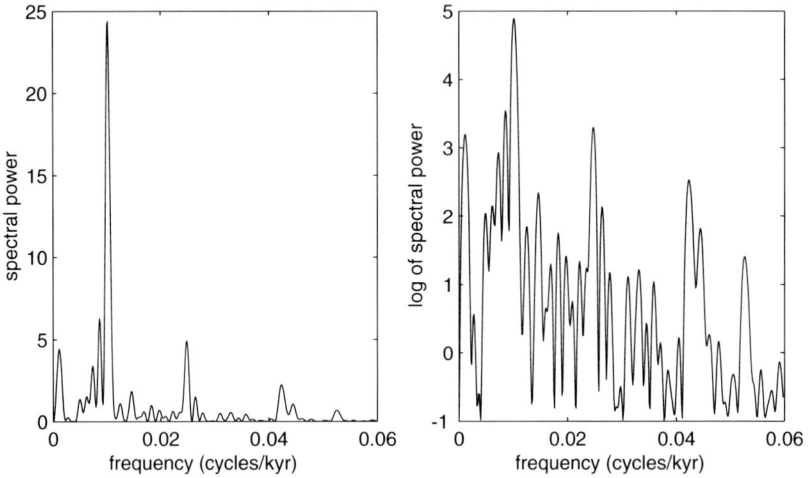

Fig. 5.19. Spectrum of SPECMAP with the adjusted time scale.

the two separate solutions matched to 10% (i.e. to within 10 kyr). This is important since the 100 kyr cycle was used neither in the original SPECMAP tuning, nor in the adjustment of the time scale beyond 650 kyr (which was based on the 41 kyr cycle). Thus there was no *a priori* reason to expect any agreement in the phase, unless the cycle in both segments derives from the same physical cause. This result strengthens our confidence in the reliability of the revised time scale.

The spectral power of the SPECMAP stack, with the adjusted time scale, is shown in Fig. 5.19. Compared to the spectrum with the original time scale, Fig. 4.5, the primary difference is that spectral power has been moved out of the precession region and into the 100 kyr peak. This has a consequence that after the adjustment, there is essentially no precession power near the 23 kyr band (see Fig. 5.17), and the power that was previously in the unexplained 52 kyr region has been moved into the 100 kyr peak.

5.4.3 SPECMAP and orbital inclination

In Fig. 5.20 we show the SPECMAP $\delta^{18}O$ data plotted on the new time scale. Superimposed is the plot of orbital inclination. The only adjustable parameters used were a scale change (to convert from $\delta^{18}O$ to inclination) and the lag of global ice with respect to inclination, which achieved its best match at 33 ± 3 kyr. This delay is physically reasonable when we recognise that the $\delta^{18}O$ is a proxy for accumulated ice; i.e. it is the *integral* of precipitation. If the climate driving force is approximated by a cosine function, then the $\delta^{18}O$ response is approximately a sine, and it will lag the climate by about 1/4 cycle $\approx$ 25 kyr. We discussed the 1/4 cycle phase lag for integrative proxies in Section 4.4.

The match between the variations in the inclination and those of $\delta^{18}O$ is excellent. Not only are the major cycles accommodated, but even the variations in the

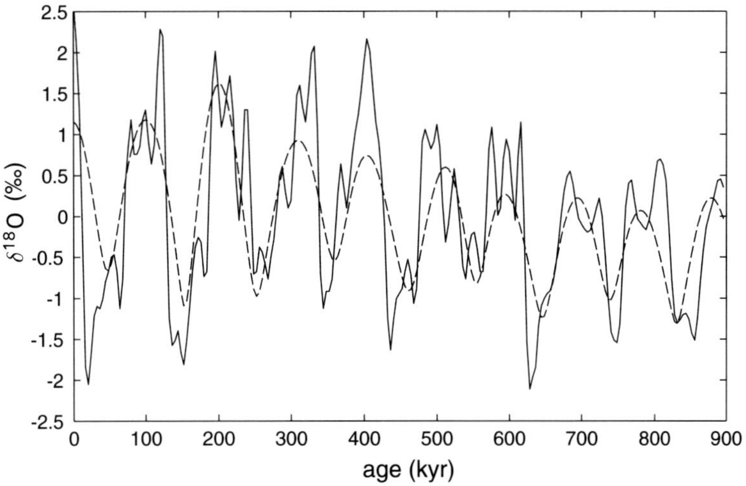

Fig. 5.20. Adjusted SPECMAP and orbital inclination (33 kyr lag).

amplitudes of the cycles and the gradual drift of the mean value seem to be accounted for. The inclination curve accounts for 49% of the variance of the adjusted SPECMAP curve. (Recall that the sum of the variances of all the components, including background, must add to the variance of the signal.) A similar fit done with eccentricity rather than inclination (see Fig. 8.4) gives a much poorer fit, with eccentricity accounting for only 15% of the variance.

5.5 TUNED AND OVERTUNED SEDIMENTATION RATES

When a record has been tuned, the result is the set of tie points, each representing a depth and a date. It has become traditional to present this as a plot of depth vs. age. In Fig. 5.21 we show such an example, from two tuning curves done for Site 607 in the North Atlantic, one by Ruddiman et al. (1986), and the other by Shackleton et al. (1990).

When plotted in this manner, the model looks very reasonable. The age appears to be linearly related to depth, at least approximately. But such a plot can hide large and discontinuous changes in sedimentation rates, since they appear only as changes in slope. We think a better sense of the tuning that has been performed can be obtained by plotting the sedimentation rate vs. age. We have done this for the same two tuning curves, and present the results in Fig. 5.22. In this plot we can see the enormous and abrupt excursions that were put into the model in order to ensure that the $\delta^{18}O$ excursions were aligned with the precession peaks in the target curve. Such detailed tuning can align not only true precession peaks, but any statistical fluctuation, and so the resulting data set is likely to contain an overestimate of the amount of precession. When this occurs, we say that the data set is "overtuned".

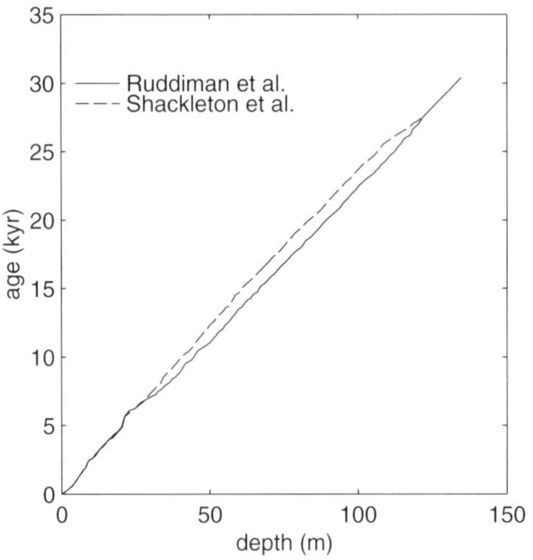

Fig. 5.21. Age vs. depth for Site 607 tuning model.

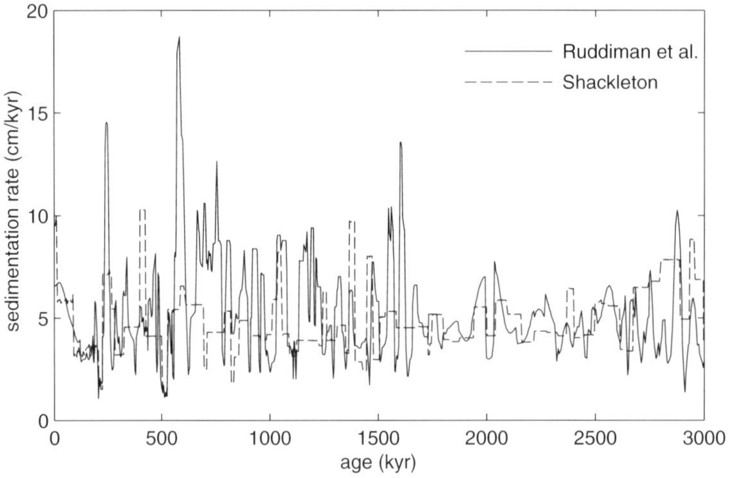

Fig. 5.22. Sedimentation rates from published tuning of Site 607.

Note in particular the region from 1500 to 2500 kyr. In Fig. 5.23 we show a plot of the spectral power in this time interval, on the left assuming absolutely constant sedimentation rate, and on the right assuming the tuned time scale of Ruddiman et al.

The tuning process does increase the spectral power in the peak, as it must. But did it artificially and incorrectly enhance the spectral power beyond that we would find present in the record if we knew the true time scale (as we hope to, someday)? We don't know. In that interval, Ruddiman et al. used 69 adjustable parameters, averaging one every 15 kyr. The ratio of maximum sedimentation rate to minimum in their model was 17. Are such variations physically accurate? A spectral analysis of

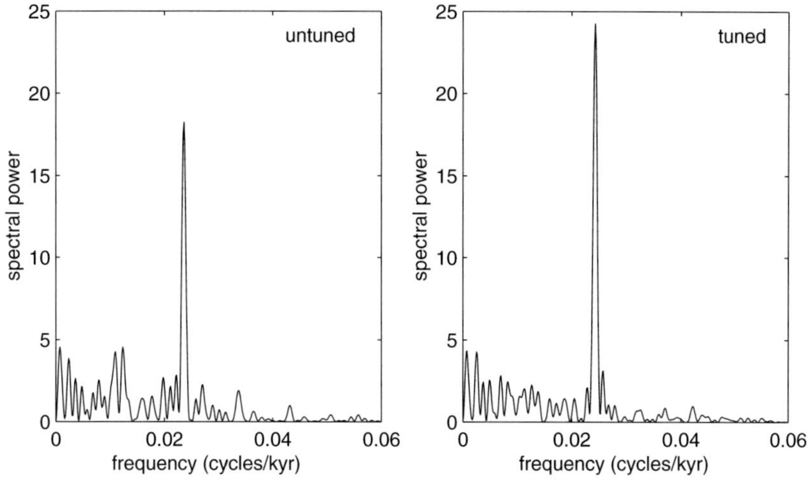

Fig. 5.23. Spectra of Site 607, untuned and tuned.

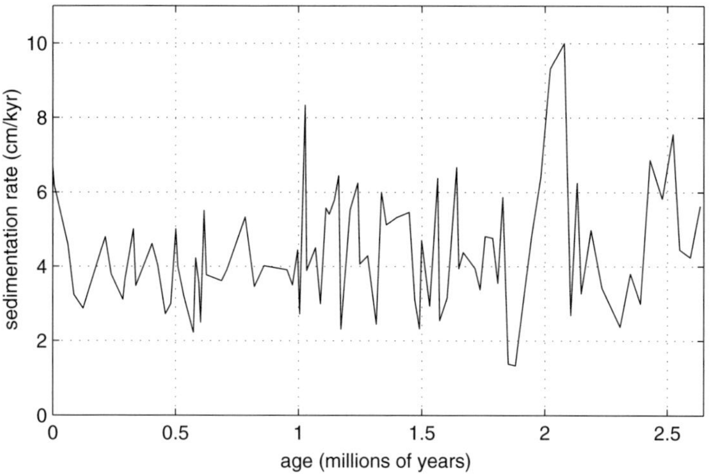

Fig. 5.24. Tuned sedimentation rate for Site 677.

these sedimentation curves shows no significant power at astronomical frequencies, so if these rates are accurate, they are not being driven by orbital forcing.

The sedimentation rate for Site 677, devised by Shackleton et al., is shown in Fig. 5.24. Once again, abrupt changes were incorporated into the tuning curve. The ratio of maximum to minimum sedimentation is 7.5, with the maximum change taking place over a time interval of only 0.2 Myr. And yet we know that the *average* sedimentation rates must be correct, since they are what led to the correct identification of the age of the Brunhes/Matuyama reversal from this data set (Shackleton et al., 1990).

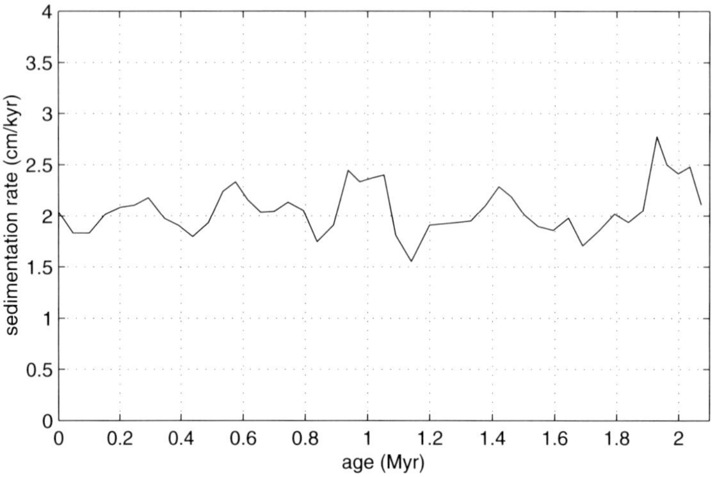

Fig. 5.25. Sedimentation rate for Site 806.

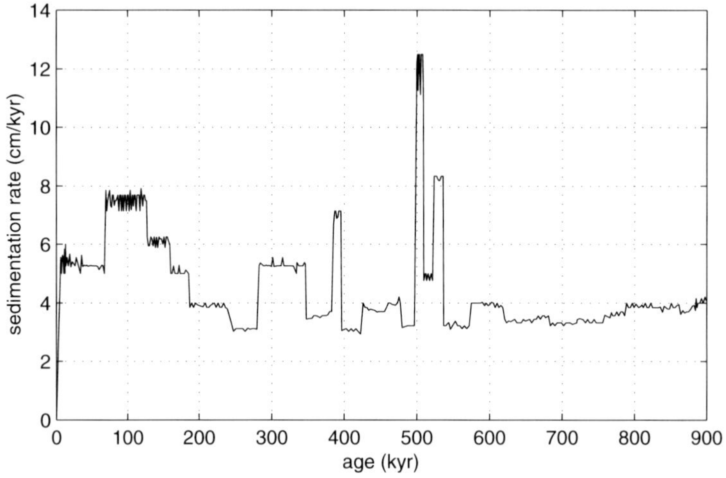

Fig. 5.26. MD900963 sedimentation rates.

Not all tuning efforts lead to rapidly varying sedimentation rates. In Fig. 5.25 we show the tuned sedimentation rates obtained by Berger et al. for Site 806 in the equatorial Pacific. The sedimentation rate for this tuning varies by only 12% RMS. And finally, in Fig. 5.26, we show the sedimentation rate for Site MD900963, used by Bassinot et al. (1994), in combination with data from Site 677, to create a new time scale to replace Specmap. For these data, there were two stages of tuning. For the first stage, only 25 control points were used. These resulted in the 25 approximately level regions in the sedimentation plot. But then there was a stage of "fine tuning" with changes of ±3 kyr; these resulted in the fine variations seen on each plateau.

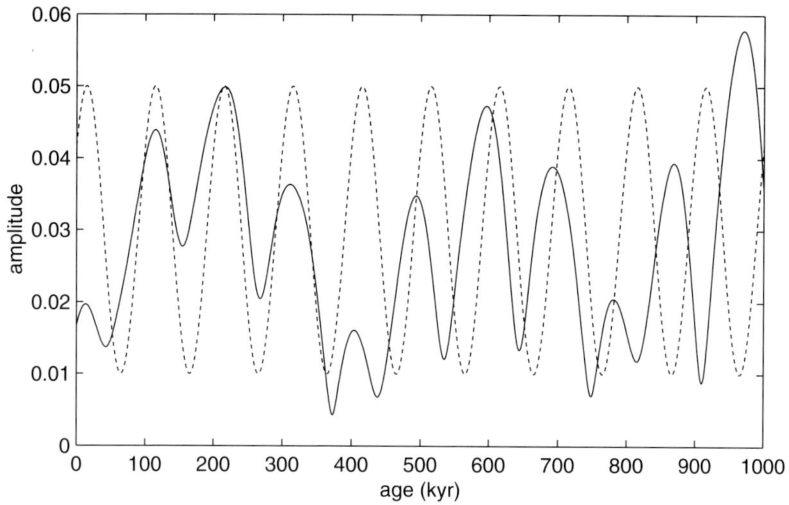

Fig. 5.27. Cosine (dashed curve) and eccentricity (solid curve).

5.6 OVERTUNING

5.6.1 Example 1: tuning a sine wave to eccentricity

What happens with tuning if the wrong target is chosen? Unfortunately, the tuning has a good chance of being "successful", and there is real danger that the scientist will think (incorrectly) that the real data match the target. We'll demonstrate this through an example. We assume that the real data are described by a pure cosine,

$$f(t) = \cos(2\pi ft + \phi) + B$$

with $f = 0.01$ cycle/kyr, and the data going from 0 to 1 Ma. The spectrum of this is, of course, a single peak.

But let us assume that we *don't* know that the time scale is correct, so we plan to adjust it. (That will, of course, distort it and make it incorrect.) Let us assume that we believe in the Milankovitch model, and expect that the main variation will follow eccentricity—not a perfect cosine. We compare our data to eccentricity in Fig. 5.27. (For that figure, the parameters in our cosine were $\phi = 0.94$ and $B = 0.03$.)

We note the first peaks match pretty well. Both the data and the eccentricity target curve reach their first maximum at 15 kyr. Good—no adjustment is necessary for the first 15 kyr. But we see that the location of the first minimum for the data occurs at 65 kyr, but for the eccentricity target curve it occurs at 45 kyr. So we tune the original time scale. We make this one of our tie points. We will now say that the first peak in the data occurs as 45. All ages between 15 kyr and 65 kyr will now be changed so that they fall between 15 kyr and 45 kyr instead.

We proceed in this manner, adjusting each peak and each minimum, and that way generate a list of tie points. They are shown in the following list:

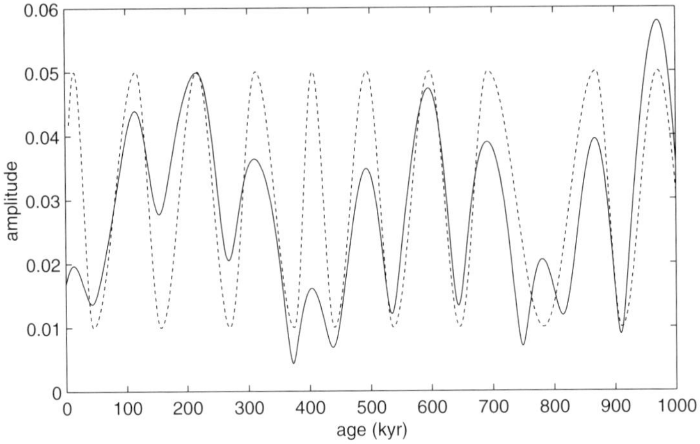

Fig. 5.28. Cosine (dashed curve) tuned to match eccentricity (solid curve).

cos = [15 65 115 165 215 265 315 365 415 465 515 565 615 665 715 765 815 865 915]′;

ecc = [15 45 117 157 218 269 313 375 406 440 495 537 598 646 693 783 870 911 971]′;

The cos points are the times of all the maxima and minima on the original time scale; the ecc points are the corresponding times for the target curve. We used one tie point for every 50,000 years. This is less than the typical usage in the tuning community. For example, the SPECMAP δ^{18}O stack used 47 tie points for one record to cover the period 0–600 kyr. That is one data point for each 12 kyr.

Using the above tie points, we tune the time scale by linear interpolation. In Matlab, this is done with one statement: t_new = interp1(data, ecc, t_old). We now replot the same data using the new time scale. The results are shown in Fig. 5.28.

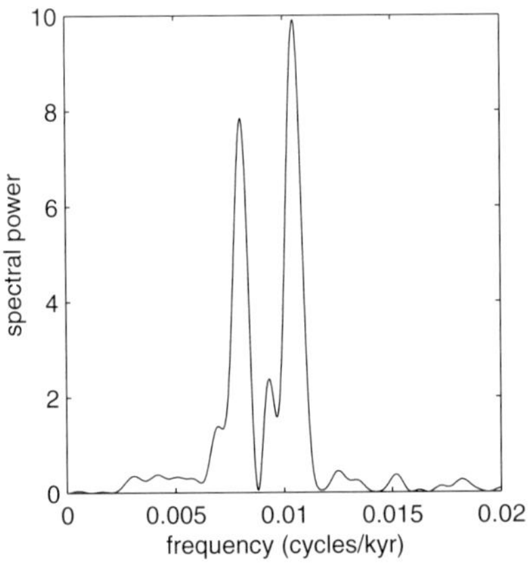

Fig. 5.29. Spectrum of the cosine after tuning to eccentricity.

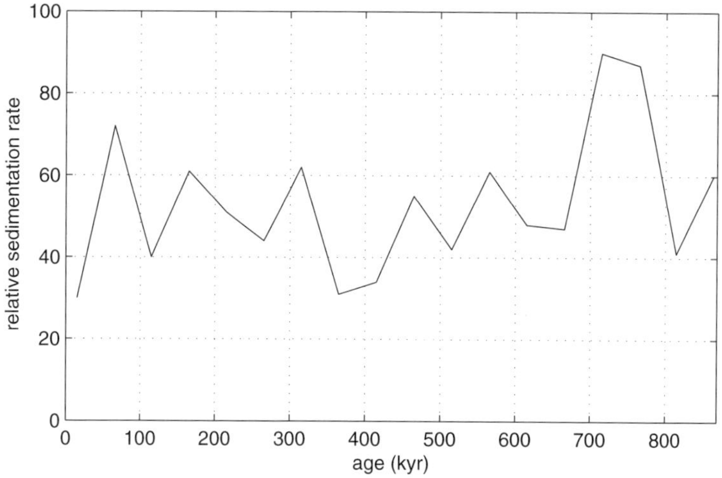

Fig. 5.30. Equivalent sedimentation rate for sine tuned to eccentricity.

The spectrum of the tuned cosine curve is shown in Fig. 5.29.

One peak in this spectrum is at a period of 95 kyr, and the other is at 125 kyr. We have managed to reproduce the double peak of the eccentricity spectrum by incorrectly tuning a cosine wave! It is interesting to note that we did not succeed in replicating the 400 kyr peak of eccentricity. That is because our data were too regular. The 400 kyr oscillation in eccentricity is responsible for the dip in the average value of eccentricity in the regions near age = 0, 400 and 800 kyr, and there was no good structure in our cosine plot to match to these dips. We would have been able to do better with random data.

From the set of tuning points, we can calculate the effective sedimentation rate variation needed to achieve the fit. This is shown in Fig. 5.30. The sedimentation rate, for this completely wrong example, has considerably less variation than was used in SPECMAP (compare with Fig. 5.12) or for Site 607 (compare with Fig. 5.22).

This is an example of *overtuned* data, in which we forced it to fit a preconceived but incorrect target. It is easy to do, if you allow yourself enough parameters. In the above example we used 19 tie points. The effective sedimentation rate that we used went from a minimum of 30 to a maximum of 90—only a factor of three in variation. The RMS deviation of the sedimentation rate is 31%. These figures could serve as a basis for caution. Any time the sedimentation rate varies as much as this, the user should recognise that the spectrum can be changed in a drastic and unphysical way.

5.6.2 Example 2: tuning inclination to insolation

In this example, we will take a pure orbital inclination curve, and see what happens when we try to tune it to a preconceived target. We take as our target the "Low Latitude Stack" of Bassinot et al. (1994), a target that is commonly used for tuning

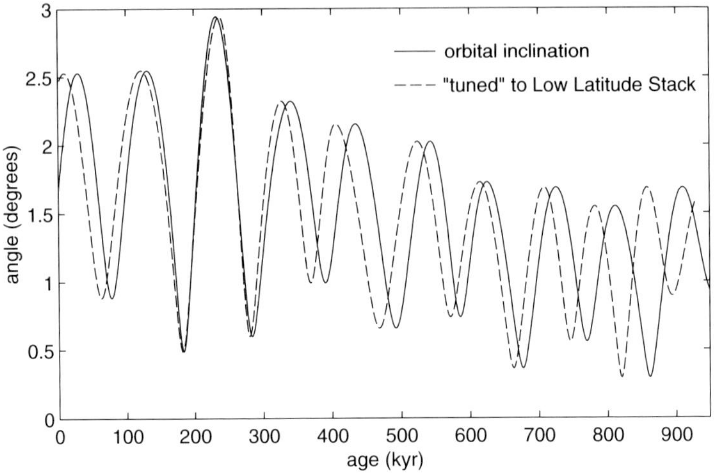

Fig. 5.31. Orbital inclination "tuned" to the Low Latitude Stack.

of other $\delta^{18}O$ data. To do the tuning, we chose the peaks of the orbital inclination curve as our control points. The ages of these maxima, in kiloyears, are:

$$t(\text{inclination max}) = [30, 131, 231, 341, 435, 543, 625, 725, 812, 910]$$

We then took each of these points on the inclination curve and reset them to the age the Low Latitude Stack assigns to the maximum of the interglacials (i.e. Stages in Bassinot et al. numbered 1, 5.5, 7.5, 9.3, 11.3, 13.3, 15.5, 17.5, 19.3, and 21.5). The ages of these control points in kiloyears, are:

$$t(\text{target}) = [10, 122, 236, 328, 406, 524, 615, 708, 782, 858]$$

The new times for the inclination curve were then calculated by interpolation between the target points. The original and the resulting "tuned" inclination are shown in Fig. 5.31. Keep in mind that it makes no sense to tune inclination; we *know* the time scale. This exercise simulates what happens when you mistakenly adjust data that are already correct, i.e. it simulates overtuning.

The spectral power of orbital inclination, before and after tuning, is shown in Fig. 5.32. We have expanded the usual scale to include only frequencies up to 0.02 cycles/kyr, since there is no significant spectral power at higher frequencies, and it is the behaviour of the peak shape that is most interesting. Note that the narrow peak in the untuned data, near period 100 kyr, has been broadened due to the inaccurate time scale. Two side lobes have appeared, one 0.008 cycles/kyr (period 125 kyr), and the other at 0.013 cycles/kyr (period 77 kyr). In Section 5.9 we will show examples in the literature of data that have been tuned in the same manner, i.e. to the Low Latitude Stack, and which show the same side lobes present in this example.

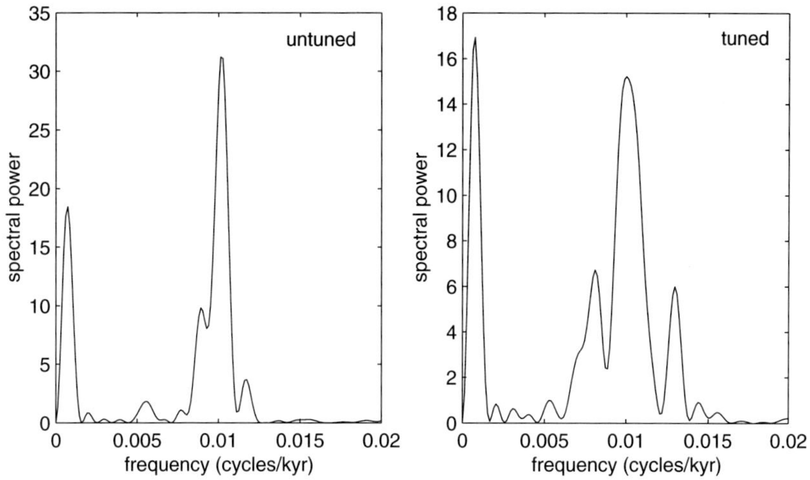

Fig. 5.32. Spectrum of orbital inclination before and after tuning.

5.7 "VALIDATION" OF TUNING—INVALID

Even when tuning was introduced, the potential for circular reasoning was recognised. However early Monte Carlo tests (Pisias, 1983) appeared to validate the method. Of course, in those days the number of tuning parameters was kept very small, and in a sense these tests were validating a procedure that we now call minimal tuning. A test was devised to give what was thought to be an independent check of whether the procedure was valid. If the tuned data passed the test, then the tuning was considered valid.

The test consisted of passing the data through a narrow-band filter centred on the precession frequencies, so that only those components would pass through. Then the filtered data were compared to the tuning target. Of course, the phases would agree; that was guaranteed by the tuning procedure. The independent check was the comparison of amplitudes. The precession amplitude varies rapidly in time, and if the filtered data showed a similar variation, then that was supposed to prove that the method truly brought out a real precession signal, and not an artificial one.

The validity of this test was largely assumed in the literature. It was put to its first explicit test by Neeman (1993) who performed extensive Monte Carlo tests on the tuning procedures. Neeman concluded that the test was not valid; that the tuning procedure, even though it explicitly adjusted only the phases, would often still reproduce the amplitude of the target signal! How could this be? The key is in the application of the narrow band filter. If you change frequencies, and then apply such a filter, the amplitude of the signal that passes through the filter is also changed.

In Fig. 5.33 we plot a simulation of a precession signal; it consists of a 24 kyr sine wave added to a 22 kyr sine wave; as the two frequencies come in and out of phase, beats in the amplitude are created. Plotted on top of this is a constant amplitude 23 kyr sine wave.

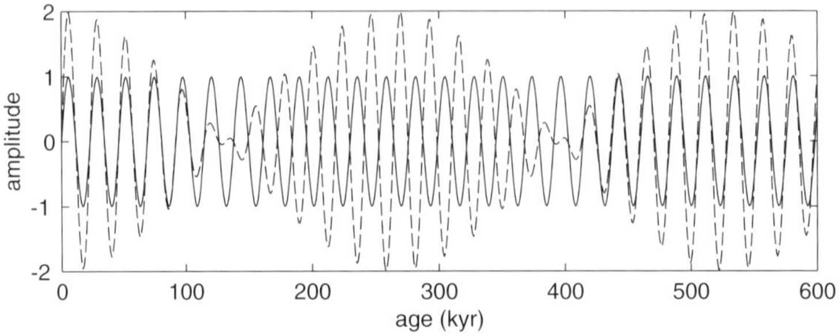

Fig. 5.33. Precession (dashed curve) and 23 kyr sine wave (solid curve).

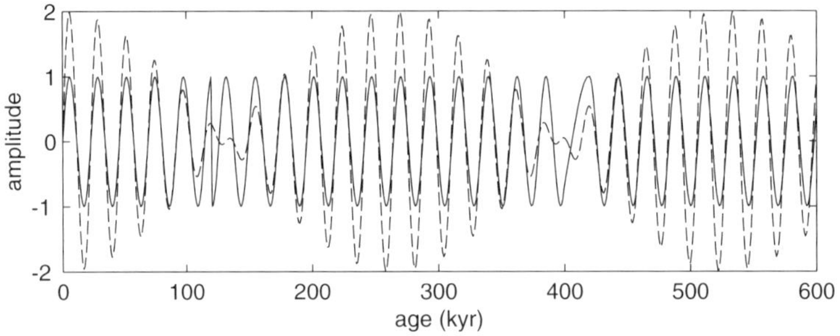

Fig. 5.34. Sine wave (solid curve) after being "tuned" to precession (dashed curve).

We will take the 23 kyr sine wave to be our "data", and the dashed "precession" signal to be our target. We will simulate the effects of false tuning by creating a time scale that forces the data to match the precession target.

Looking at Fig. 5.33, we see that the phases of the data and the target seem to match well, except in the region 120 to 400 kyr. We therefore conclude (incorrectly) that our sedimentation rate is not constant. We introduce four tuning tie points (near 120 and 400 ka), and shift all the data in this interval by a half cycle. The resulting "tuned" plot is shown in Fig. 5.34.

We have successfully tuned the data to match the phase of the target. The spectrum of the new tuned sine wave is shown in Fig. 5.35. The effect of introducing a false phase delay in one segment of the data, from 120 to 400 kyr, has been to split the single peak into a doublet, with frequencies that match those in the target.

We appear, however, not to have reproduced the amplitude of the target. Our sine wave still has constant amplitude, and does not have the slowly varying envelope characteristic of precession. But now we apply the final step. Since the precession signal is usually only a small part of the total climate signal, a band-pass band filter is applied to the data before it is plotted (see Section 5.3). This removes all the frequencies except for those in the region near precession. So let us take the tuned data (solid line in Fig. 5.34 above, a phase-modulated sine wave) and pass it through a band-pass filter. The filter we use passes all frequencies between 1/20 and 1/25 cycles/kyr. We now plot the band-passed data in the time domain shown in Fig. 5.36.

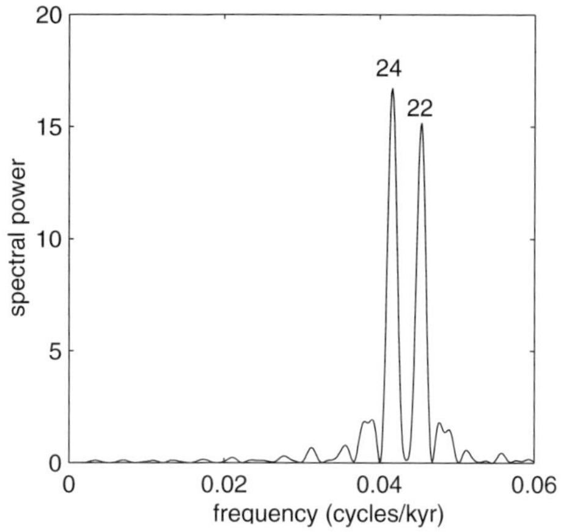

Fig. 5.35. Tuned sine wave spectrum.

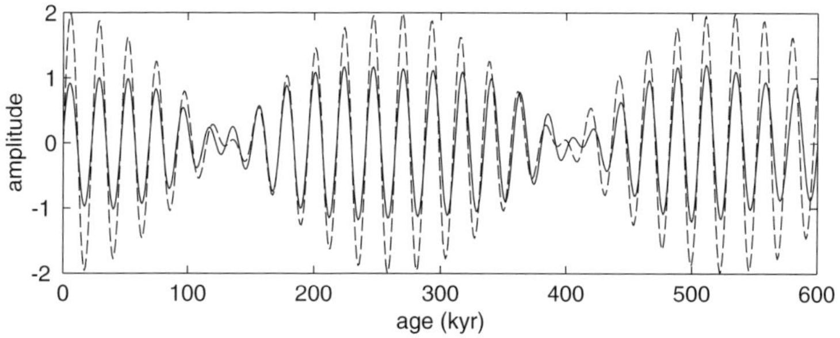

Fig. 5.36. Tuned sine wave with narrow-band filter applied (solid curve) compared to the target (dashed curve)..

The band-pass-filtered data match the target in both phase and amplitude! By the standards employed by many in the paleoclimate community, we have "proven" that our tuning was proper. Yet we know in this case (since we set it up) that the true signal is not a simulated 22 kyr/24 kyr precession, but a pure 23 kyr sine wave. This example demonstrates that the most common method used to validate tuning is itself invalid.

It is worthwhile to look at this example in more detail to see why the method failed. We begin by writing the target signal as:

$$
\begin{aligned}
h(t) &= \cos(2\pi f_1 t) + \cos(2\pi f_2 t) \\
&= 2\cos\left(2\pi\left(\frac{f_1 + f_2}{2}\right)t\right)\cos\left(2\pi\left(\frac{f_1 - f_2}{2}\right)t\right) \\
&\equiv 2\cos(2\pi f_{AVG} t)\cos\left(2\pi\frac{\Delta f}{2}t\right)
\end{aligned}
$$

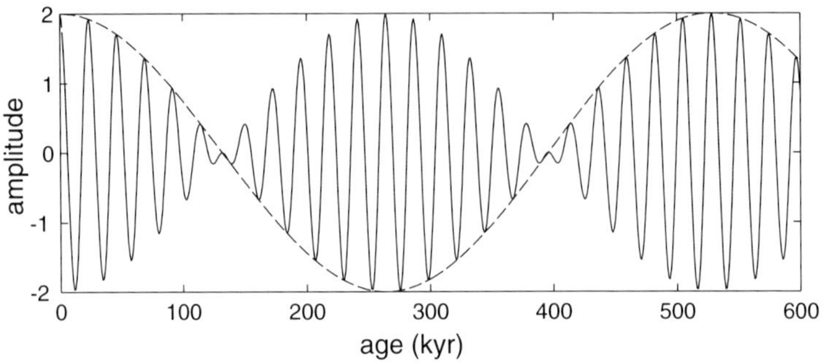

Fig. 5.37. Sum of two cosines is a modulated cosine.

where we define $f_{AVG} = (f_1 + f_2)/2$ and $\Delta f = f_1 - f_2$. This equation is illustrated in Fig. 5.37, where we have plotted $h(t)$ as the solid curve, and superimposed the modulating function $\cos(2\pi\Delta f t/2)$ as the dashed curve.

The two sine waves of slightly different frequencies fall in and out of phase. The net result is a sine wave (in this case, a cosine) with a frequency equal to the average of the two, but with an amplitude that is modulated at a frequency $\Delta f/2$. If the beats are produced as sound waves, by two tuning forks of slightly different frequencies, then the cycles of loudness are referred to as beats, and the beat frequency is Δf, not $\Delta f/2$, since the human ear is sensitive to the pulses, and not sensitive to the inversion of the envelope that takes place. (By inversion of the envelope, we mean that the smooth slow sine wave in Fig. 5.37 has a negative amplitude in the period 120 to 400 kyr, but this is not detected by the ear.)

From this plot, it is clear that the phase shift that we identified in Fig. 5.33 was really a phase *inversion*. The terms "shift" and "inversion" are often confused, since for pure sine waves there is no distinction. The inversion takes place when the modulating function goes from positive to negative. When we put a phase shift at these times, we have a small region with an anomalously high or low frequency. That drives the amplitude in these regions to zero, since the frequency falls outside of the bandpass. The fall is smooth since the bandpass filter has a bandwidth of only $\Delta f = 1/20 - 1/25 = 1/100$ cycles/kyr. A signal that contains frequencies in a band Δf can vary only with a characteristic time scale $\Delta t = 1/\Delta f = 100$ kyr. So the zeros in the band-passed signal are not sharp, but appear as low points in a much slower modulation. The problems discussed in this section do not imply that tuning is always circular. Nor do we suggest that tuning is not worth doing. There is a version of tuning that is valuable, and we discuss that in the next section.

5.8 MINIMAL TUNING

The previous sections gave examples of the dangers of overtuning. Yet, as long as we have no other general method for obtaining the ages of sedimentary samples, tuning has an useful role to play. How do we know when we are overtuning? Where do we draw the line between overtuning and minimal turning?

The answer is not simple. As we showed in the previous section (see Fig. 5.36) we can overtune with just a few data points. Nevertheless, there are a few guidelines that we recommend.

1. Work with untuned data, if possible. By this we mean data whose time scale is assumed to vary linearly between radiometrically dated tie points (e.g. the present, and the Brunhes/Matuyama reversal). If there are no established points in the record, used a tuned record to estimate the ages of the end points, and assume constant sedimentation between these points.
2. Keep the number of parameters small. In general, if you are studying a cycle that has a period T, your tuning points should be separated by at least $2T$, and it is better if the separation is greater. For example, in Section 4.9.1 we discussed an example in which we "detuned" a previously tuned data set, by using only the control points that were separated by 100 kyr, and assuming constant sedimentation for data in between. This is not sufficient detuning to study a 100 kyr cycle, but it is sufficient for the study of precession, which was the focus of our interest in that section.
3. Tune to a peak that is unrelated to the one you are studying. If data are tuned to obliquity, or to a combination of obliquity and precession, then much can still be learned about the 100 kyr peak. This was the case in the example of SPECMAP.
4. Tune to a peak in a different proxy than the one you are studying, if possible. Different proxies can show very different behaviour, as, for example, the $\delta^{18}O$ and the coarse component of Site 806 (see Section 8.4.1).
5. The best test of tuning is to have an unrelated peak significantly increase in spectral power. So, for example, if you tune to the 41 kyr peak, and see an increase in the power in the 100 kyr peak, then that provides evidence that your obliquity tuning is improving the accuracy of the time scale. This happened in the SPECMAP tuning.
6. When in doubt, use Monte Carlo simulations to see how easily you can get false time scales.

Certain records easily qualify as minimally tuned, such as the Vostok record (discussed in Section 5.2.1) and Devils Hole. We give other examples of data that are minimally tuned throughout this book. In particular, we do this when we are comparing the fits of eccentricity and orbital inclination to the 100 kyr $\delta^{18}O$ cycle, in Section 8.4.

5.9 OVERTUNING: EXAMPLES FROM THE LITERATURE

Given the difficulties inherent in unbiased tuning, it is not surprising to discover additional examples in the published literature in which there is prima facie evidence that the time scale was overtuned. We will give such examples here, with apologies to

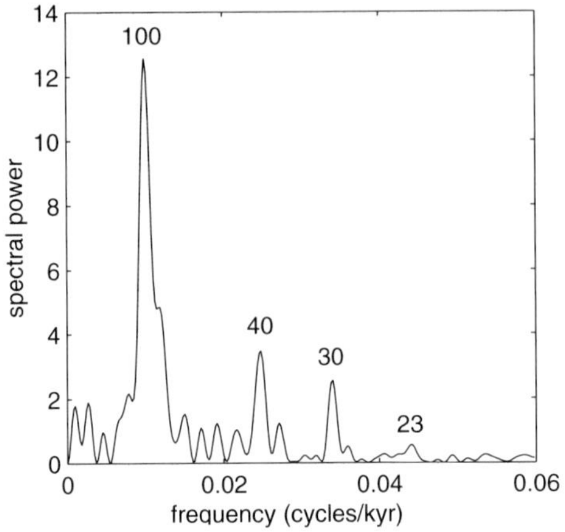

Fig. 5.38. Site 664 spectrum, prior to tuning.

the authors. The fact of overtuning does not necessarily imply that all the conclusions of the authors are wrong. Inaccurate time scales may not affect the results, but they do mean that they should be reexamined. They are very important, however, for understanding the links between climate and astronomical forcing, and this is a subject in which overtuning becomes a critical issue.

5.9.1 Site 664

Our first example comes from the $\delta^{18}O$ data from Site 664 off the coast of Africa (Raymo et al., 1997). In Fig. 5.38 we show the spectral power, using untuned data (i.e. we assumed constant sedimentation between 0 and 600 ka).

The spectrum, assuming the published tuning curve of Raymo et al., is shown in Fig. 5.39. The time scale was tuned to a target that included precession, obliquity and eccentricity.

Note that the statistical significance of the power near 0.01 (100 kyr period) has been drastically reduced. If we were to estimate the background level in the vicinity of that peak to be at 2, then the significance of the peak in the untuned data is $\exp(-6) = 0.002$. It is hard to say what the background level in Fig. 5.39 is, but the peak is now part of a structure that is consistent with shapes you see when drift and chatter are present in the time scale (see Section 5.1). Only here, the drift and chatter are probably due to improper tuning, rather than to actual sedimentation rate variations. Note that the two highest peaks in the tuned spectrum plot match the two target frequencies in the target eccentricity spectrum, near periods 125 and 95. Additional peaks have been strengthened at 41, 24 and 19, all of which were in the tuning target.

We say that this time scale shows *prima facie* evidence of being overtuned because the only peaks that are enhanced are the ones in the tuning target, and the strongest and most statistically-significant peak in the untuned spectrum is effectively

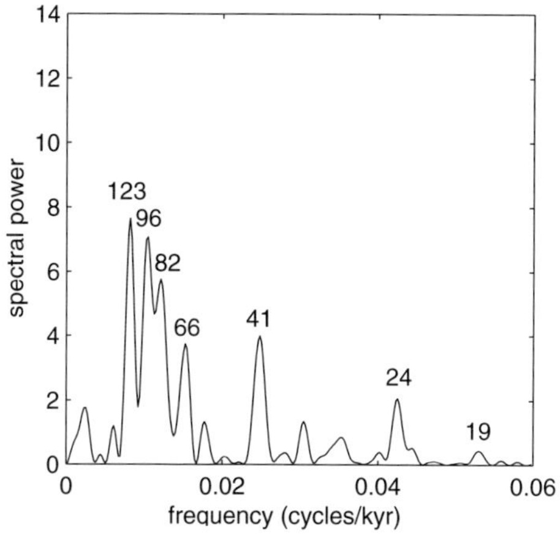

Fig. 5.39. Site 664 spectrum, after published tuning.

destroyed by the tuning process. It is very dangerous to draw any conclusions by examining data that has been so heavily adjusted to match a preconceived idea of what it should contain.

5.9.2 Site 659

As a second example, we show the spectral power for $\delta^{18}O$ in Site 659 (Tiedemann et al., 1994). In Fig. 5.40 we show the spectral power, before and after tuning. The tuned time scale was obtained by Tiedemann et al. by adjusting the control points to make the $\delta^{18}O$ pattern match the pattern of the reference stack of Bassinot et al.

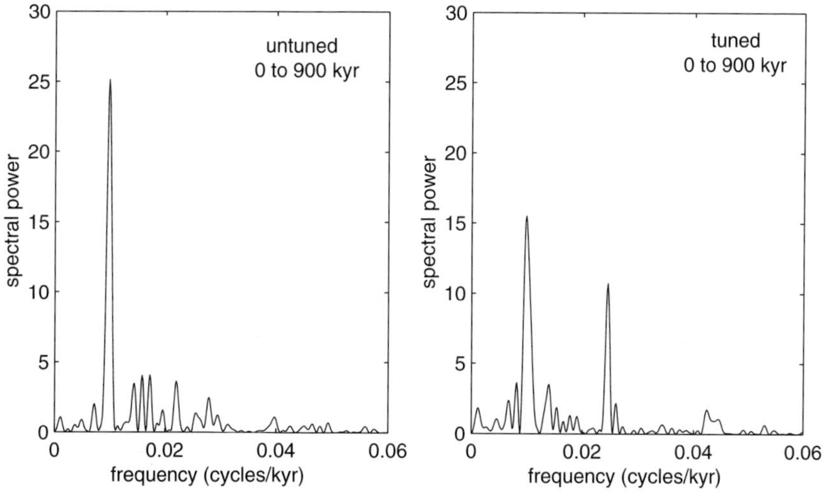

Fig. 5.40. Site 659 spectrum, before and after tuning.

(1994). Note the reduction in the statistical significance of the main peak after tuning. It is unlikely that a tuning procedure that truly resulted in an improved time scale would *reduce* the statistical significance of the main peak. We conclude that the time scale is overtuned. The tuned spectrum does have a significantly increased 41 and a 24/22/19 triplet. Of course, these frequencies were in the tuning target. It would be very interesting to know the true contributions of these frequencies to the climate measured at this site. Unfortunately, we cannot tell, because the danger of overtuning means that we must interpret the observed peaks only as an upper limit. On the other hand, the power of the peaks in the untuned spectrum may be correctly interpreted as lower limits. We will discuss this site further in Section 8.7.

5.9.3 Low Latitude Stack (Site 677 and MD900963)

In an attempt to find a more accurate replacement for SPECMAP, Bassinot et al. (1994) combined $\delta^{18}O$ data from Site 677 in the Eastern Equatorial Pacific, with $\delta^{18}O$ data from Site MD900963 in the tropical Indian Ocean, east of the Maldives. The two data sets were individually tuned to the Imbrie and Imbrie nonlinear ice model (see Section 6.2.1), and then added and smoothed. The resulting template, called the "Low Latitude Stack" or LLS, is shown in Fig. 5.41.

In Fig. 5.42 we show the LLS spectral power for the period 0 to 900 kyr, and that of the nonlinear ice model to which it was tuned. The spectra show features with common frequencies (peaks at 0.024, 0.043), although the relative amplitudes differ. Note that the peak in the LLS spectrum near 0.01 is not a single narrow peak, but that it shows complex structure. This could be due to drift in the time scale, but then the other peaks (e.g. at 0.024) should show similar structure, and they don't. The spectrum reaches its maximum power at 0.0105 cycles/kyr (period 95 kyr), and the small peak just to the left has period 121 kyr. Since these are close to the frequencies

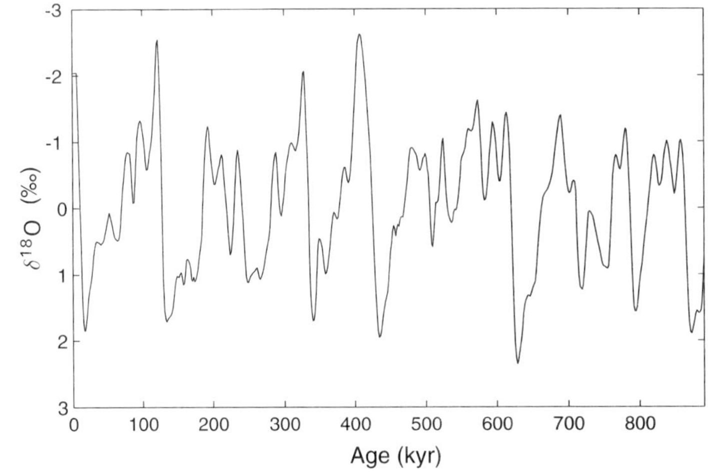

Fig. 5.41. Low Latitude $\delta^{18}O$ Stack (Sites 607 and MD900963).

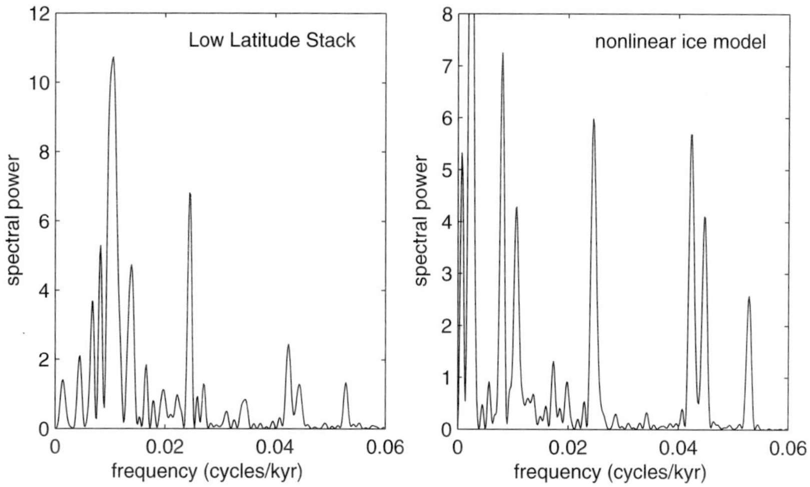

Fig. 5.42. Spectrum for LLS, and nonlinear ice model.

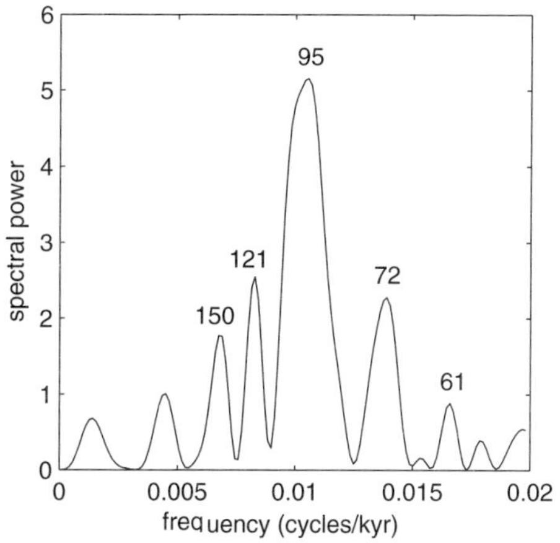

Fig. 5.43. Spectrum of LLS, 100 kyr region.

expected for eccentricity, it has been suggested that this is evidence for the presence of the split eccentricity peak in $\delta^{18}O$ sea core data. To examine this hypothesis, we show the spectrum in this region in an expanded scale in Fig. 5.43. In this plot we can clearly see the presence of the side peak at period 121 kyr. Note that there is another peak near period 72 kyr.

Does the presence of the peak near the expected 125 for eccentricity support the nonlinear ice model? For comparison, in Fig. 5.44 we have also plotted the prediction of the orbital inclination model. In this plot, the orbital inclination has been distorted by tuning it to the nonlinear ice model, to simulate what was done with the real data. It is important to note that the orbital inclination model does *not* predict a

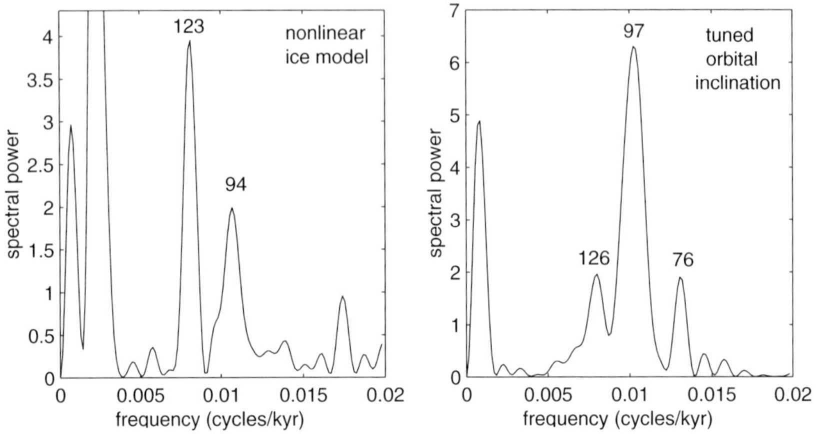

Fig. 5.44. 100 kyr region for two orbital models.

single narrow peak for these data, since the data were tuned to a multi-peak model, as we did in Section 5.6.2. So in Fig. 5.44 we reproduce the peak shape from Fig. 5.32. The similarity of the observed spectral shape to the predicted shape from the orbital inclination model leads us to conclude that the LLS data offer stronger support for the orbital inclination model than for the nonlinear ice model.

5.9.4 Site 849

In Fig. 5.45 we show the spectrum for Site 849. On the left is the spectrum for an untuned time scale, and on the right is the spectrum for the data tuned using the LLS scale. (Raymo, 1995, private communication.)

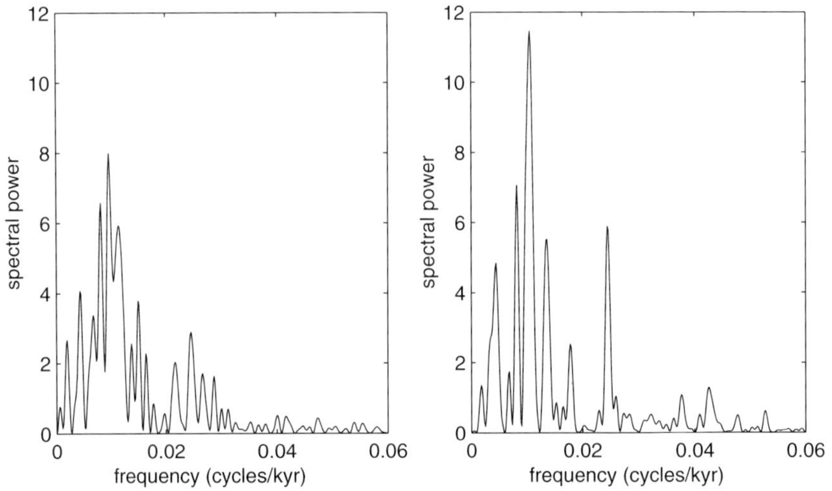

Fig. 5.45. Spectra for Site 849, untuned and tuned time scale.

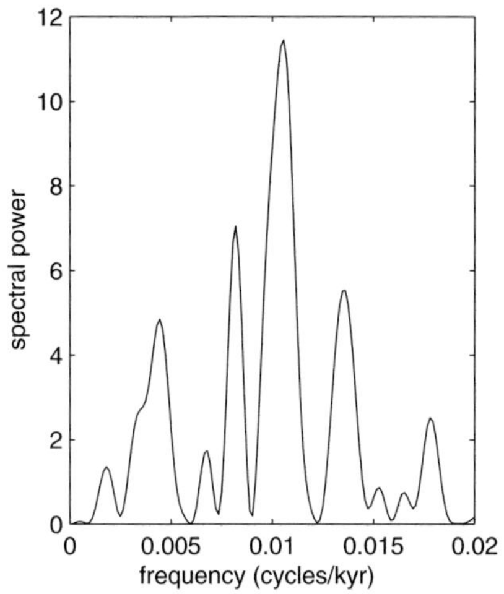

Fig. 5.46. Spectrum of Site 849 near the 100 kyr region.

As in the Low Latitude Stack, the spectral power in the vicinity of 0.01 cycles/kyr appears to show structure, using either the tuned or untuned time scale. An expanded view of this region is shown in Fig. 5.46. The maximum power appears at 0.0105 cycles/kyr (period 95 kyr). Once again, there are sidelobes at $f = 0.0082$ (period 122 kyr) and $f = 0.013$ (77 kyr). A direct comparison between this plot and the two orbital inclination models shown in Fig. 5.44 again favours the orbital inclination model. Or rather, it favours the orbital inclination model that has been mistuned to a nonlinear ice model. This is why we include Site 849 in our set of examples of overtuned data.

6

Mechanisms, models and theories

"The trouble with most folks isn't so much their ignorance. It's knowing so many things that ain't so."

Josh Billings, 1874

"In formal logic, a contradiction is the signal of defeat; but in the evolution of real knowledge it marks the first step in progress towards a victory."

Alfred North Whitehead

Anyone trying to understand climate in detail, who looks into the relevant physical processes, is likely to throw up his hands in despair. The number of important processes which must be understood to create a theory of climate is staggering. The amount of true understanding that we have of these processes is miniscule. Is there any hope?

To make things worse, weather is believed to be chaotic in nature. Small changes in the initial conditions, figuratively described as the now proverbial "butterfly flapping its wings", are believed to have an exponentially growing effect as time passes. After about a week, immeasurably small differences in the weather pattern can completely change the outcome. We may never be able to predict the path of hurricanes.

But there is hope for climate predictions. The butterfly may alter next week's weather, but the sun does rise in the morning, the seasons return every year, and the ice ages seem to come back with regularity. Even in chaos theory, there are limitations to the exponential growth—it is only a good approximation while it is still small! Eventually there are bounds to the behaviour, and these bounds determine the large-scale changes. We may never be able to predict the weather, but there is still hope that we can predict climate. The evidence for this is the regularity of climate, the precise cycles seen in the ice ages and other climate indicators. Something simple is going on, something that sets the bounds within which weather varies.

In developing an understanding of climate, we will use the terms "mechanism", "model" and "theory". Scientists sometimes reserve the term *theory* for a detailed mathematical explanation of phenomena that has sufficient rigour to make predictions from fundamental principles. Thus we talk about Maxwell's "theory" of electromagnetism, which allows us to design radio transmitters and receivers. But in the popular press, the term "theory" is often used for nothing more than a speculation: "His theory is that space aliens manipulate the climate ...". Virtually nobody sees much hope for a complete theory of climate in the near future, in the scientific sense. Rather, we all aspire to develop a *model* of climate. A model enables us to relate climate to other empirically measured phenomena, and make a mathematical prediction based on them.

By this definition, the Milankovitch theory is really a model. The Milankovitch model might assume that terminations of the ice ages are related to excursions in insolation. It may not understand in detail how they are related, so an adjustable parameter is put into the model, such as a threshold. The Milankovitch model might postulate that the ice ages terminate whenever insolation reaches this threshold; we may not understand the threshold in detail, but it should at least be plausible. (Warming plausibly melts ice.) It should be related to our understanding of climate through some mechanism that might one day be better understood. However, since the term "Milankovitch theory" is so engrained in the literature, we will continue to use it, while keeping in mind that it is not a theory in the same sense as Einstein's theory of relativity.

In this chapter we will describe mechanisms and models, the processes that affect climate, and how we can incorporate these mechanisms into a mathematical model which we can use to make predictions. We will not try to give a comprehensive summary of the published models; there are too many to allow that. In 1993, Imbrie et al. (Imbrie et al., 1993a) gave a "short list" of representative models that included nine groups of models, and references to 21 papers. This list is reproduced in Table 6.1.

Since the mid 1980s, the length of record and quality of data collected have improved remarkably, and several key issues have been recognised in the global ice proxy data. These include the Stage-11 problem, the absence of the 400 kyr cycle, the spectral narrowness of the 100 kyr oscillation, and the causality problem. Since most of these issues were not addressed by the earlier theories, our emphasis will be on those published more recently.

6.1 MECHANISMS

6.1.1 Radiative forcing and the greenhouse effect

The temperature of the Earth is determined by radiative forcing, amplified and modulated by feedback mechanisms. Sunlight hits the Earth with an average intensity of $S = 1372$ Watts per square metre. The Earth heats up until its own thermal radiation to space matches the incoming power, and that equality is what sets the temperature. The equation for radiation from the Earth is

$$P = \varepsilon \sigma A T^4$$

Table 6.1. "Short list of climate models" from Imbrie et al. (1993a).

Model	Examples
Class I. Models—with free oscillations (Interactions between orbitally forced oscillations and free oscillations of the air-sea-ice-crust system with period P.)	
Group 1. Nonlinear resonance with a damped oscillation ($P \approx 50$ kyr)	Birchfield and Ghil (1993)
Group 2. Nonlinear resonance with a self-sustaining oscillation ($P \ll 100$ kyr)	Le Treut and Ghil (1983) Ghil (1991)
Group 3. Large free oscillation ($P \approx 100$ kyr) with an arbitrary phase	Oerlemans (1982) Saltzman et al. (1982) Saltzman et al. (1987)
Group 4. Large, free oscillation ($P \approx 100$ kyr) with its phase set by orbitally forced responses at 23 kyr and 41 kyr	Oerlemans (1982) Saltzman et al. (1984) Maasch and Saltzman (1990)
Class II. Models—without free oscillations (Other nonlinear interactions between orbitally-forced responses and the dynamics of atmosphere, ocean, ice sheets, and asthenosphere.)	
Group 5. Simple rectification models (asymmetrical responses which extract power from the precession envelope)	Calder (1974) Imbrie and Imbrie (1980) Pisias and Shackleton (1984)
Group 6. Models which rectify the response to the orbital forcing by resolving regional components of the season energy budget	Short et al. (1991) Crowley et al. (1992)
Group 7. Ice sheet models in which the isostatic response and ice mass balance are explicitly calculated	
Group 7a. Snow budget prescribed by a snow line related parametrically to insolation.	Birchfield and Weertman (1978) Oerlemans (1980) Birchfield and Grumbine (1985) DeBlonde and Peltier (1991)
Group 7b. Snow budget calculated by an atmospheric model coupled to an ice sheet model	Pollard (1983) Gallée et al. (1991, 1992)

where P is the power of sunlight absorbed by the Earth, $S \times \pi R^2$, where R is the radius of the Earth, σ is the Stephan-Boltzmann constant ($\sigma = 5.67 \times 10^{-8}$ W/(m^2 s K^4)), and A is the total surface area of the Earth, $A = 4\pi R^2$. If the system were no more complex than we just described, the Earth would reach a steady-state temperature of 246°K, or −27°C.

But the Earth is neither a simple absorber, nor a simple emitter. And by climate we usually mean the climate on the surface of the Earth, which is wedged between

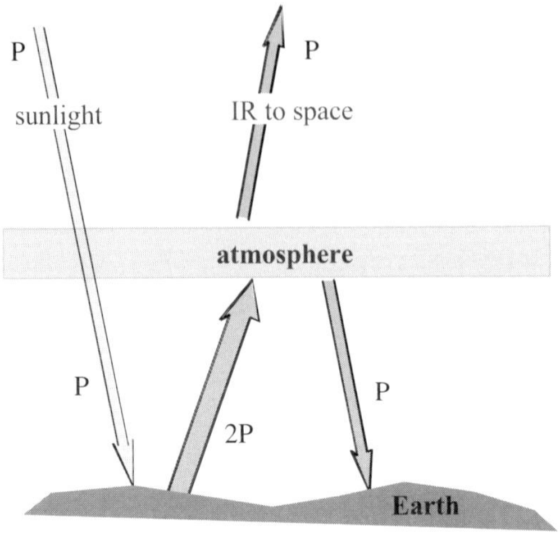

Fig. 6.1. Greenhouse effect, two-layer model.

layers of atmosphere and layers of ocean and land. Some of the Sun's radiation is reflected by clouds, ice, the surface and aerosols, and by itself this would lower the temperature of the Earth's surface. But much of the thermal radiation from the surface is absorbed by the atmosphere, and then radiated back. This trapping of radiation in the atmospheric layers increases the radiation to the surface (now a sum of this plus the original incident solar component) and causes a temperature rise known as the "greenhouse effect". (Perhaps a better name would be the "car in the parking lot" effect, since the modern student is more likely to be familiar with a hot car sitting in the sun, than with a hot greenhouse.[1])

The greenhouse effect is responsible for the fact that the average temperature of the *surface* of the Earth is over 40°C higher than it would be from the simple model—although the average temperature of the final radiating level in the upper atmosphere is close to the −27°C level. For a simple and straightforward introduction to the greenhouse effect we recommend *The Physical Universe* by F. Shu (Shu, 1982, p. 433–347). We will give an even simpler treatment here.

To illustrate the greenhouse effect, consider the following, greatly oversimplified, two-layer model. In this model, sunlight falls directly to the Earth, is completely absorbed, and then is efficiently emitted as thermal radiation. We model the atmosphere as a single layer which completely absorbs the radiation, and then re-emits it, half to space and half back to the Earth. This is shown in Fig. 6.1. Note that the incoming power from the Sun is P, but in order to have energy balance (i.e. for the system to emit P back to space) the power emitted from the Earth must be $2P$. The Earth's surface gets half its energy from the Sun, and half from the atmosphere.

[1] The greenhouse effect operates differently in the atmosphere and in the car (or greenhouse). In both, the radiation is freely admitted from outside, but then is trapped. For the atmosphere, it is the infrared absorption of the atmosphere that traps the radiation. In the car (or greenhouse), it is the disruption of the convection cells that reduces the escape of heat.

Assume that to emit P, the Earth's temperature was $T_0 = 246^\circ K$. To emit $2P$, as is required in this model, the Earth heats up to a temperature

$$T = {}^4\sqrt{2}\, T_0 = 1.189 T_0 = 293K = (246 + 47)K$$

(The factor of ${}^4\sqrt{2}$ is necessary to guarantee that σT^4 is twice as great as σT_0^4.) So in this two-layer model, the greenhouse effect causes warming by 47°K. The fact that this model yields a resulting number close to that for the observed average surface temperature (about 288°K) is due to the accidental cancellation of effects that we ignored. The measured average temperature of the upper atmosphere (the radiating region) shows that its average temperature is about 255°K.

It is clear, however, that the temperature of the Earth is determined by radiative forcing, but also is very sensitive to feedback mechanisms such as the greenhouse effect and albedo (reflection of the incoming visible sunlight). Note that this is not the case, for example, for Jupiter, since Jupiter produces more energy within its mass than it absorbs from the Sun.

6.1.2 Atmospheric processes

The primary atmospheric gases that drive the greenhouse effect are not nitrogen, oxygen or argon, which are largely transparent to infrared radiation, but water vapour and carbon dioxide. If it were the *only* greenhouse gas, atmospheric water vapour would supply 60–70% of the warming. The next most important gas is carbon dioxide, which by itself would account for about 25% of the warming. Following these, in order of importance, are methane, nitrous oxide and ozone. Their behaviour is complicated by the fact that these gases absorb very strongly at some infrared wavelengths, but are almost transparent at others. When newspaper articles refer to carbon dioxide as *causing* the greenhouse effect, what they really mean is that additional carbon dioxide in the atmosphere can effectively fill some of the wavelength holes, making the greenhouse effect a little more efficient, and thereby driving up the temperature of the Earth (if no other changes cancel this effect). Greenhouse forcing due to changes in carbon dioxide in the last century have been estimated to have an effect equivalent to increasing the incident solar power by between 0 and 0.6 W/m^2. In the last few decades, chlorofluorocarbons have also become significant greenhouse gases, but they are not important for paleoclimate.

One of the most important drivers of climate, and arguably the least understood, is the role played by cloud cover. The net effect depends on the altitude distribution of the clouds and their opacity. Thin clouds tend to cause warming (from an enhanced greenhouse effect), whereas thick clouds tend to cause cooling (by reflecting sunlight). Cloud cover also represents a potentially positive feedback mechanism that can amplify other climate drivers. To give a rough idea of the possible role of cloud cover, we show estimates from the Earth Radiation Budget Experiment (Hartmann, 1993), as cited by Kirkby (1998). They are summarised in Table 6.2.

Note that the energy budget from clouds is believed to be a negative "cooling" effect ($-27.7\,W/m^2$), which results from a difference between the more negative

Table 6.2. Estimates of cloud forcing.

Parameter	High clouds		Middle clouds		All low clouds	Total
	thin	thick	thin	thick		
Global fraction (%)	+10.1	+8.6	+10.7	+7.3	+26.6	+ 63
Forcing (relative to clear sky)						
Albedo (visible) W/m^2	−4.1	−15.6	−3.7	−9.9	−20.2	−53.5
Outgoing (infrared) W/m^2	+6.5	+8.6	+4.8	+2.4	+3.5	+25.8
Net forcing (W/m^2)	2.4	−7.0	+1.1	−7.5	−16.7	−27.7

albedo effect ($-53.5\,W/m^2$) being partially cancelled by the infrared effect ($+25.8\,W/m^2$).

Even if these estimates are only approximate, the energy budget contributions are large enough to have a key effect on climate changes. For comparison, the equivalent forcing in the nonlinear ice model amounts to a peak-to-peak variation of about $20\,W/m^2$, and the variation that is supposed to have caused the last glacial termination is only $13\,W/m^2$ (see Fig. 6.10). So anything that modulates cloud cover is going to have an important effect on the climate of the Earth.

Aerosols play an important role in the albedo and in the formation of clouds. Aerosols are particles in the size range 0.001 to 10 μm. Some are injected from fires and volcanoes; others are created from photochemical processes that involve gaseous sulfur and hydrocarbons, and still others are released from white caps of waves. Aerosols can reflect sunlight and infrared radiation. They have a short lifetime in the troposphere, where they are scavenged by rain, but they can have an important effect when they are present in the stratosphere. The concentration of aerosols is believed to affect the size distribution of cloud droplets, and that affects the optical density of the clouds. Aerosols could also play an important role in the electrical conductivity of the stratosphere, since they are efficient scavengers of electric charge; likewise, when they evaporate, they can contribute ions to the atmosphere. The aerosols in the stratosphere reach a maximum density at about 20 km, and have been strongly enhanced by upward-erupting volcanoes, such as El Chichón in 1982. (Mt. St Helens, which erupted horizontally, had little effect.) Modern-day industrial activity has significantly increased tropospheric aerosols over populated areas.

The clearest evidence that volcanic eruptions affect temperature is found in the strong correlation between temperature and volcanic events recorded in ice cores. In Fig. 6.2 we show a plot of the average yearly temperature in the Northern Hemisphere (60–90°N), with arrows indicating volcanic events detected by sulfate measured in the Greenland ice. The plot is taken from Lyons et al. (1990) and is based on work of Self et al. (1981) and Mayewski et al. (1990). Every volcanic event (except one) is associated with a sharp drop in temperature.

There is long-standing evidence that sunspot activity affects the upper atmosphere (Eddy, 1976, 1977; Herman and Goldberg, 1978), and this could give rise to a

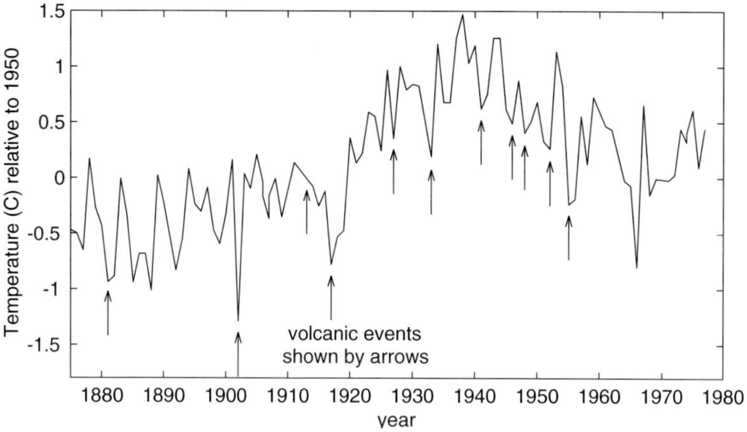

Fig. 6.2. Temperature (northern hemisphere) and volcanic events (arrows). Adapted from Lyons et al., 1990.

coupling between sunspots and climate. Labitzke and van Loon (1997) found a strong relationship between variations in atmospheric pressure and the sunspot cycle. There are several hypotheses to explain the connection, but none of them have been generally accepted. The fact that we don't understand these relationships is a reflection of the poor state of our understanding of the atmosphere. There is a known connection, however, between sunspots and cosmic rays. Solar activity strongly modulates the solar wind, and magnetic fields carried by the solar wind affect the flux and energy distribution of galactic cosmic rays that reach the Earth. Thus we might understand the sunspot/climate connection if we could understand the relationship between cosmic rays and the atmosphere. A possible mechanism is suggested by the theory of Tinsley (1996), who argues that cloud cover is strongly affected by the return electric currents in the atmosphere that are driven by equatorial thunderstorms. Thus sunspots affect cosmic rays, cosmic rays affect stratospheric conductivity, conductivity affects electrical current, and electrical current affects cloud formation. The string of hypotheses is long—but there really is a relationship between sunspots and atmospheric changes which demands explanation. It is also possible that aerosols affect atmospheric conductivity by their accretion of free electric charges.

There is other evidence linking cosmic rays to cloud formation. Svensmark and Friis-Christensen (Svensmark, 1998; Svensmark and Friis-Christensen, 1997) found a strong correlation between the cosmic ray intensity (measured with ground-based neutron monitors) and satellite measurements of mid-latitude clouds. Over a sunspot cycle, the cosmic rays varied by 15–20%, and this correlated strongly with a 3% variation (absolute) in cloud cover over that same period. Since total cloud cover is about 63% (see Table 6.2), this is about a 5% change in relative cloud cover. A 5% change in clouds would result in a variation in the radiation budget equivalent to about $1\,W/m^2$. This is roughly equal to the total radiative forcing caused by the observed human-induced increase in atmospheric carbon dioxide. It is not really surprising that clouds are important—but it is surprising that cosmic rays may be important. Of course, it is possible that something else from the sun is modulating both the cloud cover and the cosmic rays. Svensmark (1998) considered several such phenomena, and concluded that the direct cosmic ray forcing was the best fit. However some of the cosmic ray cloud correlations have been disputed by Kernthaler et al. (1999).

Another possible linkage between sunspots and climate has been reported by Friis-Christensen and Lassen (1991). They show that the change in land temperature of the Northern Hemisphere is in close agreement with the length of sunspot cycles. They define the sunspot cycle length as the time interval between successive maxima or successive minima, and show that this matches Northern Hemisphere temperature variations for the past 400 years. The cycle length is believed to be related to the magnetic activity of the Sun, with the shorter cycles associated with the more active Sun. The magnetic activity affects the solar wind, and the wind in turn modulates the intensity of galactic cosmic rays that reach the atmosphere. The sunspot cycle varies between 7 and 17 years in length, with an average of 11 years. The connection between temperature and sunspot cycles is disputed, however, by many climate modellers who attribute the variation in the twentieth century to carbon dioxide-induced greenhouse warming rather than to solar activity.

Finally, ozone is what warms the upper atmosphere, creating the layer of temperature inversion that we call the stratosphere. Changes in ozone can not only affect the greenhouse mechanism directly, but they can drive a change in the altitude of the tropopause. That, in turn, can affect such important phenomena as the position of storm tracks and the presence of thunderstorm-injected water in the stratosphere. Shifting storm tracks could increase precipitation in the northern latitudes, leading to the accumulation of ice and lower temperatures. The stress from wind drives ocean currents, and they too would change.

The ideas presented above vary from the well-established to the highly speculative. The atmosphere, especially the upper atmosphere, is not a well understood system, so even the craziest ideas may prove to be right—and maybe even important.

6.1.3 Land-based mechanisms

Historically, land-based mechanisms played *the* critical role in the original Milankovitch theory. The distribution of land as a function of latitude is shown in Fig. 6.3. Note that two-thirds of the land area occurs in the Northern Hemisphere. This was the justification for the assumption that summer insolation at northern latitudes could be the dominant driving force for the ice ages. The presence of land is necessary to allow the accumulation of ice, and so that the presence of high summer insolation could prevent ice fields from enduring. In addition, land responds more readily to seasonal changes than do the oceans. The fundamental reason for this is lack of convection. In the oceans, heat can convect between depths as well as horizontally, but on land, heat must diffuse, and that is a much slower process. Seasonal changes rarely penetrate more than about two metres. Ocean water mixes readily; cooling water contracts and sinks to the bottom until the temperature

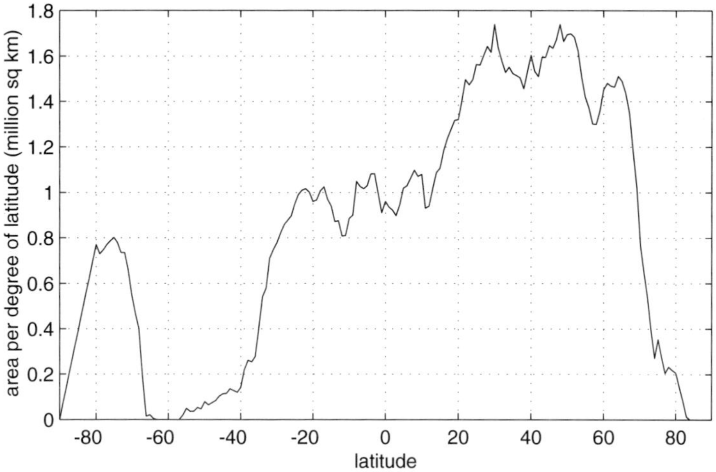

Fig. 6.3. Land area vs. latitude on the Earth. Two-thirds of the land mass is north of the equator.

of the entire depth of water drops to 4° Celsius. The uppermost 50 metres or so of the ocean, called the "mixed layer", is thoroughly mixed by wind and waves.

The importance of land in the formation of large glaciers is illustrated by the presence of glaciers in Greenland and Antarctica. These are the only land masses which extend close to the poles, and they are the only ones with extensive glaciers remaining from the ice ages. In Greenland, the glaciers extend southward almost 30 degrees from the North Pole; in Antarctica, they extend about 20 degrees from the South Pole. Large areas in Canada and Russia which are closer to the North Pole than southern Greenland but don't reach as close as northern Greenland, have no glaciation. This suggests that polar "roots" on land are necessary and, if they exist, then the glaciers can extend much further from the polar regions. Although sea ice covers the Arctic Ocean, it does not appear able to provide the same kind of roots that are provided by land.

Once the glaciers begin to form, the increased albedo from their surface may provide a positive feedback to enhance their formation, a mechanism identified by James Croll in the 1800s. We say "may" because what is needed is both cold and precipitation, and it is not obvious that increased cold alone will mean increased ice. In the California Sierra Nevada, for example, a period of extreme cold is often accompanied by a period of low precipitation; skiers know that what is really needed for snow is cold but not too much cold. Of course, it is not only ice that increases albedo, but snow.

Even when not covered with ice or snow, the properties of the land surface have strong effects on the climate. Surface vegetation and forestation affect albedo, and surface topography affects the pattern of precipitation and the availability of moisture. The amount and kind of vegetation strongly affect the humidity of the lower atmosphere. Many climate models incorporate an effect called "water vapour feedback", in which a small amount of warming causes the injection of water into the atmosphere; this enhances the greenhouse effect, and can cause substantial additional warming. This water vapour feedback is essential if the present CO_2 increase is going to account for the observed global warming of the last century. About half of the water vapour feedback in these models comes from water vapour in the upper troposphere, and the relationship of this water to surface conditions is not well understood. Modulation of the complex water vapour feedback system (e.g. by deforestation, or change in the nature of vegetation) could have an important effect on the climate.

The climate pattern affects the intensity of thunderstorm activity, which in turn probably controls the injection of moisture into the upper atmosphere. Melting ice affects the pattern of flow in the oceans. The land vegetation controls the flow of atmospheric CO_2. In addition to respiration, the transport of atmospheric carbon dioxide into sea water depends on the presence of eroding rock. When rock is exposed to rainwater, the small CO_2 component in the water makes it slightly acidic, and that dissolves the rock. The CO_2 is transformed into a bicarbonate ion, which is very soluble, and readily scavenged by calcium carbonate-producing marine organisms. Thus erosion of rocks (and their exposure to rainwater) is a major factor in the removal of CO_2 from the atmosphere.

The area of land in the form of swamps, and the grazing area for ruminant animals, determines how much methane is produced, and this is an important greenhouse gas. The methane is also affected by its destruction by bacteria in soils. Its lifetime in the atmosphere is only a few years, so it can respond very quickly to environmental changes (e.g. the drying up of swamps).

6.1.4 Ocean-based mechanisms

The oceans affect climate in many ways. The upper layer of the ocean mixes in a time measured in days, but even the deep ocean is thoroughly mixed with surface water on the time scale of a thousand years; this has been established by radiocarbon dating of deep water samples. Because of this mixing, the heat stored in the entire depth of the ocean plays a role in climate. The ocean also serves as a sink for CO_2. Knowing the circulation pattern of the oceans is essential for understanding the heat fluxes in to and out of the atmosphere. The flow includes not only surface currents, but deep flows such as the oceanic conveyor belt, also known as the thermohaline circulation. Nearly everybody who discusses the ocean currents now points out that they represent only the average flow, and that mesoscale eddies and the rotation of the Earth can be very important for local effects.

In turn, the oceanic circulation is driven by winds, salinity differences (e.g. from melting ice, which can disrupt the thermohaline circulation), and temperature (e.g. the heating of the tropics that drives the hurricane season). Subtleties in the ocean flow can have major consequences for climate. For example, only in recent years have we come to appreciate the instability of Pacific sea level which gives rise to the "El Niño Southern Oscillation". An active El Niño affects climate around the world, and yet the source is located in the equatorial Pacific.

The ocean provides many possible mechanisms for positive and negative feedback. The ice ages result in substantial changes in the depth of the oceans, by over 100 metres, and such changes could affect circulation patterns. Large areas of the current oceans, particularly in the western Pacific Ocean, are less than 100 metres deep, and would become dry during glaciation, a fact pointed out to us by Walter Alvarez. It is believed that much of the water vapour in the upper troposphere and lower stratosphere is injected by very large thunderstorms, and the drying of a major oceanic area could affect this process. It is difficult to even guess the sign of the effect, since although the absence of water area seems as though it would reduce the evaporation of water, the presence of shallow regions could enable the water to reach warmer temperatures, and thus increase evaporation. Changes in sea level (and temperature) could also affect the oceanic storage of methane clathrates, and exert a negative (or possibly positive) feedback for climate change (MacDonald, 1990). There is evidence that there was a sudden release of methane into the atmosphere from clathrates during the Paleocene, about 55 million years ago (Katz et al., 1999).

Sea ice is important both for transport of cold and non-saline water. The sudden release of melt water from ice near Greenland at the last termination is believed to be responsible for the brief return to cooling known as the "Younger Dryas". Sea ice also increases albedo, and reduces the exchange of heat between the oceans and the atmosphere. The motion of sea ice is determined by ocean currents and by wind. The sea ice floating near the source glacier might play a role in stabilising the glacier.

6.1.5 Insolation forcing

The temperature of the Earth is determined by insolation (albeit amplified by a complex greenhouse effect) so it is natural to believe that cycles of climate could be driven by changes in this insolation. But in trying to develop this theory, we immediately run into problems. Even though the orbit of the Earth does change with time, it does not change very much. The average distance from the Sun is remarkably constant, as was demonstrated by Laplace. The eccentricity of the Earth varies between 0 and 0.05. Although this changes the instantaneous distance to the Sun, during half the orbit the Earth is closer and during the other half of it is farther away; the two effects almost cancel (see Fig. 2.5). The total yearly insolation over the eccentric orbit of the Earth is proportional to:

$$I \propto \frac{1}{\sqrt{1-e^2}}$$

The fractional change in yearly insolation is

$$\frac{\delta I}{I} = \frac{e\delta e}{1-e^2} \approx \frac{1}{2}e^2 \approx \frac{(0.05)^2}{2} \approx 0.001$$

where we have taken $\delta e \approx e/2$. Thus the total insolation changes by about one part in a thousand. That makes the equivalent effect less than 1 W/m^2, much smaller than most of the other effects we will consider. But if we don't integrate over the entire

Earth—if we decide that there is a particular location and/or a particular season that is important—then there is no cancellation between the two halves of the Earth. If the obliquity of the Earth were zero, then the insolation at the poles, for example, would remain at zero (sunlight always arrives horizontally, and the power per square metre on the polar surface is zero). Croll and Milankovitch focused on the summer insolation in the Northern Hemisphere, since weak insolation during those periods might mean that snow would last through the summer, and that could give rise to enduring glaciers. Perhaps a more important reason for their choice is that by picking a single location and a single season, the computations (in the pre-computer days) became tractable. In fact, given the list of mechanisms discussed earlier in this chapter, it seems that such a choice is clearly too simple.

The matter is made even worse by the fact that in the earlier part of this chapter we listed many mechanisms which can affect climate that are not readily identified with insolation at a single location. For example, insolation at 65N is believed to have very little direct effect on the intensity of El Niño, and yet El Niño affects the climate around the world.

Fortunately, there is a viewpoint that cuts through this Gordian knot. Let us assume that the Earth responds *linearly* to insolation. So, for example, let us assume that global ice is driven by some complicated sum of summer insolation at 65N, winter insolation at 35S, and midseason insolation at the equator. We will not assume that these contributions are equal, but simply that they contribute in a linear way to the driving force F:

$$F = A_1 I_{\text{July, 65N}} + A_2 I_{\text{Dec, 35S}} + \dots$$

Secondly, we will assume that the climate responds in a linear way to the driving force. We may not know what the coefficients A_i are, but we can still make a strong prediction. This will be based on the fact that the frequencies present in the climate will be the same as those present in the insolation. Insolation, at essentially all latitudes and seasons, is dominated by just a few frequencies: those of precession (dominant periods of 24, 22 and 19 kyr), and obliquity (dominant period 41 kyr). Thus, even if we don't know precisely what combination of locations and seasons to choose in our model, we can make a prediction about what frequencies will be present in the climate. The relative amplitudes of these frequencies will depend on the locations and season, but the value of these frequencies will not.

However, we cannot likewise conclude that our favourite proxy, $\delta^{18}O$, representing ice volume, is also linearly related to climate. It is plausible that ice accumulation, or melting (negative accumulation), is linear in climate, but ice volume is the integral of the accumulation. However, a simple integration does not change the frequency, it merely changes phase (e.g. by changing a cosine into a sine wave).[2] Moreover, the phase of the insolation components also depends on latitude and longitude. When

[2] If two frequencies are present, then integration also changes the relative ratio of their amplitudes by a factor of $1/f$.

you add two sine waves which have the same frequency but different phase, you get yet another sine wave of the same frequency (but intermediate phase).

What can we conclude from all this? Something very simple: if $\delta^{18}O$ is the accumulation of a signal that is being driven by insolation at various locations around the world, then a spectral analysis of the $\delta^{18}O$ signal should show strong periods of 41, 24, 22 and 19 kyr. Moreover, an analysis of the amplitudes and phases of these frequencies will help us decide which seasons and which latitudes are most important.

This view was used to justify the tuning procedure adopted by Imbrie et al. in their SPECMAP timescale (Imbrie et al., 1984). Rather than picking a particular latitude and season, and tuning to insolation, they applied bandpass filters to their data, and tuned separately to insolation and precession. They set their phases by making the time scale match dates that had been determined radiometrically.

Unfortunately, this point of view also limits the predictive power of the insolation model. As long as the right frequencies are present, virtually any relative amplitude of their contributions can be forced into the model. Recognising this sloppiness, we will refer to any signal that has strong peaks at both obliquity and precession frequencies as showing an insolation response.

One of the best data sets with a strong insolation response is the atmospheric $\delta^{18}O$ signal from Antarctica. The spectrum is shown in Fig. 6.4, along with the spectrum for July 65N insolation. The Vostok data show the precession peaks at 0.043 cycles/kyr (period 23 kyr, consisting of unresolved 24 and 22 kyr components) and at 0.053 c/kyr (period 19 kyr) and the obliquity peak at 0.024 c/kyr (period 41 kyr). There is an additional narrow peak near 0.01, possibly due to either eccentricity or to orbital inclination.

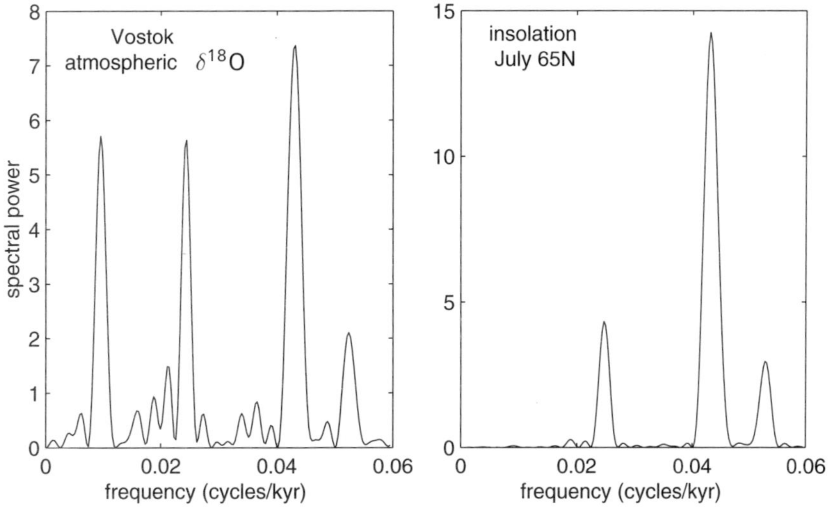

Fig. 6.4. Vostok $\delta^{18}O$ compared to insolation July 65N.

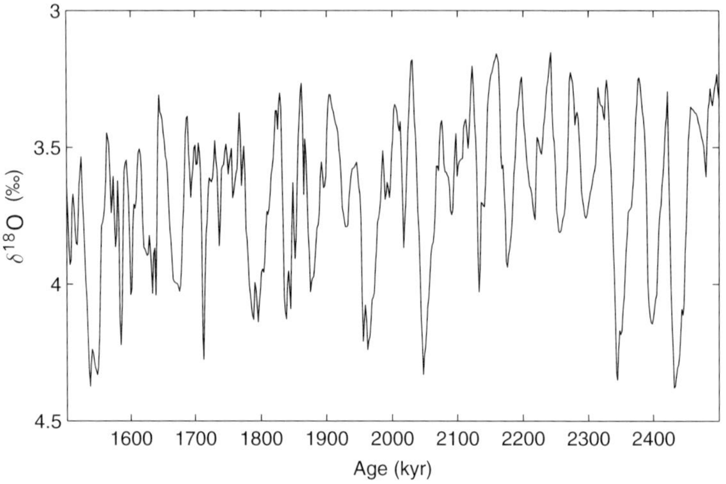

Fig. 6.5. δ^{18}O for Site 607, 2.5 Ma to 1.5 Ma.

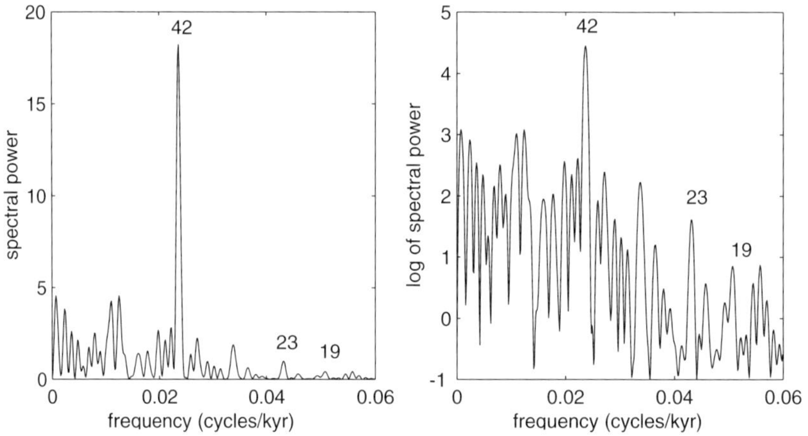

Fig. 6.6. Spectrum of δ^{18}O, Site 607, 2.5 Ma to 1.5 Ma.

A second example showing a strong insolation response is the terrigenous component from Site 721. The data are shown in Fig. 4.22, and the spectrum in Fig. 4.23. Note that this is not an ice volume proxy, but probably represents local climate of Africa.

As a third example we look at the δ^{18}O signal from Site 607 in the North Atlantic Ocean from the period 1.5 to 2.5 Myr (Raymo, 1992). The data are plotted in Fig. 6.5 using a time scale based on an assumed constant sedimentation rate with the end point ages based on those of Raymo. The spectrum for these data is shown in Fig. 6.6.

The linear plot is most useful for estimating the fraction of the variance that arises from the oscillation; the log plot is most useful for looking for small peaks. The

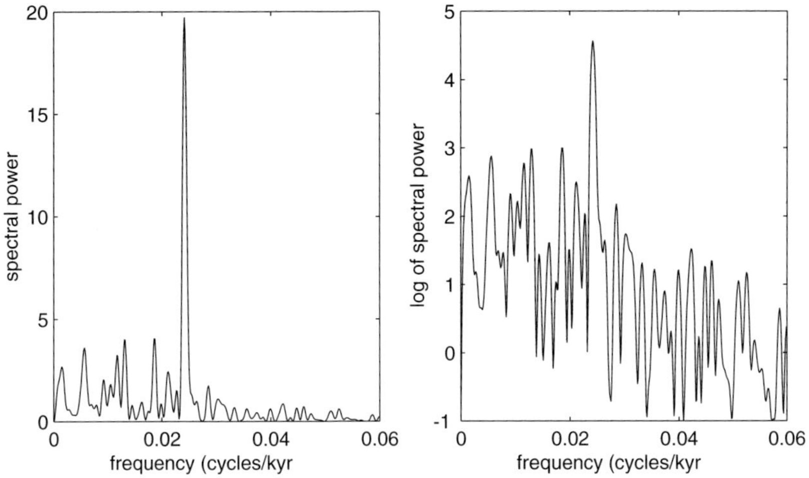

Fig. 6.7. Spectra of Site 677, 1 to 2 Ma.

spectra show a very strong 42 kyr peak, although it contains, at most, a weak precession signal. (The signal at 23 kyr is not statistically significant, and it is certainly not climatologically significant; there is nothing at 19 kyr.)

The dominance of the 41 kyr obliquity peak is also seen in the $\delta^{18}O$ data from Site 677 in the equatorial Pacific. The spectrum of this proxy, in both linear and log plots, is shown in Fig. 6.7. Note the absence of any significant signal at precession frequencies.

Further study of the insolation signal shows a clear pattern. Precession is strong in many climate proxies, but it is consistently weak in ice volume proxies (Karner and Muller, 2000b). The exception is the entrained atmospheric oxygen in the Vostok core but, as we discussed in Section 4.7, this signal does not track the ocean floor ice volume proxies. In fact, it is possible that ice volume is not driven by precession at all, but that when precession is present in a $\delta^{18}O$ signal, it is due to local climate. This implies that if the ice volume signal is being driven by insolation, that it is *not* the standard Milankovitch insolation of July 65N, but some other insolation or combination of insolations for which obliquity dominates.

6.2 NONLINEAR MODELS

The fundamental problem with insolation-based models is that although insolation is dominated by precession and obliquity, the ice volume proxies for the past million years are dominated by a 100 kyr oscillation. (There are other problems, too, as we discuss in the next chapter, but this one has been the most bothersome.) However, the envelope of the precession signal varies with a quasi-100 kyr cycle (see Fig. 2.10), equal to the variation of eccentricity. It seems reasonable, therefore, that the 100 kyr cycle might somehow be derived from this envelope.

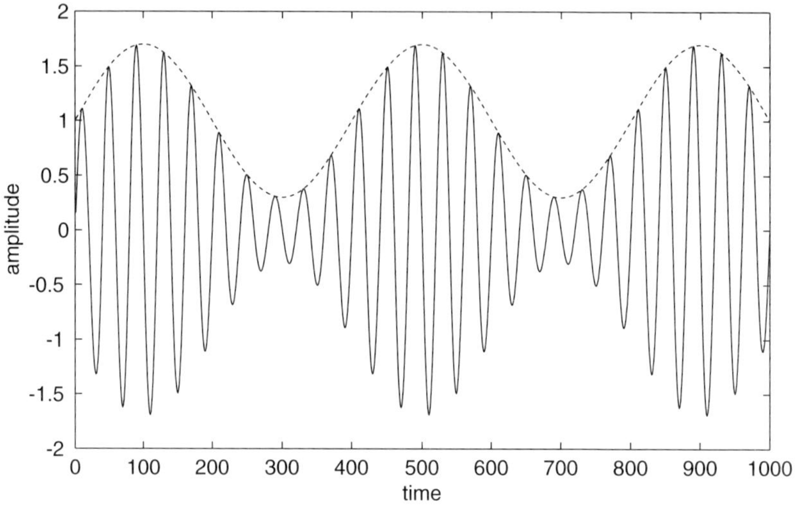

Fig. 6.8. Amplitude modulated signal. The dotted line is the envelope.

There are several ways to "extract" this 100 kyr variation. We can, for example, postulate that there is a threshold beyond which climate behaves differently—and that this threshold is related to some plausible mechanism. The list of mechanisms from which we can choose is large, so it is not hard to make this work. Milankovitch (who didn't know about the 100 kyr cycle) postulated that there was a level of summer insolation which prevented permanent ice fields from forming. This is an example of such a threshold.

But, in the end, a threshold is just one example of a nonlinear response to a forcing function. There are an infinite number of nonlinear operations that will create an eccentricity signal out of precession. In fact, it is difficult to construct a nonlinear operator that *doesn't* do this! To see why this is so, consider the example of an AM radio. AM stands for "Amplitude Modulated"; it means that the carrier (which may have a frequency of 100 kHz) has its amplitude slowly modulated by the audio signal (which may include frequencies near 1 kHz), as illustrated in Fig. 6.8. In the original AM radio receivers, the AM signal is first rectified (made positive) and then smoothed. The result is the extraction of the envelope from the carrier. This process is illustrated in Fig. 6.9. This process is so easily performed that some people pick up and hear AM radio signals from fillings in their teeth.

We take the signal from Fig. 6.8 and square it, and then smooth (by taking a running average); the result is shown in Fig. 6.9. We have succeeded in extracting the envelope of the carrier. This is how an AM radio works. The original signal will have no spectral component at the envelope frequency; the demodulated signal will be dominated by the envelope frequency. Note that there is a slight phase delay in the extracted signal. This arose from the method of smoothing that we applied; at each time point we plotted the average from the *previous* 40 time points. We did that because that is how an AM radio works: in real time.

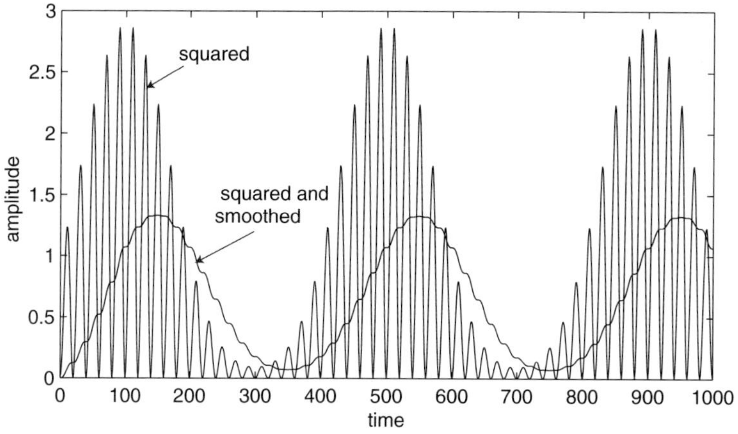

Fig. 6.9. AM signal after squaring and smoothing.

For a nonlinear insolation model, the carrier is the precession signal, and the modulation is the eccentricity. The nonlinear operator used in an AM radio (square then smooth—or rectify and smooth) will create a strong eccentricity signal out of precession, as we will show in Section 6.2.2.

There are an infinite number of nonlinear models that will perform this task. That means that we cannot judge which one is correct by its ability to create a strong eccentricity signal. We must use other criteria, such as whether the spectrum created matches that of the $\delta^{18}O$ data, and whether it can reproduce the variation in the time domain (which depends not only on the spectrum, but on the relative phases of the components). In addition, the model should be physically plausible. One of the simplest and most plausible such models is called the "nonlinear ice model".

6.2.1 Nonlinear ice model

The nonlinear ice model of Imbrie and Imbrie (1980) is based on a simple nonlinear equation that was inspired by the fact that the advance and decline of glaciers could easily occur at different rates. We represent the size of the ice sheet by the symbol S. Then the behaviour of this sheet is assumed to follow the differential equation:

$$\frac{dS}{dt} = \frac{(I - S)}{T}$$

I is the insolation, and in these units, S will have the same dimensions as I. S is the value the insolation would have to have to mimic the nonlinear ice model. We call it the "equivalent insolation". T is a time constant which represents changes in the growth or contraction of the ice sheets. This is a linear equation; the equation is made nonlinear by putting in different values for T depending on whether dS/dt is positive or negative, i.e. on whether the ice sheets are increasing or decreasing. The equation says that the rate of ice accumulation dS/dt depends on the difference between the insolation and the state of the ice. So, for example, if insolation suddenly

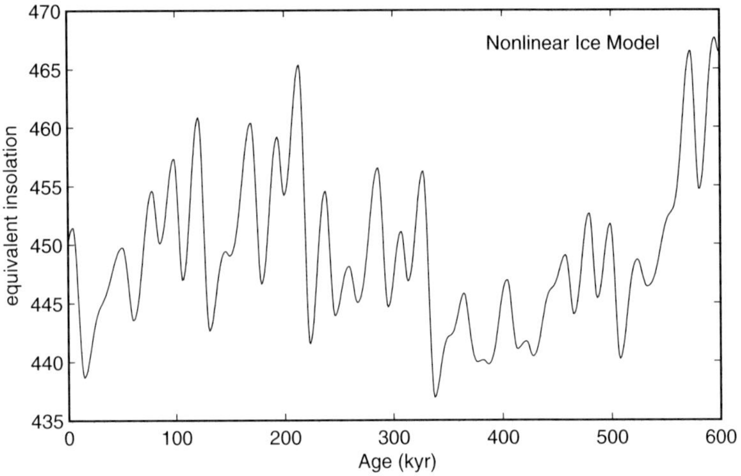

Fig. 6.10. Nonlinear ice model, 0 to 600 ka.

rose to a high level, and there was still lots of ice left, then $(I - S)$ would be large, and the rate of ice change, dS/dt, would be large.

If we integrate the nonlinear ice equation, we get $S(t)$. We have done this using the values for T suggested by Imbrie and Imbrie (1980); the results are shown in Fig. 6.10.

In effect, the model replaces ordinary insolation with an effective insolation which is supposed to represent more accurately the true driving force and response. The model does not predict the level of ice to be found. It does not explain how the changes in insolation drive changes in ice level. It simply assumes that the ice volume responds linearly to the new driving force. Two arbitrary parameters can be introduced to convert insolation to ice. The fact that the model does not really connect the revised insolation to expected ice levels in a numeric way is not widely appreciated by the paleoclimate community.

It is worthwhile noting that the nonlinear ice model does not solve the Stage-11 problem discussed in Section 8.2. The large glacial termination that takes place at about 420 kyr does not correspond to any strong behaviour in Fig. 6.10. And in the time variation seen in Fig. 6.10, the presence of the 400 kyr cycle can be deduced. The spectrum of the data for this nonlinear ice model is shown in Fig. 6.11.

Precession is still higher than obliquity, but the eccentricity peaks have been enormously enlarged. The split character of the 95/125 peak remains. Another peak has appeared just above 0.02.

6.2.2 The AM radio nonlinear model

To illustrate the fact that almost any nonlinear model produces eccentricity, let us perform the "AM radio nonlinear model" operation described in Section 6.2, used by AM radio receivers to extract the audio from the carrier. To do this to the

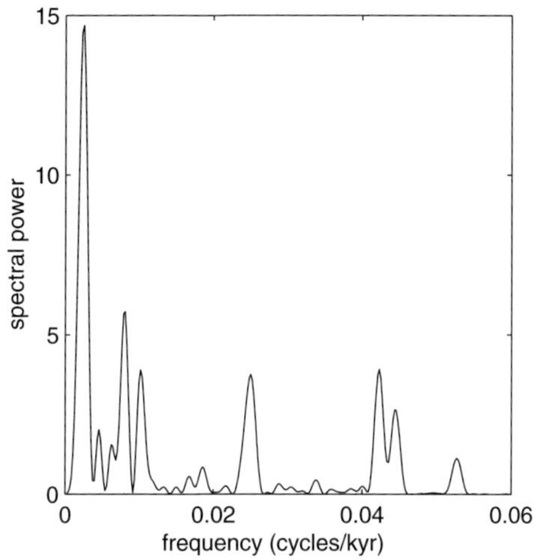

Fig. 6.11. Nonlinear ice model spectrum, 0 to 600 kyr.

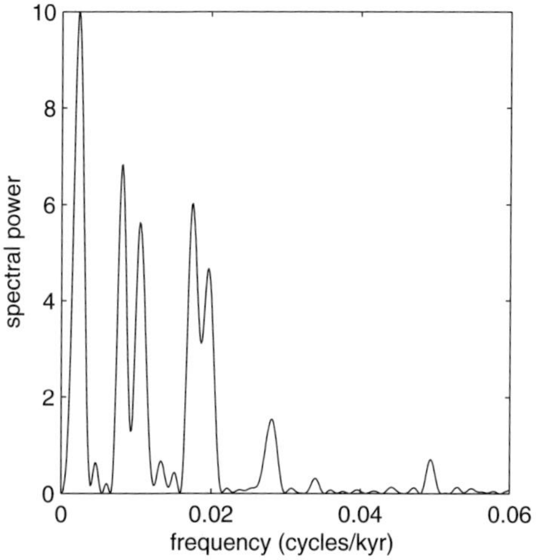

Fig. 6.12. Nonlinear operator applied to eccentricity.

insolation, we subtract the mean and square the signal (just rectifying it will do about the same thing). The spectrum of the resulting function is given in Fig. 6.12.

This spectrum has peaks at 95, 125 and 400 kyr, just like the spectrum for the nonlinear ice model. We have suppressed the inclination cycles even more (which, actually, makes it a better fit to benthic $\delta^{18}O$ data), and have produced one large extra peak near 0.02, which is a bit too large. But the nonlinear ice model had an extra peak too, and given one parameter we could suppress this. For example, if we smooth the output by taking a running average over 40 kyr (smoothing is also done for AM radio signals), the resulting spectrum is shown in Fig. 6.13.

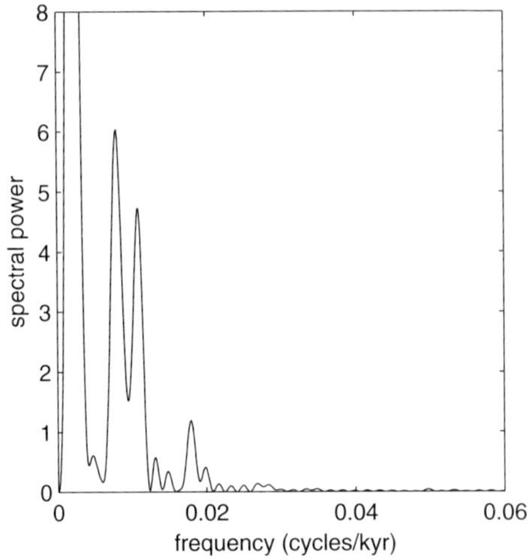

Fig. 6.13. Nonlinear operator with smoothing applied to eccentricity.

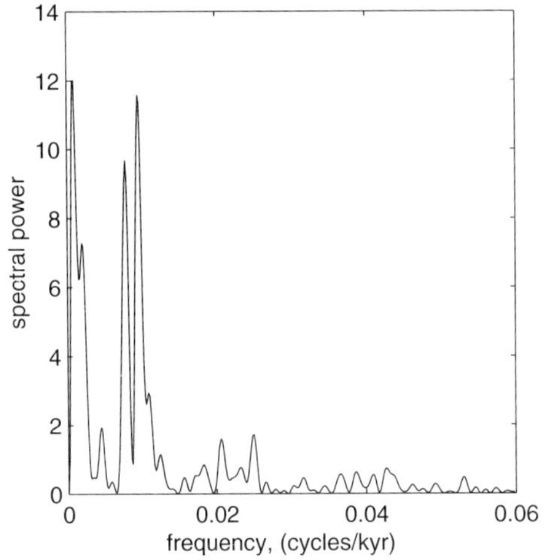

Fig. 6.14. Site 806 sand spectrum.

This is actually a closer match to the spectrum of eccentricity peaks seen in the data (e.g. the foram size spectrum of Site 806) than that given by the nonlinear ice model. We reproduce that spectrum in Fig. 6.14 for comparison.

Note that the smoothed square function even "reproduced" the little bump near 0.02. Yet this spectrum was produced without using *any* physics of ice sheets; we simply assumed that the Earth listens to insolation in a fashion similar to the way humans listen to an electromagnetic wave using an AM radio. We would get a similar behaviour if we rectified the signal and then took the squareroot instead of the square. As we said earlier, the problem with nonlinear models is that they enhance the eccentricity power regardless of the details, so the success of the model does not demonstrate that the details of the model bear any relationship to the actual nonlinear physical mechanism. The danger of modelling is that, given a few parameters and a little insight about the math, it is often easy to fit the data with a model that has no bearing on the real problem.

6.3 PROBLEMS FOR MODELS TO ADDRESS

There have been too many models proposed for us to discuss them all (or for us to *want* to discuss them all). So we will choose to give a close look at a subset of recent models which have the following features: (1) they directly address significant problems that have been discovered in the standard Croll/Milankovitch insolation models (including the nonlinear ice model); and (2) they illustrate how the methods of analysis which we have presented in the previous sections can be applied to reach a conclusion about the correctness of the model.

In the earlier sections of this book, several problems have appeared that challenge the standard model. They will be addressed in detail in Chapter 8. Here we summarise them, so that they can be used as a focus for the discussion of other models. The problems are as follows:

1. The 100 kyr problem. For the past million years, the global ice proxy $\delta^{18}O$ is dominated by a 100 kyr oscillation. The standard model predicts that obliquity (41 kyr), and precession (24, 22 and 19 kyr) should be stronger.
2. The "Stage-11 Problem" articulated by Imbrie et al. We combine this with the related "Stage-1" problem. This is the fact that strong global ice oscillations were taking place 400 thousand years ago (and 15 thousand years ago) when variations in insolation were very small.
3. The causality problem. The sudden terminations of the ice ages appear to precede the warming that was supposed to cause them.
4. The unsplit peak problem. According to the standard theory, the 100 kyr "eccentricity" peak should be split into two components, with periods 95 and 125 kyr; yet the data indicate it is a single narrow peak.
5. The absence of a 400 kyr cycle. This is predicted to be the dominant oscillation in the nonlinear ice model, and yet it is virtually absent in the $\delta^{18}O$ data.
6. The bispectrum problem. The bispectrum of the $\delta^{18}O$ data shows peaks in areas which do not match the peaks from the standard model. The bispectrum tests aspects of coherence in the data that are not tested in the standard spectrum or coherence plots.
7. The transition problem. The global ice proxy underwent a shift about a million years ago, from dominance by the obliquity signal, to dominance by the 100 kyr cycle. A good theory of the ice ages should be able to accommodate this shift, or—better still—explain it.

8. The 41 kyr problem. In the global ice proxy, $\delta^{18}O$ data, the precession component is consistently very small. In the period between 1 Ma and 2.5 Ma it is essentially negligible. The signal is completely dominated by obliquity. Why?
9. The sudden termination problem. Why do the ice ages end so abruptly? The terminations appear to be more sudden than the initiations of the glacial periods, and that gives rise to the sawtooth shape seen in the $\delta^{18}O$ data, although this shape seems more evident in the last 400 kyr than in the earlier data.

6.4 ALTERNATIVE ASTRONOMICAL MODELS

We will now describe a series of "alternative models" that have been published in order to address some of the "problems" listed in the previous section.

6.4.1 Orbital inclination model

This model postulates that it is orbital inclination, not eccentricity, which is responsible for the 100 kyr cycle of glaciation. The Earth's orbital inclination, as explained in Section 2.2.6, is tilted with respect to the average orbits of the other planets, and this variation undergoes a 100 kyr oscillation. The oscillation has a spectrum and bispectrum that are very good matches to those of the global ice proxy $\delta^{18}O$. It has no causality problem; it has no Stage-11 problem, and no Stage-1 problem. It is claimed to have other advantages: it gives a natural (no adjustable parameters) explanation of the transition from the 41 kyr cycle to the 100 kyr cycle. Moreover, the theory correctly predicted that the accretion of interplanetary dust on the Earth would show a 100 kyr cycle locked in phase with that of the glacial oscillation, and subsequently that prediction was verified. It accounts for all of the problems listed in the last section, with the exception of the sudden terminations (for which it must postulate an ad hoc cause, such as positive feedback that occurs when the ice begins to melt).

The physical mechanism behind the model is the linkage between dust, cloud cover and climate. Cloud cover exerts a driving force on climate that is comparable to that of insolation variations. As shown in Table 6.2, the net forcing of clouds in the present climate is 28 W/m^2. For comparison, the RMS deviation in 65N summer insolation for the past 600 kyr (Fig. 4.38) is only 20 W/m^2. Clearly variations in cloud cover is important. Unfortunately, cloud cover is the least well understood aspect of climate, not only in the past, but also in present climate models (see, for example, IPCC, 1996). Moreover the link between dust and cloud cover, although reported in diverse papers, is not firmly established (see Chapter 7) and must be handled through parameterisation. The only defense for using this approach is that a similar parameterisation is used in *all* present-day climate models, including those based on insolation.

It is important to note that there is not enough extraterrestrial dust to affect climate by the direct scattering of sunlight. One year's accumulation, if spread evenly over the Earth, would create a layer that is less than an Angstrom thick (i.e. less than

0.0001 micron). An amplifying mechanism, such as cloud cover, is required. Dust can modulate clouds both by supplying condensation centres (a role they play prominently for noctilucent clouds), and by its effect on stratospheric electric currents (which can affect clouds of the lower stratosphere and upper troposphere). But clouds are not the only possible amplification mechanism. Bromine, which is enriched in accreting dust, destroys ozone. Heterogeneous chemical reactions (those that take place simultaneously in two phases) already play an important role in the destruction of the ozone over Antarctica, and the dust provides a surface for such reactions. Ozone modulation can have important climate consequences, not only because of its direct greenhouse contribution, but because of the role it plays in determining the configuration of the upper troposphere. None of these mechanisms can yet be ruled out.

As we mentioned, an attractive aspect of the orbital inclination model is that it offers a solution to most of the problems listed in the last section. Based on this success, we devote Chapter 7 to a detailed study of this model. We happen to be particular fans of this theory, although that may be, in part, because the authors of the theory are the same as the authors of this book.

6.4.2 Statistical fluctuation model

Let us concentrate, for a moment, on just one of the problems affecting the insolation/eccentricity theories: the lack of a split 100 kyr peak. This problem is essentially the same as the Stage-11 problem, since it is the interference between the 95 kyr and 125 kyr frequencies which causes the cancellation at the age 420 kyr. Could this disagreement be a statistical fluke? That is, is it possible that variations in background levels managed to accidentally turn the split peak into a single peak?

This idea is not as farfetched as it might appear at first. Although eccentricity predicts a 95/125 doublet, the estimates of the frequency of the "100 kyr peak" have uncertainties that can be as large as 5 kyr. As we showed in the section on frequency determination (Section 3.14), in the presence of 10% noise, the frequency of the peak cannot be determined to better than a full-width of the peak. So perhaps the observed 100 kyr peak is really the expected 95 kyr peak of eccentricity. But why is the 125 kyr peak missing?

There are several possibilities. One is that there is a sharp resonance in the Earth's response that strongly enhances the 95, but not the 125. We will discuss this in the next section. Another possible solution to the absence of the split peak could be the following: most climate records contain significant background, at about the 10% level in power. Moreover, this background could be coherent around the world, that is, it could be global in character. Is it possible that this background accidentally cancels, say, the 125 kyr peak, while leaving the 95 kyr peak untouched (or maybe slightly amplified)? If this were so, then the 95 kyr peak alone could be what we are seeing.

To test this hypothesis, we simulated 100 spectra consisting of eccentricity plus red noise (generated by taking a random number every 10 kyr and linear interpolation). The RMS (root-mean-square deviation) of the background was set equal to one-third the RMS of the eccentricity signal, to simulate a signal to background ratio of roughly 9 : 1, similar to that seen in $\delta^{18}O$ records. In all of these records, the clear split in the peak was clearly visible. Only at higher background levels, with the RMS of the background approaching that of eccentricity, was the split nature of the peak often cancelled, and replaced by a single peak. We conclude that the absence of the split in the peak cannot be attributed to background cancellation when the background is at 10% or less. This is the case in many records, e.g. Site 806, Site 659, Site 664, Site 552 and the SPECMAP stack, all of which show a single narrow peak in the untuned or minimally-tuned records.

6.4.3 Resonant enhancement of the 95 kyr peak

The eccentricity has spectral power at periods 400, 125 and 95 kyr. Is it possible that the observed "100 kyr peak" is actually the 95, strongly enhanced by some resonant mechanism within the Earth's climate system? The resonance would have to have a high Q, so that the 125 kyr peak is not similarly amplified. The resonance might consist of a natural oscillation period for glacial formation and melting.

A theoretical difficulty with this approach is that resonance enhances a signal by taking energy from it and putting it into the natural oscillation over many cycles. That is, a resonance depends on the presence of internal memory that can store energy. It takes many cycles for the resonance to reach its full amplitude. It is conceivable that such a memory phenomena occurs in the glacial system, and that it is sufficiently narrow in response that it does not amplify the 125 kyr signal. This solution would also address the Stage-11 problem, since elimination of the response to the 125 kyr cycle would eliminate the cancellation that takes place at this stage.

Ignoring this conceptual difficulty ("if it happens, it must be possible"), we plot the 95 kyr component of eccentricity on top of the SPECMAP data in Fig. 6.15. The

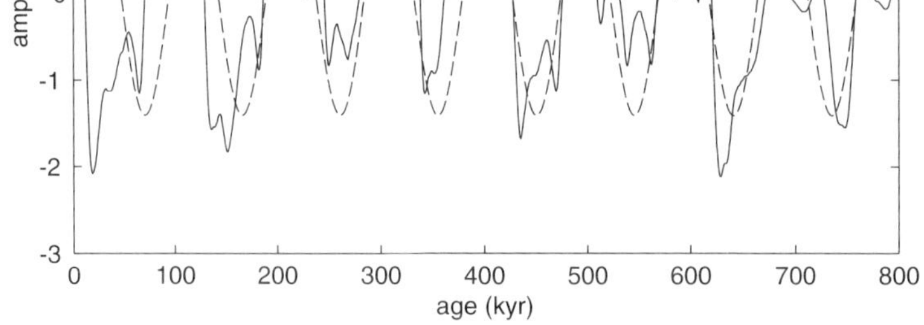

Fig. 6.15. 95 kyr component and SPECMAP.

agreement appears to be excellent, except for the last termination (near age = 12 kyr). Unlike the orbital inclination solution, there appears to be no phase lag between the signal and the ice volume response.

But this absence of lag is actually a serious argument *against* this model. As we discussed in Section 2.1.3, the phase lag between a driving force and the response, at resonance, is $\delta = \pi/2$ radians, i.e. a quarter of a cycle, or 95/4 = 24 kyr. The near zero phase lag between eccentricity and the $\delta^{18}O$ signal had always been considered a plus for the eccentricity model; the cycles appeared to agree. But that very agreement implies that resonance could not be taking place and, without resonance, the response should include both the 125 and the 95 kyr peaks.

6.4.4 Precession triggering

If the 100 kyr glacial cycles are measured from termination to termination, then they do not appear to be very regular at all. Maureen Raymo postulated that the termination requires two phenomena: a relatively long period of cold, so a large amount of ice ("100 kyr ice" in her terminology) can accumulate, and then the appearance of a sudden warming, associated with a peak in the precession parameter (Raymo, 1997). Based on this concept, she attempted to explain the exact timing of the terminations. An important prediction of this model is that the terminations were synchronous with the precession peaks.

The essence of the model can be seen in Fig. 6.16. The upper plot shows the $\delta^{18}O$ data from Site 849. The horizontal line passing through these data represents the "100 kyr ice" line. Only long periods of cold enable ice to accumulate to the point that the $\delta^{18}O$ data fall below this line. The lower curve is the calculated northern summer insolation. The sloping line passing through the insolation represents the gradual change in the average level of carbon dioxide in the atmosphere. As this line rises (towards the left), the level of insolation required to trigger a glacial termination is higher. The blackened areas represent periods when the insolation was sufficiently low (typically for two precession periods) that sufficient 100 kyr ice could accumulate. Without this accumulated ice, the phenomena of a sudden termination could not happen. The arrows at the top of the plot signify the terminations. Raymo argued that the bulk of the terminations were accounted for by this model.

This model was criticised by Karner and Muller (1999). They argued that the model was not as successful as claimed. At the bottom of the plot we have indicated with the letters "S" and "F" those locations when the model made a successful (S) or a failed (F) prediction. There are five successes and six failures. In this model there were approximately nine free parameters that had been adjusted. They include the location of the two reference lines (three parameters), and several choices of data sets, insolation curves and termination criteria. Given this many free parameters, it is not hard to create entirely nonphysical models that would show comparable success. The model has other severe problems. It predicts that the dominant signal in benthic $\delta^{18}O$ should be at 23 kyr, which is false. It assumes that carbon dioxide levels should be gradually changing, at variance with the observed levels in glacial ice; the straight line in Fig. 6.16 bears no resemblance to the actual measured atmospheric CO_2

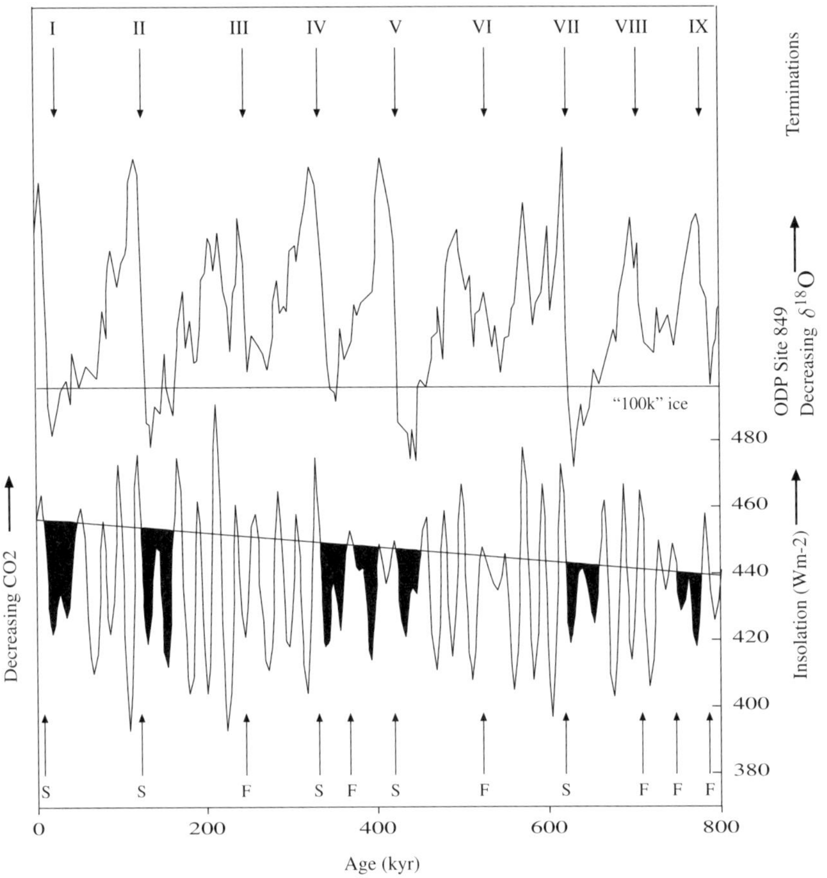

Fig. 6.16. Precession triggering model of Raymo.

plotted in Fig. 4.21. The fact that a termination always lies close to a precession peak is simply a consequence of the fact that precession peaks occur every 23 kyr, so there is always one within 12 kyr, and this is approximately the precision of the time scale.

The fundamental prediction of this model, that the terminations are coincident with peaks in precession, was contradicted by new measurements of the timing of the terminations. These measurements show that for Termination II, the melting of the ice (indicated by the rise in sea level) preceded the increase in precession by about 10 kyr. We discuss the data that led to this conclusion in the section on the causality problem, Section 8.3.

6.4.5 Multi-state climate model

A "Perspective" published in *Science* magazine (Raymo, 1998) discussed the 100 kyr glacial "puzzle", and drew attention to a model by Paillard (1998) that it said "elegantly avoids some of the shortfalls of earlier models". Paillard had published a simple multiple-state climate model based on calculated changes in solar insolation, and then stated that it could account for the timing of Pleistocene glaciation. In particular, he compared his calculations with measured $\delta^{18}O$ from a stack of sea-floor cores, and found good agreement. He concluded that this contradicts recent claims made by Muller and MacDonald (Muller and MacDonald, 1997a), that the classical Milankovitch theory is inadequate to account for the dominant 100 kyr cycle of glaciation.

We will not describe the model in detail, because it is more valuable as an example of the need to pay close attention to the data and, in particular, its time scale. In his paper, Paillard did not mention the fact that the time scale for the data had previously been modified to bring the data into agreement with a theoretical insolation target. This critical fact was also not mentioned in the *Science* Perspective. The $\delta^{18}O$ data consisted of the low-latitude sea-floor stack discussed in Section 5.6.2. The time scale had been constructed to make the data match the nonlinear ice model (Section 6.2.1). If Paillard's model didn't match insolation, it only would have indicated that a poor mathematical job had been done in the time-scale tuning. To use agreement between the new model and previously tuned data as confirmation of the multi-state climate model amounts to circular reasoning.

Even ignoring the circularity of the model, a spectral analysis of the Paillard model shows that it does not successfully account for the $\delta^{18}O$ data. In Fig. 6.17 we show the results of a spectral analysis. Plot A shows the $\delta^{18}O$ spectrum from Site 659 (Sarnthein and Tiedemann, 1989; Tiedemann et al., 1994) for the time span 0 to 600 kyr, using a time scale calculated by assuming a constant sedimentation rate. The presence of a single narrow line in the spectrum validates the assumption of constant sedimentation (Muller and MacDonald, 1997a). The spectrum shows the single narrow peak that is typical of $\delta^{18}O$ spectra. In Fig. 6.17B we show the spectrum of the Paillard insolation model estimate of ice volume. It is not a good match to the spectrum of the data in A. For contrast, we show in C the expected spectrum from the orbital inclination model.

Paillard concluded his paper with the statement that, "contrary to recent claims" (with reference the orbital inclination theory), his model "clearly demonstrates that the geological record can easily be explained in the framework of the classical astronomical theory", i.e. insolation. Comparison of spectra in Fig. 6.17 shows that this is not the case, and that the orbital inclination theory is a better fit when unbiased data are used for the comparison.

6.4.6 Derivative of eccentricity

Rial (1995) suggested that the driving force for the 100 kyr cycle is not eccentricity, but the derivative of eccentricity. In other words, it is not the envelope of precession that drives climate, but the *rate of change* of that envelope. This is a very intriguing idea that bears close scrutiny. Whether or not it proves to be right, the concept is one that might find a place in the ultimate climate model.

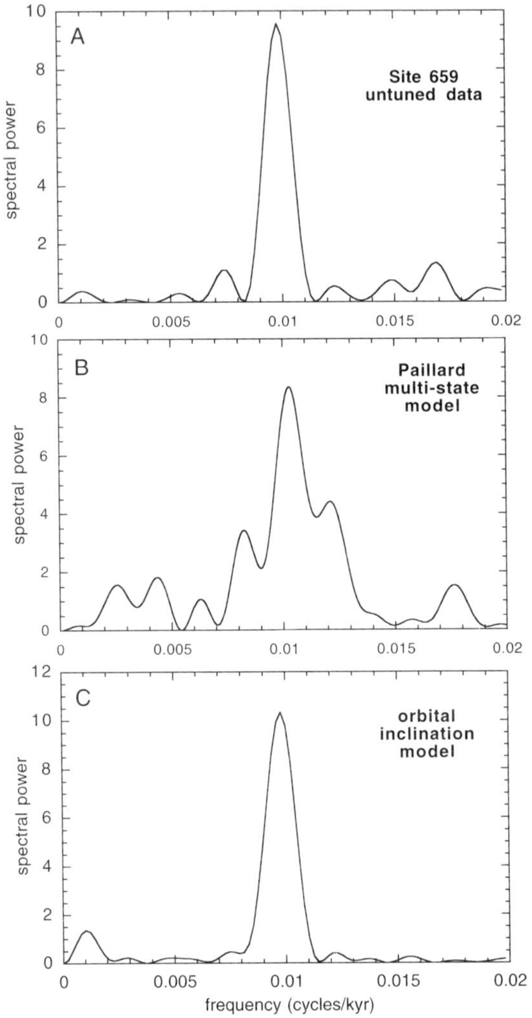

Fig. 6.17. Spectra of Paillard model. (A) $\delta^{18}O$ ice volume proxy for Site 659, (B) ice volume according to the insolation-driven multiple-state climate model of Paillard, and (C) linear response to orbital inclination, calculated with respect to the invariable plane.

Taking the derivative of a function actually has a very simple effect on the spectrum of the function: it multiplies it by frequency squared, as we showed in Section 3.13.2. The spectrum of the derivative of eccentricity for the last 600 kyr is plotted using a linear and then a log scale, in Fig. 6.18.

This model has two important strengths. The first is that it provides excellent suppression for the low frequency 400 kyr signal. This is a result of the fact that the eccentricity spectrum is multiplied by f^2, strongly suppressing the low frequency region. Another plus for the theory is that it involves no adjustable parameters, other than the usual two that provide scale and phase lag.

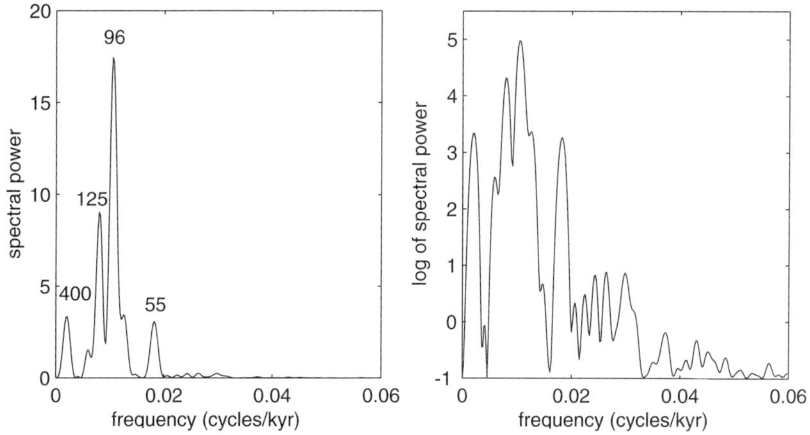

Fig. 6.18. Spectrum of the derivative of eccentricity.

The model does not have a ready physical explanation. Remember that insolation does not contain an eccentricity term. Eccentricity is the envelope of precession, and that term is present in insolation, but a spectral analysis of insolation shows no spectral power near period 100 kyr. You might think this problem could be solved by taking the derivative, not of eccentricity, but of something that contains eccentricity, such as the nonlinear ice model. But this approach does not work, because the frequency term (the factor of f^2) greatly enhances the 23 kyr peak which already was too strong in the nonlinear ice model. The spectrum of the derivative of the nonlinear ice model is shown in Fig. 6.19. Unfortunately, this plot looks very much like the spectrum for insolation. We have undone the advantage of the nonlinear ice model,

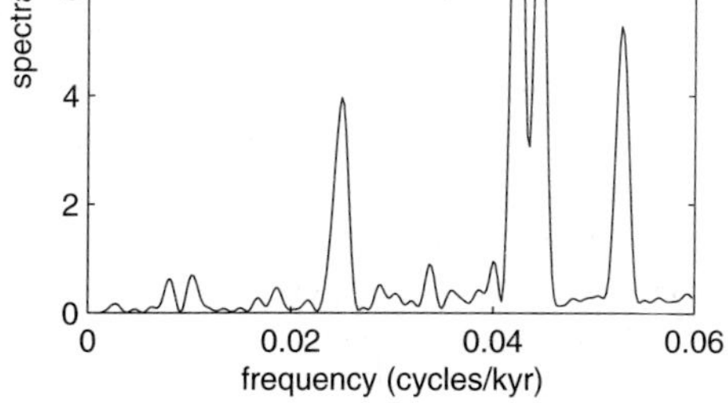

Fig. 6.19. Spectrum of the derivative of the nonlinear ice model.

which was to enhance the 100 kyr region with respect to the 23 kyr region. We can go further by introducing additional parameters, but we lose the simplicity of the model, and that was what made it so potentially attractive.

6.4.7 Frequency modulation

The narrowness of the 100 kyr peak was addressed directly by Rial in an article titled "Pacemaking the Ice Ages by Frequency Modulation of Earth's Orbital Eccentricity" (Rial, 1999). He proposed that the peak, despite its narrow width, could still be attributed to eccentricity if the nonlinear linking mechanism involved frequency modulation. He presents a mathematical model which he states does this. He concludes that orbital inclination is therefore "unnecessary". This conclusion has been referred to by subsequent papers, e.g. Clark et al. (1999a).

Mathematically, frequency modulation means that the frequency varies with time, i.e. $f = f(t)$. Frequency modulation is equivalent to phase modulation:

$$\begin{aligned} \sin(2\pi f(t)t) &= \sin(2\pi(f_0 + \delta f(t)t)) \\ &\equiv \sin(2\pi f_0 t + \phi(t)) \end{aligned}$$

where we define $\phi(t) = \delta f(t)t$ as the phase modulation function. This means that frequency modulation can do for a theoretical model what tuning does for a time scale: if you use enough parameters, you can make any phrases match. Thus, in principle, if treated as a mathematical exercise, frequency modulation of *any* model should be able to match *any* data. For example, in Section 5.6.1 we tuned a sine wave to match eccentricity. The inverse transformation turns eccentricity into a sine wave. To do that took one free parameter every 50 kyr. It would be impressive if such theoretical tuning could be done with few adjustable parameters. Rial's mathematical model appears to do this. It consists of the following equation (equation 3, from his paper):

$$\begin{aligned} \Phi_{\mathrm{FM}}(t) = {} & a\sin[2\pi t/95 + \varepsilon\sin(2\pi t/413) + \varepsilon'\sin(2\pi t/826)] \\ & + b\sin[2\pi t/100 + \varepsilon\sin(2\pi t/413)] \\ & + c\sin[2\pi t/125 + \varepsilon\sin(2\pi t/413)] \end{aligned}$$

The adjustable parameters in this equation are a, b, c, ε, ε'. The periods that appear in the above equation are 95, 100, 125, 413 and 826. Three of these are the periods found in eccentricity (95, 125 and 413). Rial justifies the 826 as a *subharmonic* of 413. He gives no explanation for the inclusion of the period 100. Rial states that "despite the simplicity" of this equation, its spectrum "reproduces every spectral peak in the power spectra of the tuned and untuned records ..."

But Rial has not showed that orbital inclination was unnecessary, since he included it in his model, implicitly, unnamed, by using a sine wave with period 100 kyr. To include this period is virtually equivalent to including orbital inclination.

6.4.8 Nonlinear memory model of Berger

W. Berger invented a nonlinear model insolation-based that gives a remarkably good fit to the data (Berger, 1999); additional analysis was done by Paul and Berger (1999). Berger begins the analysis by examining some of the issues that had been raised by Muller and MacDonald (1977), and concludes that the problems with the standard insolation-based models are substantial. In particular, he confirms that the observed $\delta^{18}O$ spectrum for the 100 kyr cycle contains only a single narrow peak, and not the triplet that is expected from simple nonlinear models. But he does not consider the orbital inclination model an adequate solution to the problem. He is bothered, in particular, by the 33 kyr delay between the forcing function (inclination) and the responding $\delta^{18}O$. He argues that this delay is unreasonably long, and that such a delay requires that the climate system have some mechanism for remembering the past. (We disagree with this assessment. As we argue in Section 4.4, no "memory" other than the presence of accumulated ice is required to provide a 25 kyr or more delay.)

Yet Berger does not want to rule out the theory based on a perceived absence of a good physical mechanism for such memory. He cites "Kelvin's fallacy"—rejecting an observed phenomena because of the lack of a known mechanism. It is a law of science that if something happens, it must be possible. So instead, he explores the consequences of the existence of such memory. Using physical reasoning, he invents a nonlinear model containing such memory, based on 65N summer insolation, that does not require any input from extraterrestrial dust.

The model has the following equation describing the rate of ice formation and melting:

$$\frac{d(ice)}{dt} = const - ins(t)^A \times ice(t) \times mem(t)^C$$

The constant *const* represents the natural tendency of glaciers to grow. Berger found a best value for this parameter of 0.03. The function *ins* is the July insolation at 65N, transformed such that its maximum value is 1 and its minimum value is zero. The function *ice* is the integral of $d(ice)/dt$. If we take $dt = 1$ kyr, then $ice(t)$ is just the sum of the previous values of $d(ice)/dt$. Berger does the sum using a simple spreadsheet. The *mem* function is the average value of $ice(t)$ for the previous 57 kyr. Berger finds that the best values for the constants $A = C = 4$. By raising *mem* to such a high power, he makes it very difficult to sustain an ice age for much longer than 57 kyr. By raising *ins* to a high power, he makes a high level of insolation into a triggering event that terminates an ice age. Once the ice drops, then the factor $ice(t)$ prevents the level from going negative. It holds it near zero, while the constant term begins the slow rebuilding.

This equation can be integrated to yield $ice(t)$. After 100 kyr, the result is insensitive to the starting value of *ice*. In Fig. 6.20, we show the resulting function $ice(t)$, plotted with the SPECMAP $\delta^{18}O$ data. On the right-hand side is the spectrum. The model has succeeded in suppressing the 400 kyr cycle, and it has suppressed the 125 kyr cycle.

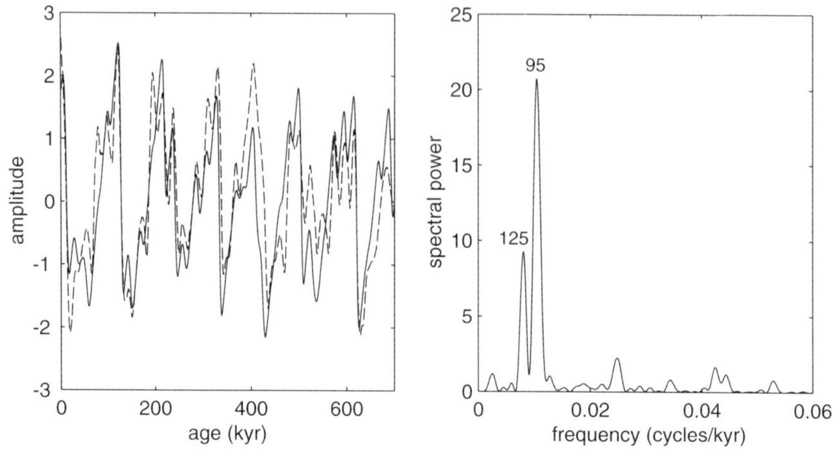

Fig. 6.20. Berger's memory model (solid line) compared with SPECMAP (dashed line). The spectrum of the memory model is shown on the right.

The fit between the Berger model and the SPECMAP data is excellent. The sudden terminations are reproduced, and the Stage-11 problem has gone away. Of course, there were a several adjustable parameters, including the constants *const*, A, and B, the 57-year averaging time, the linear transformation that turned insolation into $ins(t)$ (involving two parameters). The fact that the small oscillations are well matched is primarily a consequence of the fact that the SPECMAP data were tuned to precession, so the precession built into the Berger model reproduces them. Likewise, the fact that the model reproduces the sudden terminations is a consequence of the fact that it was designed to create terminations when insolation is high—and SPECMAP was tuned to the same criterion. Thus, by design, the sudden terminations were forced to appear at the times of high precession in both the Berger model and in the SPECMAP data set. Nevertheless, the match between theory and data is impressive. The spectrum shows that although the 125 kyr peak is suppressed, it is still present.

To try to reduce the 125 kyr peak, we reran the Berger model using different parameters. The results, for $constant = 0.02$ and $C = 8$, are shown in Fig. 6.21. With this high coefficient for the memory factor, the 125 kyr line is almost completely suppressed.

The fact that the termination is driven by insolation to the 8th power may appear unphysical, but this mathematical result is largely reflecting the need to have a very strong nonlinear response to insolation. Recall that the insolation was transformed so that it fell between zero and 1. Unless the value of *insol* is close to 1, the factor of $insol^8 \approx 0$. Thus the negative term in Berger's equation contributes strongly only when the insolation is near its peak. In Raymo's precession triggering model (Section 6.4.4), a similar effect was obtained by setting a threshold above which the terminations were triggered. Her model also had a "memory effect" in that she required there to be at least two precession cycles of "100 kyr ice" before any termination would take place. This plays a similar role to the Berger condition that the negative

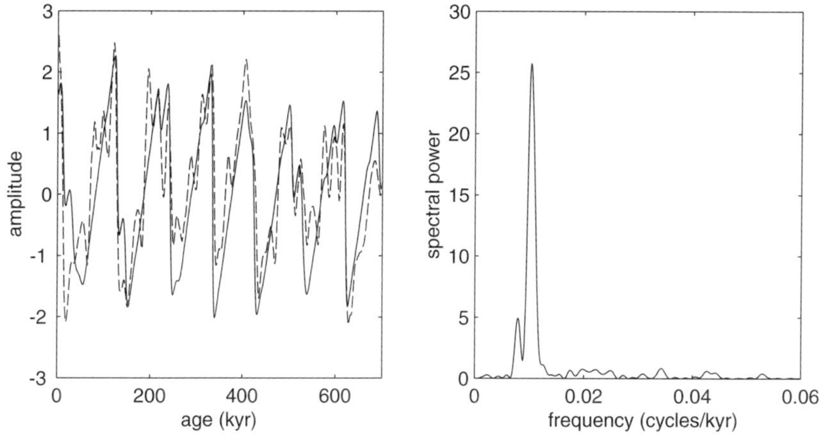

Fig. 6.21. Berger model with $C = 8$. The left plot shows the model (solid line) plotted on top of the SPECMAP $\delta^{18}O$ data (dashed line), and the right plot is the spectral power.

term in the $d(ice)/dt$ equation was multiplied by the 4th (or 8th) power of the average ice over 57 kyr. Unless ice had been around for a long time, there is no strong negative term, and so there is no termination.

Thus we can see that the effect of the mathematics in the Berger model is to create slow ice growth (through the *const* term). Terminations (sudden decreases in ice) occur when two conditions are satisfied: (a) there has been substantial ice for at least 57 kyr, (b) precession is near its peak value. Thus Berger's model creates a sawtooth shape. The fact that precession is a multiplying factor means that the model will also reproduce the 23 kyr oscillations. By design, the sudden terminations are forced to be synchronous with the precession cycles.

Could the Berger model be correct? It suffers from the same weakness as does the Raymo precession model: it ties the terminations to precession cycles, and this creates a causality problem. We will discuss this in more detail in Section 8.3. Briefly, the problem is as follows: the most accurately determined age for termination II is 135 ± 2 kyr, and this occurs well before the insolation increase. Although consequences can follow the causes, they may not precede them. Unless the radioisotopic ages measured in the corals turn out to be incorrect (and there are several independent measurements, and these agree with the Devils Hole data—see Section 8.3), the Berger model fails the causality test.

7

Accretion climate models

"Science is always wrong. It never solves a problem without creating ten more."

George Bernard Shaw

"... and to dust shalt thou return."

Genesis, 3:19

In this chapter we begin with a discussion of the nature of interplanetary dust, how it reaches the Earth, and what form it takes when it enters the atmosphere. Then we describe the effects it might exert on climate, many of which were briefly mentioned in Chapter 6. A good general reference to the science of interplanetary dust is *Physics, Chemistry, and Dynamics of Interplanetary Dust* (Gustafson and Hanner, 1996).

7.1 ORIGIN AND DISTRIBUTION OF INTERPLANETARY DUST

Extraterrestrial dust originates in gradual evaporation and disintegration of comets, and in the collisions of comets and asteroids. Much of this dust is spread out in a broad region along the ecliptic. At a dark location it is visible at night as the "zodiacal light". Under ideal conditions, the zodiacal light can be as bright as the Milky Way. It is about 10 degrees wide—but that is the full-width at half-maximum; the band is visible over a width of about 30 degrees. It is also known as "the false dawn", since it occurs above the horizon at the same location where the sun will rise, which leads some people to guess, incorrectly, that sunrise is imminent.

It was shown by the Infrared Astronomical Satellite "IRAS" that this dust has a very uneven distribution (Low et al., 1984; Neugebauer et al., 1984). In particular, some of it is concentrated into very narrow bands, as shown in Fig. 7.1.

These narrow dust bands were a surprise, although they were quickly explained by Dermott et al. (1984). They reflect the uneven distribution of asteroids. Hirayama (1918) had discovered that asteroids tend to cluster into groups he called "families"

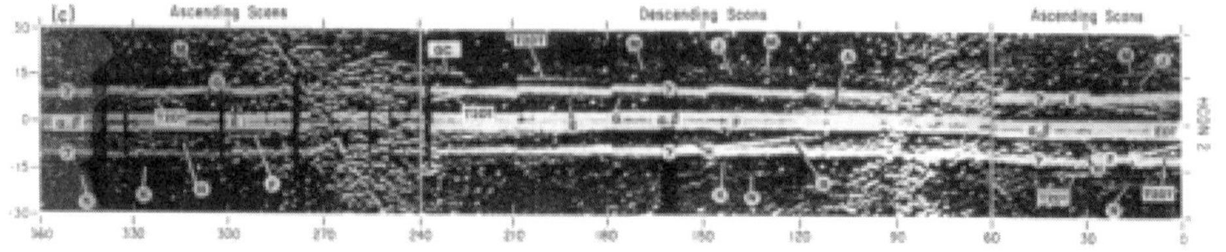

Fig. 7.1. IRAS dust bands.

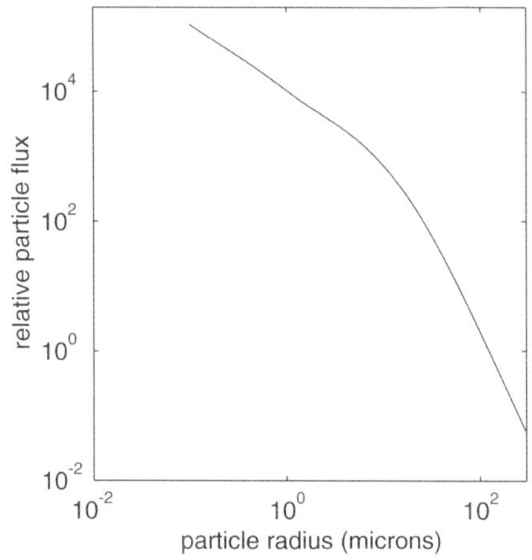

Fig. 7.2. Interplanetary dust particle size distribution.

that have similar orbits. The reason for this clustering is unknown, although it is plausible that each family was once a larger asteroid that was broken into many pieces by a collision. Since the asteroids within a family are in similar orbits, they tend to collide with each other at a greater rate than unassociated asteroids. These collisions give rise to the bands of dust seen in the IRAS image. Some of the dust is probably created in the collision, but much of it may simply be dust accumulating on the surface from micro-impacts, which is then set free of the asteroid's gravity by the energy of the collision.

Most of the collisions between family members are relatively gentle, since they proceed around the sun at nearly identical speeds (they are in similar orbits), so the initial orbits of the dust particles will be similar to those of the asteroids themselves. But the dust particles are small. Their size distribution has been estimated by Grün et al. (1985) based on a combination of measurement and dust collision theory. The distribution of particle size is shown in Fig. 7.2.

Based on this distribution, we can estimate the mass distribution. If we let the distribution in Fig. 7.2 be called $N(r)$, then the mass distribution is proportional to $r^3N(r)$. This mass distribution is plotted in Fig. 7.3. Note that although the peak in the number of particles occurs at small radii (Fig. 7.2), the bulk of the mass is carried by much larger particles (Fig. 7.3), between 30 and 200 microns in size.

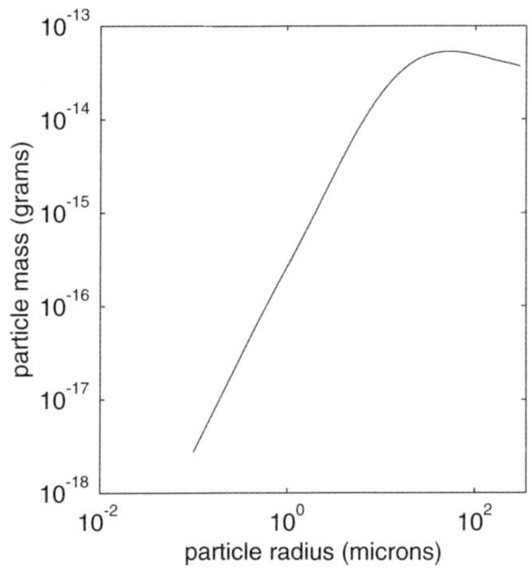

Fig. 7.3. IDP particle mass distribution.

7.2 TRANSPORT TO THE EARTH

The dust particles are sufficiently small that their orbits depend not only on gravity, but also on the effects of pressure from solar radiation and from solar wind. (These effects are negligible for more massive objects such as the asteroids.) The pressure from sunlight dominates. It has two effects on the particles. The first is that it exerts a force directed away from the Sun. Since the intensity of the sunlight decreases as $1/r^2$, as does that of gravity, this has the same consequences that you would get if the gravity of the Sun were reduced slightly. So the dust particle orbits are still elliptical, but for the same size orbit, their periods are slightly longer. The particles do not move around the Sun quite as quickly as their parent asteroids.

The second effect is more interesting. The collisions of the particles with photons from the Sun also cause them to lose energy, and transverse momentum. This is called the "Poynting–Robertson effect" (Poynting, 1904; Robertson, 1937). Sunlight acts as friction; it makes the space of the solar system viscous. This viscosity tends to circularise the orbit (since it has its greatest effect when the particle is moving fastest). As the particle loses energy (and transverse momentum) the orbit decays, i.e. the particle "falls" into the Sun with a spiral orbit. The effect is analogous to what happens to an artificial satellite orbiting the Earth when it begins to feel friction from the upper reaches of the Earth's atmosphere, and quickly spirals into the Earth.

The rate of decay of the orbit depends on the ratio of frictional force to the momentum of the particle. Momentum is proportional to mass (i.e. to ρr^3), whereas the force of sunlight is proportional to the area (i.e. to r^2); the net result is that smaller particles spiral into the Sun more rapidly than larger particles. For a good review of the Poynting–Robertson effect, see Kortenkamp and Dermott (1998). The time that it takes a particle of radius r (cm) and density ρ (gm/cm^3) to move into the sun from a distance R in AU (1 astronomical unit = 1 AU = average Earth–Sun distance = 150 million km) is given by

$$t(\text{years}) = (7.0 \times 10^6) r \rho R^2$$

So, for example, a particle in the asteroid belt at $R = 2.7$ AU, with radius $r = 1$ micron (10^{-4} cm) and density $\rho = 3$, will take 15,300 years to reach the Sun. A particle with radius 100 microns will take 1.5 million years for the same journey. This results in an interesting filtering of the particles. If there is a particularly catastrophic collision (and we will discuss later evidence that such a collision took place about a million years ago) then the smaller particles pass by the Earth first, followed by the larger ones.

The Poynting–Robertson effect means that the dust, originally created in a ring (at the orbits of the asteroids), evolves into a disk around the Sun. The bands of dust seen in the IRAS image (Fig. 7.1), are actually bands that extend all the way past the Earth and to the Sun. Thus, just as Saturn has rings, so does the Sun.

The dust is perturbed by the gravity of Mars and the Earth when it passes close to these planets. Some very interesting effects result. The presence of "resonant dust bands" was predicted by Thomas Gold (Gold, 1975). Jackson and Zook (1989) identified evidence for these resonances in the IRAS dust bands; additional analysis was carried out by Dermott et al. (1994). Resonant capture refers to the fact that some of the dust is captured into an orbit just outside that of the Earth. The process by which this happens is that the dust, as it approaches the Earth, is given a slight kick by the faster Earth; this gives it back some of the energy it lost from Poynting–Robertson drag, so the particle stays in orbit for a slightly longer time. If the orbital speed is just right, when the Earth catches up to it a second time (the Earth is closer to the Sun, and therefore moving faster) it will receive a second kick. The dust can stay in the same orbit for many cycles. The particles that have the correct orbit are said to be in "resonant lock" with the Earth. The resonant bands can be seen in IRAS images (Neugebauer et al., 1984; Sykes et al., 1989), and were later confirmed in great detail with the DIRBE satellite (Kelsall et al., 1998).

When the dust comes close, the gravitational attraction of the Earth becomes important. Particles that would miss the Earth by a distance b with velocity v are sucked in towards the Earth. They will hit the Earth if their distance of closest approach is less than or equal to the radius of the Earth R. From angular momentum conservation (valid in the approximation that we can ignore the acceleration due to the Sun), we conclude that $vb = v_e R$, where v_e is the escape velocity of the Earth (and is roughly equal to the velocity of the particle when the particle hits the surface). Thus, the Earth pulls in all particles which would come within a distance $b = R(v_e/v)$. This means that the effective cross-sectional area of the Earth for particle capture is enhanced by the factor $(b/R)^2 = (v_e/v)^2$. Since the velocity of the dust relative to the Earth can be small, e.g. 1 km/sec, and the escape velocity of the Earth is large, 11 km/sec, this factor can easily be greater than 100.

As the above example shows, the accretion is extremely sensitive to the relative velocity of the particle and the Earth. Since the particle is moving in a circular orbit (circularised by Poynting–Robertson drag), the relative velocity depends on both the eccentricity of the Earth's orbit, and on the location of that part of the Earth's orbit where the encounter takes place. Thus we expect accretion to depend on the eccentricity of the Earth's orbit. A second potential relationship between eccentricity comes from the fact that the location of the Earth's orbit (the regions that the Earth passes through) also depends on eccentricity. This could also lead to an eccentricity dependence. Kortenkamp and Dermott (1998) studied these accretion effects in detail. Unfortunately, the calculations are greatly complicated by the presence of the resonant bands, so simple ideas must yield to computer simulations. Early calculations indicated that the "bulk" of the accretion would come from the dust bands that came from the Themis and Koronis asteroid families (Dermott et al., 1996), although later calculations suggested that the Eos family and others might be more important (Kortenkamp and Dermott, 1998).

In 1993, when we first hypothesised that the accretion of interplanetary dust could cause a 100 kyr cycle in the ice ages (Muller, 1994), we assumed that the cyclicity would arise from the variations in the eccentricity of the Earth's orbit. However, we recognised that the orbital inclination could also lead to a modulation. At that time, we did not know what the frequency of this modulation would be. Berger had published a perturbation theory calculation of the Earth's orbital elements, and his work showed that the orbital inclination varied with a 70 kyr frequency (Berger, 1978; Berger and Loutre, 1992). These calculations had been done using the *present* orbit of the Earth as a reference, not the frame that is symmetric with respect to the dust. Nevertheless, the presence of the 70 kyr cycle was encouraging, since a cycle with this period had been identified (MacDonald, 1990) in the $\delta^{18}O$ record as the most significant cycle in the data that had not been previously recognised (see Section 8.9).

To make a better estimate of the expected accretion from orbital inclination, it was necessary to transform the calculations to the symmetry plane of the Earth. The procedure for doing this is outlined in Section 2.2.6. Once the transformation was made, it was clear that the dominant cycle of oscillation was 100 kyr, not 70. The 70 kyr cycle was still present, but it represented the precession of the inclination axis, not the variation in inclination itself. When measured in a tilted reference frame (i.e. the present ecliptic of the Earth) this precession appeared as large changes in inclination. But these changes were not changes with respect to the dust bands.

The fact that the orbital inclination was dominated by a 100 kyr oscillation came as a surprise. It meant that both dust and insolation could be contributing to the 100 kyr spectral power. It was to estimate the relative contributions of these two effects that we first applied spectral analysis to the problem (Muller and MacDonald, 1994b).

7.3 HITTING THE EARTH'S ATMOSPHERE

The interplanetary dust particles are accelerated by the Earth's gravitational field, and finally hit the atmosphere with a velocity equal to or greater than the escape velocity of 11 km/sec. What happens next is largely unknown. Large, apparently

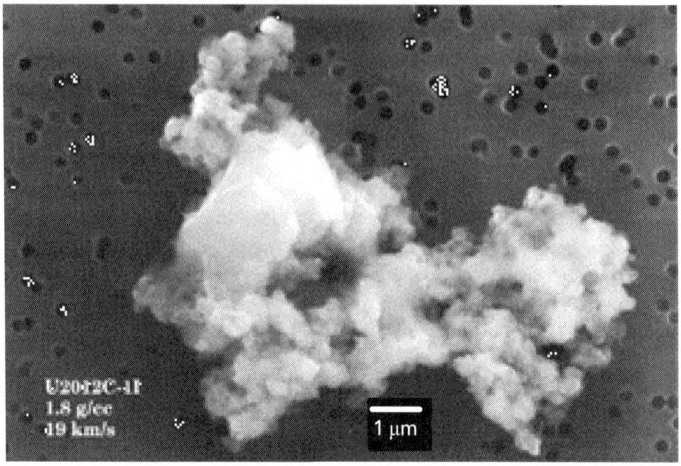

Fig. 7.4. Interplanetry dust particle ("IDP").

unmelted particles have been collected in the Earth's atmosphere by Brownlee (Brownlee, 1985). A photo of one such particle is shown in Fig. 7.4. The surface is complex and the average density is low.

Most particles hit the Earth's atmosphere at a velocity near 14 km/sec, and about two-thirds of their mass is vapourised. This occurs at altitude of about 90 km. The vapour is believed to recondense into "smoke" particles with an average radius calculated to be 0.01 microns (Hunten et al., 1980). The lifetime of these small particles in the stratosphere is estimated to be about 15 years, although this value is not well established (Toon and Farlow, 1981). They are expected to coagulate in the stratosphere, and this reduces their concentration from as large as $10^5/cm^3$ in the upper stratosphere, to 10^2–10^3 at 40 km. These numbers are all highly uncertain, since the particles are too small to be easily captured and measured. They do not represent the bulk of the particulate mass in the stratosphere; that comes from aerosols and terrestrial particles. However, they could easily represent the largest contribution to particle number in the stratosphere and, as such, could play an important role in such stratospheric parameters as electrical conductivity. Once particles fall to an altitude of 20 km, near the tropopause, they are flushed out by rain; their remaining lifetime is estimated to be just one to two years.

7.4 MEASUREMENTS OF ACCRETION RATES

Measurements of the accretion rate of dust on the Earth have been made from micrometeorite measurements, from radar, and by using the elements iridium and osmium as proxies. These elements are relatively rare in the crust of the Earth, because they were drawn into the molten core when the Earth was still liquid. Accretion measurements have been reviewed by Rocchia et al. (1990) and by

Table 7.1. Estimates of extraterrestrial accretion.

Measurements	Extraterrestrial flux (kiloton/yr)	References
Antarctic & Greenland snow	10 to 20	Takahashi et al. (1979)
Atmospheric particulates	6 to 11	Tuncel and Zoller (1987)
Greenland cosmic spherules	4	Maurette et al. (1987)
Antarctic snow	100	Ganapathy (1988)
	10	Rocchia et al. (1990)
Satellites & radar	16	Hughes (1978)
	15	Grün et al. (1985)
	40 ± 20	Love and Brownlee (1993)
Iridium in oceanic clay	72 ± 36	Barker and Anders (1968)
	60 to 120	Kyte and Wasson (1986)
Osmium in oceanic clay	49 to 56	Esser and Turekian (1988)

Peucker-Ehrenbrink (1996). A sample of these measurements is shown in Table 7.1, adapted from Rocchia et al. (1990). In each case, the extraterrestrial flux is calculated by assuming uniform dispersion over the surface of the Earth. For the iridium and osmium proxy estimates, it was assumed that these elements were present with the abundance typical for chondritic meteorites.

The disagreement in the estimates could reflect changes in the flux with time, uneven distribution on the Earth, and systematic errors. The measurements in snow and ice have time resolutions, for the recent past, of better than a year. The measurements of Kyte and Wasson (Kyte and Wasson, 1986) represent the time interval of 33–67 million years ago. Thus, although they are the most extensive in period covered, they do not overlap with the period for which we have the best climate records.

The older rates, measured in oceanic clay, are not particularly sensitive to rapid variations (100 kyr time scale). Kyte and Wasson (1986) made their measurements in a location with low sedimentation rate; this gave a relatively high level of measurable iridium, but it meant that they could not see pulses or variations on short time scales. The measurements by Rocchia et al. (1990) have the opposite problem: their time resolution is excellent, but the range that they cover is only a few years. They were searching for a pulse from the Tunguska impact of 1908, and they saw no such pulse.

Farley et al. have used ^3He to track variations of accretion during the past 70 million years (Farley, 1995), and for high time resolution measurements during the past one million years (Farley and Patterson, 1995; Farley et al., 1996). ^{3}Helium originates in the solar wind, and is absorbed on interplanetary dust particles before they hit the Earth. The ^3He level is not directly proportional to the total accretion flux, since it is carried primarily by particles in the 7–20 micron size range, whereas the bulk of the accreted mass comes in particles 50–200 microns in size (Farley et al., 1997). Thus although ^3He measures accretion, it does not necessarily track the bulk of the accreted mass.

We summarise these measurements by emphasising not their disagreement, but their agreement. We can say that the accretion rate is known to be in the range between a few kilotons and 120 kilotons per year. The uncertainty in the level may reflect actual variations, but all the numbers represent relatively low yearly amounts. How can accretion in this range affect climate?

7.5 SCATTERING OF SUNLIGHT MODEL

The most obvious way that dust could affect climate is by scattering sunlight. Light that does not reach the Earth cannot warm it. In particular, scattering in the stratosphere and above interferes with the greenhouse amplification process. This process is seen in the climate effects of the great volcanic eruptions, which tend to cause a cooling on the Earth. However, in this section we will show that direct scattering of sunlight is *not* a likely mechanism to link interplanetary dust to climate, since the accretion levels are so low.

To estimate the climate effects of accreted dust, assume that the accretion is in the middle of the range of measurements described in Table 7.1, i.e. about 60 kilotons per year, over the entire Earth. This is equivalent to an average flux of 4×10^{-9} gm/cm^2/yr. For $\rho = 3\,\text{g/cm}^3$, in one year this is equivalent to a layer only $1.3 \times 10^{-9}\,\text{cm} = 0.13$ Angstroms thick. We do not know how long the dust stays in the atmosphere, but the best estimates indicate about 15 years (Toon and Farlow, 1981). This implies that the total equivalent thickness of dust in the atmosphere at any one time is $15\,\text{yr} \times 0.13$ Angstroms/yr $= 2$ Angstroms, or about 1/2500 of the wavelength of light. Such a small amount of material is not an effective scattering source for visible light. We conclude that direct scattering of sunlight in the atmosphere is probably negligible.

Next we will consider the dust that lies between us and the Sun. Such dust could, in principle, reduce the sunlight hitting the Earth. Most of the mass of this dust is in particles with a radius of about 100 microns. From the Poynting–Robertson effect (Section 7.2), and assuming a typical density of 1.3, we can use the Poynting–Robertson equation to estimate the time it takes the particles to reach the Sun: $t = (7 \times 10^6) \times 0.01 \times 1.3 = 90\,\text{kyr}$. If their flux is similar to that hitting the Earth, the column density is 0.13 Angstroms/yr $\times$ 90,000 yr $= 1$ micron. This is a high enough column density that it should cause considerable scattering.

However, in doing the calculation we overestimated the flux. The flux on the Earth is increased by gravitational focusing, i.e. particles that would not otherwise hit the geometrical cross section of the Earth are drawn to it by gravity. We estimated that this could have enhanced the flux at the Earth by a factor of 100. Thus, the column density of particles between us and the Sun may have been overestimated by this same factor, and may be only 0.01 microns.

Our gravitational focusing calculation was quite crude, and could easily be wrong by a large factor. Fortunately, we are rescued by experimental results. The solar irradiance has been measured with great precision by the Earth Radiation Budget

Satellite. The Earth moves up and down through the Sun's dust cloud during the year, and yet the satellite measurements show that the irradiance varies by less than two parts per thousand over the year (Lee et al., 1995). Thus the variability, as a function of height above the plane of symmetry of the solar system is negligible. This agrees with the conclusion we draw from the crude gravitational focusing calculation.

7.6 CLOUD COVER MODEL

In this model, we assume that extraterrestrial dust modulates cloud cover. We showed in Table 6.2 that cloud cover exerts a very strong driving force on the climate. In Section 6.1.2, we discussed mechanisms that could modulate cloud cover. Hunten et al. (1980) argued that smoke from extraterrestrial accretion is a significant source of solid surface in the middle and upper stratosphere. To the extent that extraterrestrial smoke absorbs ions and reduces their mobility, the return path of the electric currents is changed. Tinsley (Tinsley, 1996) has argued that the conductivity of the stratosphere is primary in cloud formation. If the results of Svensmark and Friis-Christensen (Svensmark, 1998; Svensmark and Friis-Christensen, 1997) are correct, cloud cover is strongly modulated by cosmic rays, presumably through their effect on stratospheric conductivity. As a major absorber of free charges, it is plausible that accreted dust, through its modulation of stratospheric electric currents, can also affect cloud cover, although it is not clear whether additional dust would cause an increase or a decrease. This model awaits confirmation that the cosmic rays and cloud cover are truly linked, and a better understanding of the role of electric currents in cloud formation.

Hunten et al. estimate that the residence time in the stratosphere for these very small particles is about 15 years. If they coagulate into more massive particles, then they will fall out with the lifetime closer to one year. The material is eventually washed out when it reaches the tropopause. If we take the average incoming mass per year to be 60 kton per year, then the stratosphere holds between 60 and 900 kton at any one time. The mass of the stratosphere is about 5×10^{20} gm $= 5 \times 10^{11}$ kton. Taking the smoke to be 500 kton, the accreted matter, at any one time, is about one part in 10^9, by mass. If the accreting material is similar to carbonaceous chondrites (and almost everybody assumes it is) then it should be 18% iron, 10% silicon, 10% magnesium and 6% sulfur. Most of the rest is oxygen.

This material is likely to be the primary source of heavy atoms in the upper and middle stratosphere. Note that heavy material coming from the surface of the Earth would probably have a much shorter lifetime in the stratosphere, since such particles were not recondensed smoke from vapour, and would presumably be larger in size.

What effect does the extraterrestrial smoke have on the atmospheric electrical properties? The number density of air molecules at 15 km altitude is approximately 2×10^{18} per cm^3; for these heavy molecules, the number density is 2×10^{11} per cm^3. This density is substantially larger than the ion density and, therefore, the attachment to heavy molecules could be the most important parameter in the determination of the lifetime of free ions. The conductivity of the stratosphere is proportional to this lifetime. Therefore, in this model, the conductivity of the stratosphere is inversely proportional to the accretion of extraterrestrial dust. From Tinsley's association of conductivity with cloud formation, we consider it plausible that extraterrestrial dust plays a major role in this process.

Our cloud cover model then consists of the following mechanism: modulation of dust linearly affects currents in the stratosphere, and these linearly affect the cloud formation. This model leads to the following equations:

$$I(t) = \alpha A(i(t)) + \beta$$

$$\delta^{18}\mathrm{O}(t) = \gamma \int_{t-T}^{t} I(t')dt'$$

Here $A(i)$ is the accretion rate as a function of orbital inclination i. The constants α, β, γ are adjustable, but they can be eliminated from the fit by normalising both the model and the data to zero mean and unit variance. The function A is presently unknown; estimates have been made by Kortenkamp and Dermott (1998), but are not yet able to account for the measured variations of Farley and collaborators (Farley and Patterson, 1995, Farley et al., 1996). In order to test our model, however, it will only be necessary to know A. The orbital inclination is dominated by a single frequency, $f = 0.01$ cycles/kyr. It will be sufficient to assume that $A(i)$ likewise is dominated by this same frequency. This will certainly occur if A is linear, but it will also be true for a large range of monotonic or nearly monotonic functions. The integral in the second equation represents the fact that the δ^{18}O represents total ice volume rather than accumulation rate, and so it depends on the integral of the driving force from the present back to a time T. A reasonable starting value for T is negative infinity; this is equivalent to a quarter wave phase shift in the frequency (i.e. it turns a cosine wave into a sine wave). To the extent that I is cyclical, this integral can be represented by putting a time delay into the equation. Given these simplifying assumptions, the equation becomes

$$\delta^{18}\hat{\mathrm{O}}(\text{theoretical}) = \hat{i}(t - \tau)$$

In this equation, the hat over the variables indicates that the mean has been subtracted, and the functions normalised to unit variance. The variable τ represents the time delay that comes from the integral. For a signal dominated by a single frequency, we would expect τ to be approximately a quarter period (equivalent to the phase delay of integrating a sine wave). We discussed this issue at length in Section 4.4. For the 100 kyr cycle of orbital inclination, this would be 25 kyr. However, we leave this constant as a variable (actually, the only remaining variable) to take into account the possibility of additional delays in the response of the climate, which could come about (for example) by ice sheet dynamics, or from the positive feedback of the decreased albedo when extensive ice sheets are formed.

If this equation appears to be distressingly simple compared to the equations found in other models, it is because of the absence of adjustable parameters. This is not a defect, of course, but a virtue. The fact that the equation is linear is not an

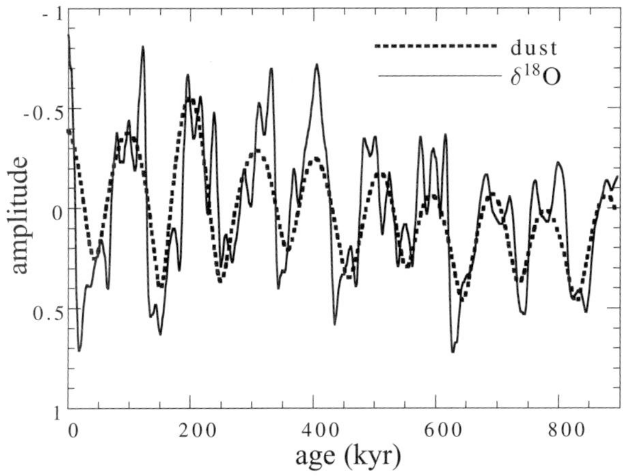

Fig. 7.5. The dust model fit to δ^{18}O from SPECMAP (no insolation term).

essential assumption (why should climate response be linear?) but we will now show that it appears to be sufficient to account for about half of the variance of the observed signal.

A numerical fit (minimising the variance of the difference) of this "dust model" to the adjusted SPECMAP data is shown in Fig. 7.5. Two parameters were used: the amplitude of the dust signal and the time delay. The best fit was obtained with a delay of 33 kyr, with an amplitude that accounts for 49% of the variance, as shown. The delay of $\tau = 33$ kyr is close to the nominal quarter cycle value of 25 kyr, and it suggests that (if the model is correct) another mechanism is active (such as ice sheet dynamics) that gives an additional 8 kyr. Such dynamics have been discussed by Weertman (1964). But another possibility is that the age assigned to the terminations is incorrect. The SPECMAP time scale was tuned to precession and obliquity, and this forced the termination to be near a precession peak. However, recent determination of the age of this termination from radioisotopic measurements of corals (see Section 8.3) disagrees with the tuning values, and places the termination 5–10 kyr earlier than SPECMAP. If such an effect is systematic across the entire time scale, then the required time delay for the fit is between 23 and 28 kyr, and that is an excellent match to the canonical 1/4 cycle = 25 kyr.

The model successfully accounts for the behaviour of the data, although it misses some of the details. It is very likely that the small time-scale variations in the SPECMAP data are due to the presence of a small precession component (perhaps due to insolation); we also know that a significant 41 kyr cycle is present, and not accounted for in the dust model. Yet this is a remarkable fit, given the fact that only one parameter was used.

The absence of the 41 and 23 kyr components can be rectified by adding a small insolation component (see Fig. 8.5). To account for the sudden terminations, we make the *ad hoc* assumption that the sudden terminations are not precisely timed by astronomy, but are instabilities that occur within time slots set by the climate forces. (As we stated earlier, often the first day of spring is not spring-like—but that doesn't disprove the astronomical theory of the seasons.) The sudden terminations could come about from an instability in feedback. For example, in many northern regions where rivers freeze over during the winter, there is a phenomena every spring when the ice suddenly melts. The sudden cracking can be heard for miles, and it sounds like a gunshot. Once it happens at one location, the increased water flow triggers the crack to propagate along the river and, within a few hours, the river is clear. This is a "catastrophic event" (in the mathematical sense) based on positive feedback. The exact day when it will occur is chaotic; it cannot be predicted based on knowledge of the driving function. Every year the date changes. If the sudden terminations come about from a similarly catastrophic change, then no model will be able to predict the precise timing.

7.7 NOCTILUCENT CLOUD MODEL

The most clearly established connection between clouds and extraterrestrial dust occurs for noctilucent clouds. These are extremely high clouds, typically 90 km high, observed after sunset when the sun lies just under the horizon, but which can still illuminate the upper stratosphere. Noctilucent clouds consist of water droplets, but most models required the existence of extraterrestrial dust to provide nucleation sites (Fiocco and Grams, 1971; Keesee, 1989; Toon and Farlow, 1981; Turco et al., 1982). Other than the connection with extraterrestrial accretion, the details are unknown. For example, we don't know whether increased accretion will increase the opacity of noctilucent clouds (by increasing the number of droplets), decrease the opacity (by reducing the average size of droplets), or have no effect whatsoever (since the formation of clouds may be insensitive to the number of condensation nuclei).

Intriguing data relating accretion to noctilucent cloud formation are found in the yearly occurrence of clouds, shown in Fig. 7.6, as first pointed out by Muller and MacDonald (1997c). A strong peak in the occurrence of these clouds occurs on about 9 July in the northern hemisphere (Foyle and Haurwitz, 1966; Thomas, 1991). This is within a day of the date when the Earth passes through the invariable plane. In the southern hemisphere, the maximum in observation of noctilucent clouds occurs on 9 January, also consistent with the invariable plane passage. At about the same time, there is also an increase in the number of other high-altitude clouds known as "polar mesospheric clouds" (Thomas and Olivero, 1989) and there is a broad peak in total meteoric flux (Whipple and Hawkins, 1956). Note that the peak in the clouds is not on the first day of summer, when observational conditions are best, but nearly three weeks later.

Measurements of accretion made with He-3 by Farley et al. (see Section 8.8) suggest that the ice ages occur when the accretion of small (7–20 micron) dust particles is at its minimum, and this occurs when the Earth is closest to the plane

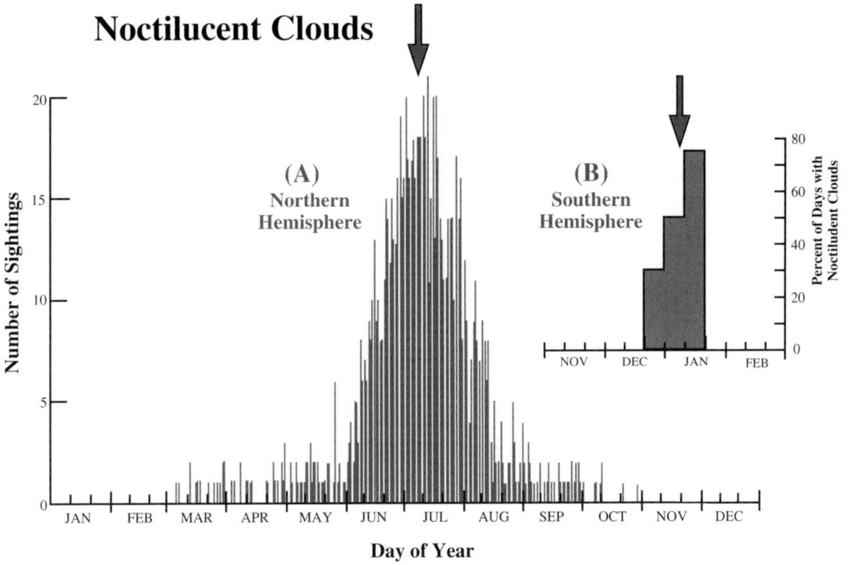

Fig. 7.6. Noctilucent clouds vs. day of year. The arrows indicate the Earth's passage through the invariable plane.

of symmetry of the dust. If the linkage is through noctilucent clouds, and if the primary role of the noctilucent clouds is to reflect visible sunlight (i.e. cause cooling), then this would imply that the presence of dust particles tends to decrease the opacity of the clouds. Unfortunately, the number of "ifs" that appear in that prediction are too great to allow it to be testable.

The present level of opacity of noctilucent clouds is such that their effect on the climate is small. However, the optical thickness of these clouds depends on the water vapour available at high altitude, and this depends in turn on other variables, such as the level of methane (which breaks down into free hydrogen, which forms water). So it is difficult to know whether noctilucent clouds are the key link between accretion and climate or whether they play just a minor role.

To relate the dust to climate, it is sufficient to postulate a linear relationship between the noctilucent cloud opacity and the orbital inclination. Mathematically, this model then become indistinguishable from the previous "dust model" shown in Fig. 7.5.

7.8 OZONE MODEL

In addition to protecting surface life from ultraviolet radiation, atmospheric ozone plays an important and complex role in climate. This is remarkable, given its low concentration. In the stratosphere, where it plays its crucial role, its number density reaches its maximum, but its concentration is still only $\approx$10 parts per million. If all the ozone were collected and compressed to a pressure of 1 atmosphere, it would form a layer only 3 mm thick. If it were solidified, it would be only a few microns thick. We now appreciate that this layer is not only remarkably thin, but also remarkably fragile. Small amounts of chlorofluorocarbons released at ground level have resulted in enough chlorine diffusing into the stratosphere that there is now a significant depletion of ozone over Antarctica every spring.

Ozone is important in climate for several reasons. Ozone is a greenhouse gas, and so its depletion directly affects the surface temperature. But a more intriguing connection comes from its role as the primary absorber of ultraviolet radiation in the upper atmosphere. By doing so, it provides the heat that causes the temperature at high altitudes to increase with altitude, creating a permanent "inversion" layer (analogous to the inversion layers that cause smog to accumulate on hot summer days). This process makes this region of the atmosphere adiabatically stable. The resulting absence of convection cells, and the consequential stratification, are what give the stratosphere its name and its character. The stratification of this region creates a barrier that effectively puts a cap on the troposphere, the region of the atmosphere that contains our weather and our climate. When you watch a thunderhead building up on a hot summer day, it finally reaches what appears to be a glass ceiling which causes it to spread out into its famous "anvil" shape. This glass ceiling is the tropopause, the boundary of the troposphere and the stratosphere. It is the limit to which convection will rise, and it is created by the ozone layer.

The thickness of the weather layer, the tropopause, is thus set by the presence of ozone. The thickness of this layer is important in determining not only the general circulation pattern of the atmosphere (Hadley and Walker circulations), which controls the flow of heat from the equator to the poles, but also as the locations of the Atlantic and Pacific storm tracks. For a general discussion of these atmospheric phenomena, see Salby (1996). Anything that affects ozone could, in principle, affect the entire climate pattern of the Earth. In the last few decades we have become increasingly aware of how "small local effects", from the monsoons to El Niño, have global consequences. Our climate models are not sufficiently reliable to enable us to know with certainty what the global effects of ozone depletion would be, but they are likely to be first order.

It is important to remember that the primary property of the ice ages is not that they were cool—but that they had more ice. Changes in global ice could be triggered by an alteration of wind patterns, even in the absence of any direct cause of global cooling. If the winds are such that they carry more moisture to the northern land masses, then ice accumulation could grow, and the increased reflectivity of the surface when covered with snow could provide the feedback necessary to make sure that the snow remained through the summer. If this were to occur, then the cooling associated with the ice ages could be an effect, rather than a cause. Since the cooling and the accumulation of ice occur virtually simultaneously, it is difficult to be certain, from the data, which is the cause and which is the effect.

Ozone provides a natural amplifier for what would otherwise be the tiny effects of accreting dust. We have learned from the Antarctic ozone hole that ozone is very sensitive to several atmospheric phenomena. The ozone hole is directly attributable to the presence of chlorine, but the effect of chlorine was greatly amplified by a second phenomenon: the presence of particles that provided a solid surface on which catalytic chemistry could proceed. This was provided by polar stratospheric clouds, or PSCs, which consist of frozen droplets of nitric acid trihydrate that are submicron in size. These clouds provide a surface for heterogeneous chemistry, i.e. chemistry which involves several phases operating simultaneously. The solid particles are essential in keeping the chlorine present in a form that can be released when the sun rises in the Antarctic spring. These reactions were not understood until the pioneering work of M. J. Molina and T. Peter in the early 1990s.

Calculations show that most of the extraterrestrial dust vaporises upon entry into the atmosphere, and recondenses into fine smoke particles. As mentioned earlier, the smoke that is predicted to form from recondensed dust vapour could be the primary source of solid surface in the middle and upper stratosphere (Hunten et al., 1980). A photograph of typical particles (see Fig. 7.4) shows the kind of complex convoluted surface that is typical of those that provide catalytic amplification. Thus modulation of such particles, provided by variations in the orbit of the Earth, could directly modulate the chemistry. There have been several discussions of these and related possibilities, including Aiken and McPeters (1986) who discussed the possible relation between extraterrestrial accretion and ozone variations in the Antarctic, and Prather and Rodriguez (1988) who discussed the role that meteoric material could play in affecting nitric acid, with an ultimate impact on Antarctic ozone. Metallic meteoric elements may have other effects. For example, according to Turco et al. (1981), they may neutralise sulfuric acid vapour at altitudes greater than 20 km, and this could significantly reduce ambient sulfuric acid.

Other than the heterogeneous chemical reactions, the dust could also play a role in ozone modulation from the chemicals that they inject into the upper atmosphere. Extraterrestrial dust has been measured to contain an unusual and unexplained enrichment in bromine (Flynn, 1994; Flynn et al., 1993). Bromine is known to be over 100 times as effective in destroying atmospheric ozone than chlorine. Detailed calculations of such an effect are not currently possible, because we don't have sufficient understanding of particle size distributions and other aspects of relevant upper atmosphere chemistry. The safest conclusion that we can draw from all this is that, whatever the processes involved, they are probably complicated, and amplified by feedback mechanisms.

8

Evidence

"The trouble about arguments is they ain't nothin' but theories, after all, and theories don't prove nothin'. They only give you a place to rest on a spell when you are tuckered out buttin' around trying to find out somethin' there ain't no way to find out. And there's another thing about theories: there's always a hole in 'em sure, if you look close enuf."

Huckleberry Finn, in *Tom Sawyer Abroad* by Mark Twain

In the previous chapters, we have shown many of the records of the glacial cycles, and we have discussed how the most prominent models address them. We have introduced the analytical methods that can be used to help evaluate the models and, in the process of doing this, we have given many examples of aspects of the data which can and cannot be explained by various models. In this chapter, we would like to bring this all together with the goal of seeing what we do understand, and what we don't. We will review much of the evidence again, this time with a more systematic focus on using the analytical tools to evaluate the models. With a few exceptions (e.g. concerning the causality problem) little new material will be introduced in this chapter. Insolation certainly plays a role in climate, but how important is it in the glacial cycles? What is the evidence for nonlinear mechanisms? Have we been misled by tuning, with all its potential for circular reasoning? Is orbital inclination, with accreting dust as the only plausible linkage, a viable alternative to insolation?

Based on our previous arguments, we assume that the reader is convinced that the cycles of glaciation are driven by astronomical causes. The important questions are: what aspects of astronomy drive the cycles, and how do they do it. In Section 6.3 we listed the "problems" that a climate theorist faces. In this chapter we will address each one of these, and discuss how they might be solved.

8.1 THE 100 KYR PROBLEM

James Croll and Milutin Milankovitch, the inventors of the insolation theory, did not know that the major cycle of the ice ages for the last million years had a period of

100 kyr. If they did, they might never have attributed the ice ages to insolation, since spectral analysis of insolation shows negligible 100 kyr variation. Insolation is dominated by precession and obliquity.

Even the numbering of the early stages of the cycles (shown in Fig. 4.4) reflects this prejudice. At the 1999 autumn meeting of the American Geophysical Union, Nicholas Shackleton recounted his memories of the early 1960s when Cesare Emiliani was attempting to understand the chronology of data from the early sea floor cores. Emiliani named the interglacials with odd numbers, counting backwards from the present (Emiliani, 1966). By this scheme, the interglacials were Stages 1, 3, 5, 7, 9, 11, etc. To the present day paleoclimate student, one of these stands out as strange: Stage 3 is not an interglacial, but only a slight warming that took place about 55,000 years ago. Shackleton recalled that Emiliani expected the period of glaciation to be 41 kyr, based on the importance of this period in the Milankovitch theory. So, although it was weak in the record, it occurred at about the "right" time (about 41 kyr prior to the last glacial termination), so it was given the odd number that he reserved for interglacials.

When Broecker and van Donk (1970) discovered that the major cycle was 100 kyr, it was a mystery. In many ways, it still is a mystery, the biggest mystery of the glacial cycles—although many of us have our favourite explanations.

As we discussed in the introduction, immediately after the existence of the 100 kyr cycle was identified, many people associated it with the eccentricity cycle. Eccentricity appears in the insolation theory as the envelope of the precession parameter; it is the only variable in the theory that has a 100 kyr oscillation. But the amplitude of the eccentricity term in a spectral analysis of insolation is exceedingly small, as is shown in Fig. 8.1. The spectral power at $f = 0.01$, period $T = 100$ kyr, is smaller than that at the precession frequencies by a factor of $\exp(-6.4) = 1/600$. We show this in a log plot, since in a linear plot the height of the eccentricity peak is less than the thickness of the line used to draw the axis.

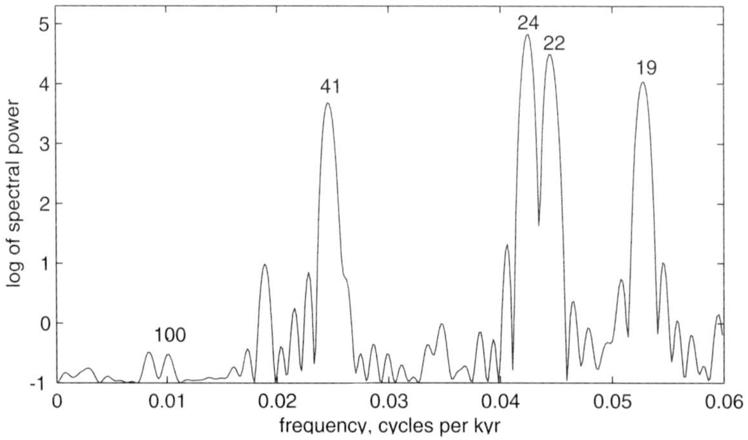

Fig. 8.1. Spectral power of July 65N insolation, log plot.

One way to solve this problem is by postulating a nonlinear response to the driving force. For example, if the climate system of the Earth has a resonant response to a 100 kyr driving force, then this amplitude could be strongly enhanced. Resonant response is common in geophysics. It is what makes the tides in the Bay of Fundy so high—the bay is resonant at the tidal period of two cycles per day. Resonant response is what makes the sound in a sea shell; when we listen we hear the ambient noise at the sea shell resonant frequency amplified.

However, the amplification of the small eccentricity signal in the glacial record presents a more difficult problem. In the actual data (e.g. the SPECMAP spectral power, shown in Fig. 4.5) the 100 kyr peak is nearly five times greater than the precession peaks. Thus the total discrepancy of the ratio of the two peaks is about a factor of 1000. No plausible resonant amplifier has been identified on the Earth that can enhance the eccentricity peak by a factor of 1000. Moreover, such an amplifier is not allowed to amplify the nearby 41 kyr peak. We don't mean to imply that a *mathematical* amplifier is difficult to construct. W. Berger was able to construct such an amplifier, by assuming that the Earth responded to the 4th power of the insolation and to the 4th power of the average ice volume averaged over the previous 57 kyr (see Section 6.4.8). If we find that the measurements of the age of Termination II are in error (they disagree with the Berger model), then we should look for mechanisms which could make physical sense of such a response.

Another obvious possibility is to *suppress* the precession and obliquity peaks, so that the only remaining spectral power is in the 100 kyr region. This is easier to do; one way, for example, is to postulate that the Earth responds not to northern hemisphere insolation, but to a combination of insolations, north–south symmetric, in such a way that the obliquity and precession components cancel. But this approach does not amplify the 100 kyr signal, so it means that the ice ages are being driven by a very small insolation change.

The problem of creating a larger 100 kyr cycle has been solved by the nonlinear ice model, as discussed in Section 6.2.1. As shown there, virtually any nonlinear operator will transfer energy from the precession region to the eccentricity region, by putting it into the envelope of the precession curve. As shown in Fig. 6.11, the nonlinear ice model by itself does not increase the amplitude of the 100 kyr signal as high as necessary to account for its dominance over precession in the $\delta^{18}O$ data. However, it does amplify it enough to make it plausible that a final amplification can be accomplished by a weak resonant response from the climate mechanisms of the Earth.

A totally different solution is found in the orbital inclination theory, discussed in Chapter 7. In this approach, the obliquity and precession cycles are attributed to insolation, but the 100 kyr cycle is attributed to cloud modulation by accreting dust. Since there is no *a priori* way to estimate the relative contributions of dust and insolation, the relative amplitudes cannot be predicted by the theory.

8.2 THE STAGE-11 PROBLEM

The "Stage-11 Problem" (Imbrie et al., 1993a) refers to the fact that there is a very large glacial termination that leads to Stage-11, and yet there is only a minor change

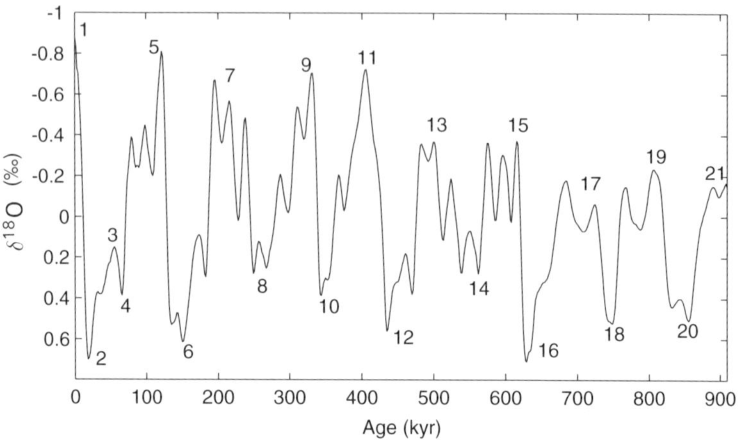

Fig. 8.2. SPECMAP $\delta^{18}O$ stack (revised time scale) with stages.

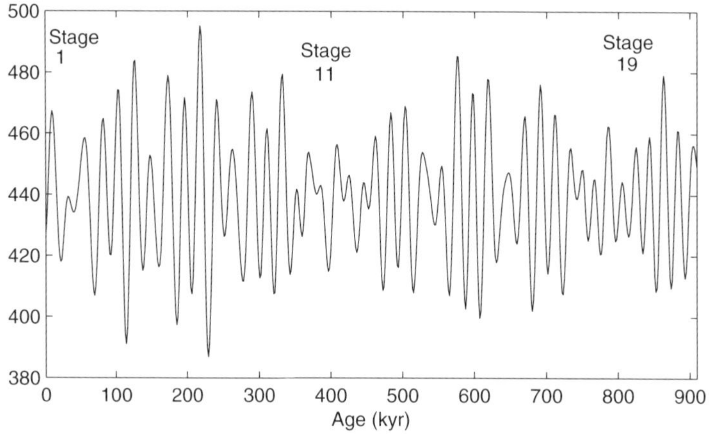

Fig. 8.3. July 65N insolation, with problematical stages indicated.

in insolation at that time. This is illustrated in Fig. 8.2, where we have indicated the stages on the revised SPECMAP time scale. In Fig. 8.3 we have plotted the July 65N insolation, and marked several stages. Note, in particular, that at the beginning of Stage-11 the insolation is undergoing very little variation—and yet according to the insolation theory, this tiny variation is supposed to be responsible for the large termination that occurs at the same time. There are similar problems at Stages 1 and 19, as can be seen from the plots, although they are not as severe.

There is no good answer to these problems in the insolation theory. Raymo attempted to account for this problem by developing a complex criterion by which even small changes in insolation could trigger a termination if enough ice had accumulated, but that approach has serious problems (see Section 6.4.4). Another approach is to postulate that the same resonant system that enhances the 100 kyr

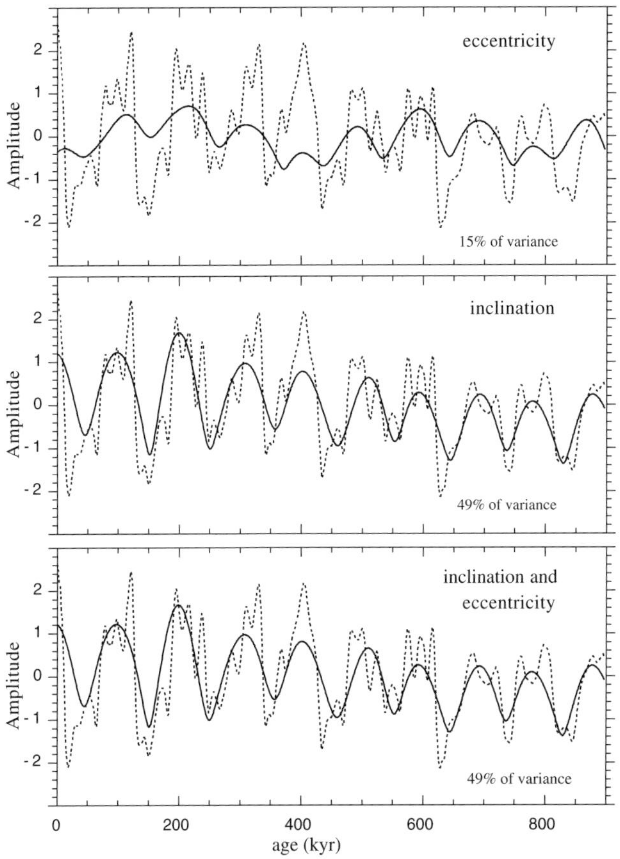

Fig. 8.4. Best fit of orbital inclination and eccentricity to the δ^{18}O data from the SPECMAP stack. The top plot shows the best fit to eccentricity alone, in which eccentricity accounts for 15% of the variance. The middle plot is the best fit for inclination alone, which can account for 49% of the variance; the best-fit lag is 33 kyr. When the data are fit to eccentricity and inclination simultaneously, the best fit is obtained with virtually no contribution from eccentricity. Of the variance, 49% is accounted for by orbital inclination.

cycle also acts as a flywheel to keep the cycle oscillating when the driving force has become negligible. Thus, you don't have to push on a swing every cycle to keep it high. W. Berger did this with his resonance memory model. The orbital inclination theory solves the Stage-11 problem immediately, since no such minimum in dust accretion occurs at Stage-11 (or at Stage-1 or Stage-19). This is evident from the fit of the theory to the SPECMAP data, shown in Fig. 7.5.

We show the results of least-squares-fits to eccentricity and inclination to the SPECMAP δ^{18}O data in Fig. 8.4. (We use the corrected SPECMAP timescale, shown in Fig. 5.18.) In the upper plot, the fit was made to eccentricity alone. Two parameters were used, one for the amplitude and the other for time delay. The time delay is included because the driving force (e.g. eccentricity) is presumed to drive the instantaneous rate of ice accumulation or melting, whereas δ^{18}O represents the total amount of ice accumulated, i.e. the integral of the rate. The best fit (minimum root-mean-square difference) for eccentricity was obtained with a lag of 1 kyr. The best fit amplitude had a variance equal to 15% of the variance of the SPECMAP signal.

The fact that only 15% of the variance was matched is a consequence of the different shapes of eccentricity and SPECMAP. Eccentricity has a long 400 kyr modulation that is not seen in the data. This is seen in the fact that the local average of the eccentricity signal is minimum at age = 0, rises to a maximum at about 200 kyr, falls again to a minimum near 400 kyr, and so on. Such behaviour is not present in the SPECMAP data. The programme also has difficulty fitting the data because of the very low amplitude of the eccentricity oscillation near age = 0 (the Stage-1 problem), 400 (the Stage-11 problem), and 800 kyr (Stage-19).

In the middle plot of Fig. 8.4, we show a fit between the data and the linear orbital inclination model. The fit is surprisingly good. Orbital inclination fits the overall behaviour of the data, and accounts for 49% of the variance. The best fit lag was 33 kyr, i.e. the changes in inclination precede the changes in $\delta^{18}O$ by this much. When both orbital inclination and eccentricity are used in combination (lower plot), the numerical fit chooses to fit the data with inclination alone. The amplitude of inclination is essentially the same as in the middle plot, and the amplitude of eccentricity is close to zero. The fit is not improved by the inclusion of eccentricity.

Orbital inclination cannot fit the sharp changes and narrow features present in the data, since it does not contain any rapid oscillations itself. Of course, orbital inclination is not supposed to account for the entire behaviour of the climate; we know that obliquity and precession are also present. To take these into account, we did a fit to the SPECMAP data (again, with the revised time scale), to a linear combination of orbital inclination and insolation. There were three parameters: the relative amplitude of the two terms, and the time delay for each. We chose summer insolation at 65N, and we did not try varying the latitude or longitude. The results of the best fit are shown in Fig. 8.5.

The combined inclination plus insolation model gives a remarkably good fit to the SPECMAP data. Note that even the shape of Stage-11 (around 400) kyr is well represented by the combination. The broad shape is a result of the fact that insolation variations have a low amplitude at this time.

We could try to improve the fit by trying other latitudes or longitudes, but there is a serious danger of circular reasoning. The small peaks in the SPECMAP stack were already adjusted to match an insolation-based model, and the success of the fit shown in Fig. 8.5 may already reflect this.

8.3 THE CAUSALITY PROBLEM

"The great tragedy of Science: the slaying of a beautiful hypothesis by an ugly fact."

Thomas Huxley

Climate theories would be much easier to test if our paleoclimate records had accurate age scales. The methods of tuning to which we devoted Chapter 5, were

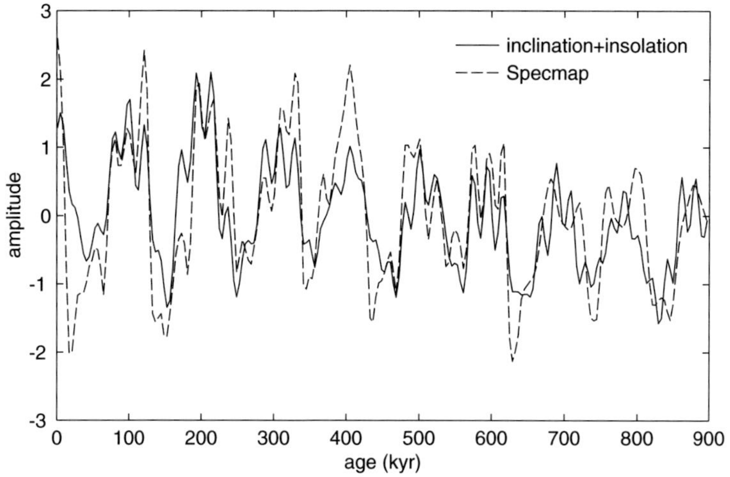

Fig. 8.5. SPECMAP fit to inclination and insolation. The lag for insolation is zero; the lag for inclination is 33 kyr (i.e. $\delta^{18}O$ lags inclination).

designed to improve this time scale. Unfortunately, when the time scale is based on a theory, the data are corrupted, and great caution must be used if they are to be used as a test of that theory or any other theory, lest they post the danger of circular reasoning.

The paleoclimate record with arguably the best time scale is the $\delta^{18}O$ record from the Nevada cave known as Devils Hole (Winograd et al., 1992), as we mentioned in Section 4.10.1. The full $\delta^{18}O$ data set is shown in Fig. 4.27. The spectrum will be shown in the next section (Fig. 8.10); the fact that the spectrum is dominated by a 100 kyr peak is evidence that the climate at Devils Hole is not entirely local, but reflects the global changes of the glacial cycles.

In Fig. 8.6 we show the Devils Hole $\delta^{18}O$ data for the age near Termination II, superimposed on a plot of the 65N July insolation (the basis for the standard Milankovitch theory). Pay particular attention to the timing of the last termination shown in the data. At 135 kyr (dashed line) two things are occurring:

1. Insolation is near its lowest point in 45 kyr. It is at such a low level (418 watts/m^2) that it is lower than it had been for 150 out of the prior 180 kyr.
2. $\delta^{18}O$, probably reflecting temperature, is higher than it had been for over 180 kyr. It is almost two-thirds of its way to the peak that it will reach 10 kyr later.

The fact that the rise in temperature precedes the rise in insolation is what we call the causality problem, since the effect precedes the cause. If the interpretation of the data is correct, then this comparison represents strong evidence against the insolation hypothesis. The ice age is terminating prior to the onset of high insolation; it seems to be well into its termination when the insolation is very low.

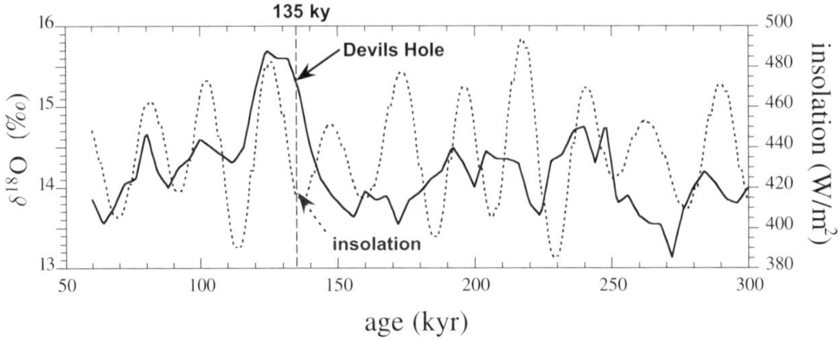

Fig. 8.6. Devils Hole δ^{18}O data and insolation.

Could the causality problem be a consequence of error in the time scale? The time scale for insolation is beyond dispute. The discrepancy is not removed by picking a different date for the insolation calculation. August insolation reaches its minimum at 134 ka; June at 139 kyr, May at 141 kyr. So even shifting the date for the insolation into spring (rather than summer) doesn't solve the causality problem. Changing the latitude makes very little difference. But could the time scale for the Devils Hole data be off by 10–15 kyr? If we slide the temperature data to the left, we are left with striking *confirmation* of the Croll/Milankovitch theory, rather than a refutation. This issue was raised soon after the original publication of the Devils Hole data. One of the most important possibilities was that there was a systematic error in the uranium-thorium method, as applied to Devils Hole. Was it possible, for example, that excess thorium was attracted to the cave walls, distorting the U/Th ratio, and making the ages appear older than they were?

Not surprisingly, these results have been hotly debated. The following references make for very interesting reading: Johnson and Wright, 1989; Winograd and Copland, 1989; Ludwig et al., 1992; Shackleton, 1993; Ludwig et al., 1993a, b; Edwards and Gallup, 1993; Imbrie et al., 1993b; Crowley, 1994; Winograd and Landwehr, 1993; Winograd et al., 1997.

Detailed answers to the objections were summarised in the 1997 paper in *Quaternary Research* (Winograd et al., 1997). Additional measurements made at different levels in the cave, where the thorium systematic would have been different, confirmed the original time scale.

The insolation theory could still be salvaged if Devils Hole δ^{18}O were reflecting a local climate effect, and that the rise in the proxy at the Devils Hole site simply reflected a local temperature rise, a possible one-time event unrelated to the cycles of glaciation. However, new data have subsequently become available that suggest that the causality problem is real. For a review, see Karner and Muller (2000b). The data consist of sea level records from sites from opposite sides of the globe. Coral terraces have long been used to obtain measurements of the ages of the rise and fall of sea level. For example, *Acropora palmata*, found in reefs, is known to grow only within two metres of the surface. The presence of such fossils indicates the ancient sea level and, combined with ^{230}Th/^{234}U dating, can yield the sea level vs. age (Bard et al.,

1990c). In January 1999, Tezer Esat and collaborators published measurements of coral terraces of the Huon Peninsula in Papua New Guinea (Esat et al., 1999). These results imply that the sea level rose by 135 kyr to within 14 metres of present levels. This implies that the glacial termination was nearly complete by this time, in confirmation of the causality problem.

Soon, additional results from other groups measuring sea level on the other side of the world, confirmed the result. At the Fall 1999 Meeting of the American Geophysical Union in San Francisco, Gideon Henderson and Niall Slowey reported an age of 135 ± 2.5 kyr for the termination, based on U-Th dates of aragonite-rich sediments from the slopes of the Bahamas (Henderson and Slowey, 1999). Christina Gallup and collaborators (Gallup et al., 1999) reported their preliminary results that sea level had risen to within 18 metres of the present level by 135.8 ± 0.8 kyr. Their work was based on new U-Th and U-Pa ages of fossil corals in Barbados. Thus, they too found that most of the rise in sea level took place prior to the expected insolation warming. There is no plausible reason to think that the low level of insolation could have triggered an ice age termination that was, by that time, essentially complete.

Violation of causality is a very serious problem. Can the insolation model be salvaged? Crowley and Kim (1994) suggest that a more complex insolation model might do the job. In their model, the warming at 135 kyr is "sufficient" to cause the termination. But by "sufficient", they only mean that the insolation is as high as the insolation associated with Termination I. In a sense, they have linked two problems (the low level of insolation at Terminations I and II) rather than explaining either one.

Another possibility discussed at the American Geophysical Union meeting where the new coral results were presented, is to attribute terminations to southern hemisphere insolation. But if we allow every discordant measurement to have its own explanation, we do not have a theory that can make predictions, and that means it really isn't a theory at all. Remember, it was *northern* hemisphere insolation that best accounts for the atmospheric δ^{18}O at the *south* pole.

The resurgence of the causality problem is fatal for theories that base the terminations on northern latitude insolation, such as the precession triggering model (Section 6.4.4). In addition, the coral results show that the time scales created by tuning to insolation (including both SPECMAP and the low latitude stack of Bassinot (Bassinot et al., 1994), are incorrect by at least 5–10 kyr at Termination 2. There is no reason to think that they are not incorrect by similar amounts at the other terminations.

The orbital inclination theory has no causality problem. In it, the termination dates determined from SPECMAP followed the orbital inclination by 33 kyr. If the SPECMAP ages are incorrect by 10 kyr, as suggested by the new data, then the lag between inclination and the termination is reduced to 23 kyr—closer to the canonical 25 kyr (quarter cycle) that the theory uses as its natural delay. More importantly, the effect (the termination) still follows the cause (orbital inclination increase) rather than preceding it.

8.4 THE UNSPLIT PEAK PROBLEM

According to the standard theory and most others based on insolation, the 100 kyr oscillation is driven by eccentricity, either directly or though its presence as the envelope of the precession parameter. In virtually all these theories, the spectrum should show a triplet of peaks, with periods of 95, 125 and 400 kyr. The absence of the 400 kyr peak in $\delta^{18}O$ data has been recognised for a long time, and we will treat that separately in the next section. But it was only in 1994 that it was pointed out (Muller and MacDonald, 1994b) that the observed 100 kyr spectral peak was not split, in contradiction to the prediction of these theories.

A more extensive analysis was published in 1997 (Muller and MacDonald, 1997a), and we show the results in Fig. 8.7. The frequency scale is expanded to draw attention to the shape of the spectral power near frequency $f = 0.01$ cycles/kyr, i.e. period 100 kyr. The predicted spectra for eccentricity, the nonlinear ice model and for the inclination model are all shown in column A. The spectra of the data are shown in the other columns: data sets that were untuned are in column B, minimally-tuned in column C, and fully-tuned in column D. The data were all chosen to cover the age interval 0–600 kyr. This was done so that the spectra could easily be compared (spectral widths would be similar) and because the time scale in this range is fairly well established.

It is clear in these spectra that the insolation-based models predict a split peak, and that the data show only a single, narrow peak, in agreement with the orbital inclination model.

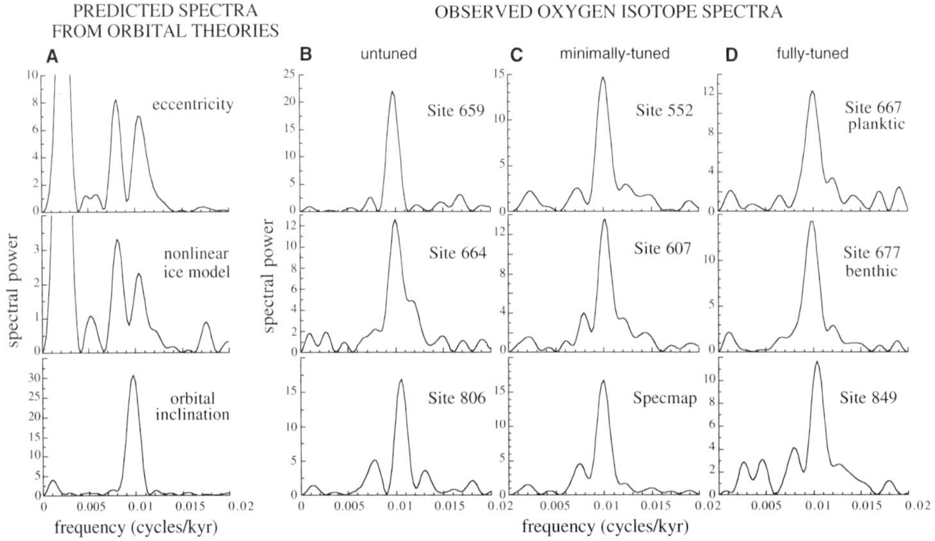

Fig. 8.7. Spectra near 100 kyr peak. Predicted spectra of the theories are shown in column A. Columns B, C and D show the observed spectra from $\delta^{18}O$ at various sites, for the period 0–600 kyr. The theoretical spectra for eccentricity and the nonlinear ice model do not match the data; the spectrum of the orbital inclination model is a good match.

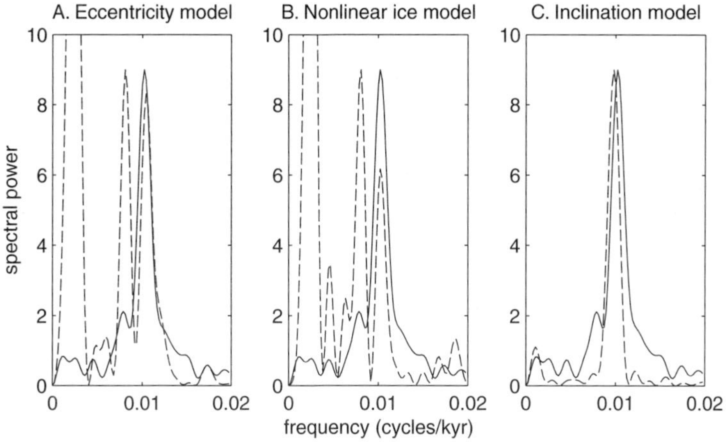

Fig. 8.8. Spectra of models (dashed curves) compared to the average spectrum of the data (solid curve). Only the inclination model is a good match to the data.

To make the comparison clearer, we have calculated the average of all the $\delta^{18}O$ spectra that appear in Fig. 8.7, and we show that average as a line in Fig. 8.8 A, B and C. Also in Fig. 8.8, we have shown the predicted spectra for the eccentricity model, for the nonlinear ice model, and for the orbital inclination model. These plots were normalised to make the maximum peak in the 100 kyr region have the same power.

In these plots, the problems with the eccentricity and nonlinear ice models are apparent. Both of these models show excessive power at low frequency (0.0025 cycles/kyr, corresponding to a period of 400 kyr). In addition, they both predict a split 100 kyr peak. Only inclination predicts a spectrum close to that observed in $\delta^{18}O$. The spectra of the data are truly incompatible with the insolation-based models. The summary of theories described by Imbrie et al. in 1993 (Table 6.1), does not contain a single theory that predicted the single narrow peak.

The "Unsplit Peak Problem" is mathematically related to the Stage-11 problem in the following interesting way. The Stage-11 problem arises because of the destructive interference of the 24 kyr cycle with the 22 kyr cycle. The cancellation occurs every 400 kyr, because the beat frequency of the interference is given by $f = 1/22 - 1/24 = 1/(400\,\text{kyr})$. But exactly the same interference occurs for eccentricity. The minimum variations in eccentricity take place every 400 kyr because of the relationship f = 1/95 - 1/125 = 1/(400 kyr). The agreement of these two results is not accidental, but occurs because the Earth's rate of precessional, which plays a dominant important role in insolation, depends on the advance of the Earth's perihelion, which in turn depends on the eccentricity of the Earth's orbit. Note that the Stage-11 problem is *not* the same as the absence of the 400 kyr peak, although both have a characteristic 400 kyr period.

Can the single peak be an artifact of variation in the sedimentation rate? Drift in the time scale can make a single peak appear to be a split peak, but although the

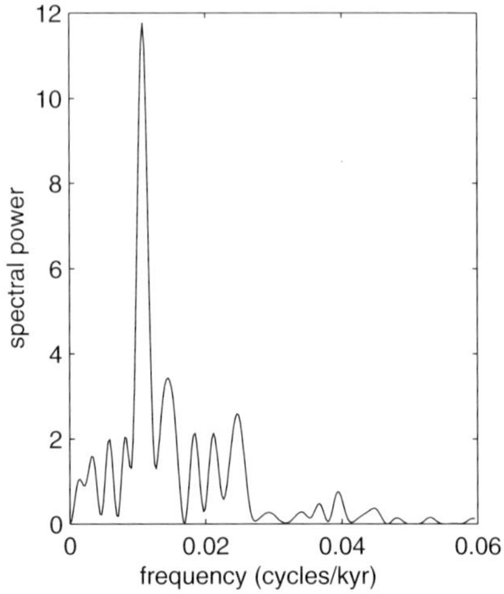

Fig. 8.9. Devils Hole δ^{18}O spectrum.

opposite is possible, it is extremely unlikely: that drift in the time scale would turn a split peak (from eccentricity) into a single narrow peak. Still, the best data for analysis are always the data that have the best time scale. Once again, the Devils Hole δ^{18}O data are attractive, since its time scale is determined by uranium-thorium dating, and has been verified by uranium-protactinium dating (Edwards et al., 1997). In Fig. 8.9 we show the spectral power for the Devils Hole data. In Fig. 8.10 we show the corresponding spectra for eccentricity and orbital inclination, covering the identical time period covered by the Devils Hole data, 60 to 568 kyr. (It is important to choose the same time periods, since the spectra of the orbital parameters do change with time.)

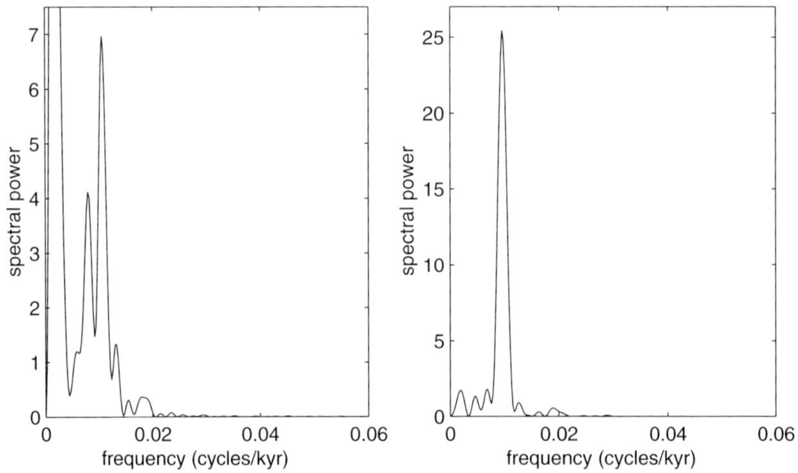

Fig. 8.10. Devils Hole time interval: spectra of inclination and eccentricity.

The single narrow peak in the Devils Hole spectrum is better matched by orbital inclination than by eccentricity. It may surprise some people that both the split peak and the 400 kyr cycle show up so well in the eccentricity spectrum for this relatively short time interval. The effect of a 400 kyr cycle is very noticeable in data lasting more than 400 kyr, so it should be expected to show up in the data. Likewise, the beats between the 95 and the 125 kyr peaks run through a complete cycle in 400 kyr. Their effect in the time domain is to produce the Stage-11 Problem, and so it is not surprising that both frequencies are visible in the frequency domain.

8.4.1 Simultaneous presence of eccentricity and inclination

Both eccentricity and orbital inclination appear to be present in data from Site 806 in the western Pacific Ocean (Muller and MacDonald, 1997b). The coarse component of the sediment (see Section 4.9.2) consists of the larger particles (primarily large foraminifera). The spectrum of the coarse component has peaks which are near the eccentricity triplet of 95, 125 and 400 kyr periods. Yet, in the same core, using the same time scale, the $\delta^{18}O$ signal shows a strong single narrow peak that is a good match to orbital inclination, shown in Fig. 8.11. The 70 kyr peak in the $\delta^{18}O$ plot may be due to the oscillations of another orbital parameter called Ω; for further discussion of this possibility, see Section 8.9.

An expanded view of the low frequency region for the coarse component is shown in Fig. 8.12, along with the eccentricity spectrum from the same period (0–862 ka). The match isn't perfect; although the peaks of the coarse component are close to those of eccentricity, they are not quite within the one standard deviation uncertainties. A possible explanation for this discrepancy will be made in Section 8.6, in which we examine the bispectra of the two signals. As we discuss in that section, the bispectra show that the coarse component is actually a superposition of eccentricity

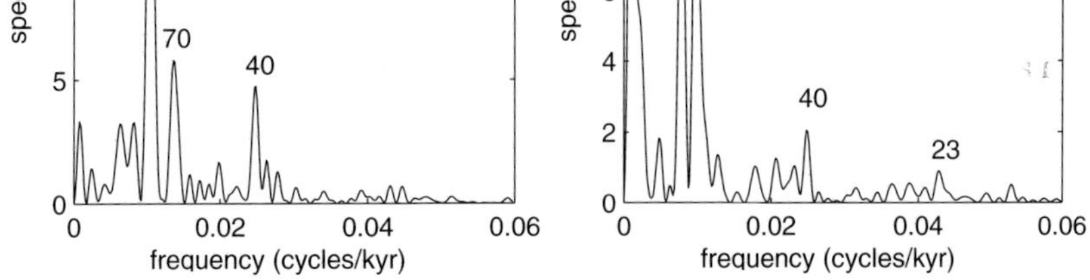

Fig. 8.11. Spectra for Site 806.

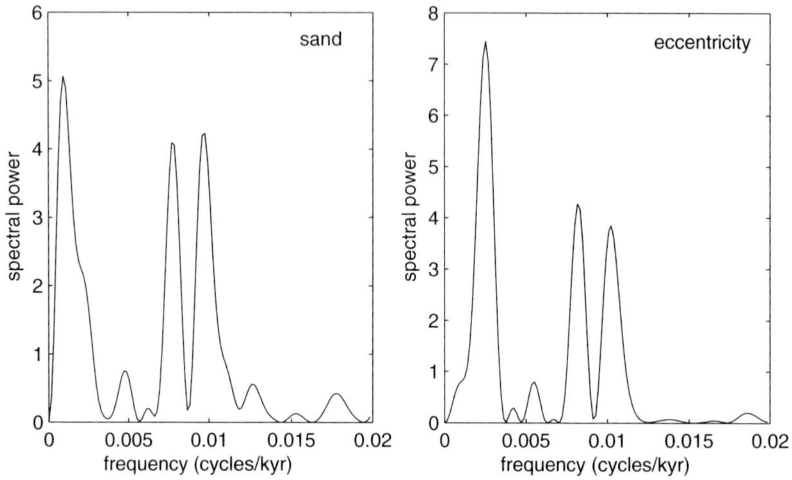

Fig. 8.12. Sand spectrum compared with eccentricity.

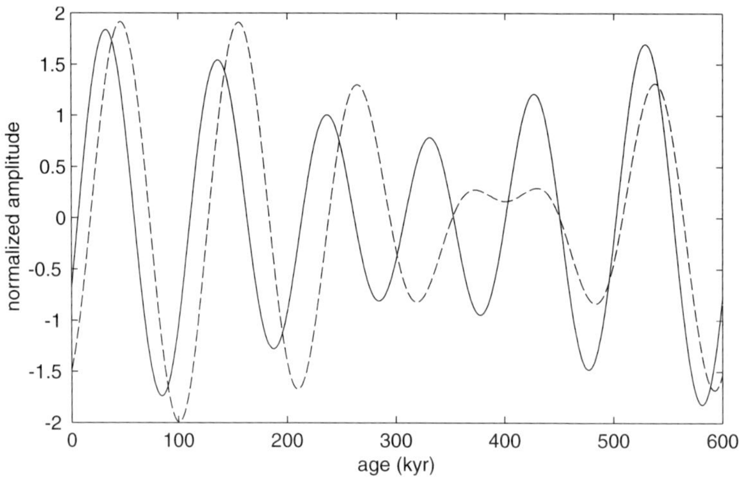

Fig. 8.13. Narrow-banded coarse component from Site 806 (solid line), and eccentricity (dashed line).

and orbital inclination. Interference between these two signals could distort the spectrum sufficiently to explain the slight disagreement found in Fig. 8.12.

We can also compare the behaviour of the sand component in the time domain. In Fig. 8.13 we have plotted the eccentricity and coarse component; both were filtered with a band pass filter with sharp cutoffs at frequencies 1/135 and 1/90 cycles/kyr, and the eccentricity plot was inverted (i.e. we plotted $-e$). The behaviour of sand and eccentricity in the time domain is very similar (as should be expected, from the good spectrum match). The eccentricity changes appear to precede the sand by an average of 15 kyr. It is plausible to believe that eccentricity is forcing the sand, and that the response is rapid rather than integrative (so the delay should be less than 1/4 cycle).

It is possible that the presence of eccentricity in the sand signal is a local climate effect brought about by variations in the strength of the monsoons (although the weakness of the 23 kyr precession signal in the sand spectrum is then a mystery). Another signal that appears to show the 95/125/400 triplet of eccentricity is the New Jersey Lake Bed depth rank data of Olsen and Kent (1996), described in Section 8.5 and plotted in Fig. 8.16. These lake beds were equatorial (as is Site 806) at the time that they were laid down.

8.4.2 Picking narrow peaks: circular reasoning?

Not all $\delta^{18}O$ data show a single narrow peak in untuned data. For example, in Fig. 8.14, we plot the $\delta^{18}O$ spectrum of Site 607 for the period 0–600 kyr. The spectrum is complex, with many peaks. It resembles neither inclination, eccentricity nor the nonlinear ice model. (The highest peak has period 87 kyr, and its full-width at half-maximum does not include the 95 kyr period of eccentricity; see description of frequency accuracy.) The presence of these complex spectra raises an important issue. In the comparison done in the previous section, did we display a bias in the selection process, by picking only those data that were in agreement with the inclination theory? Our answer is "no"—but it requires some justification.

Ever since the method of tuning was invented in the 1970s, essentially every seafloor proxy was published along with a tuning curve to show the relationship between sedimentation rate and age. Is there a way to tell whether these tuning curves were necessary?

The answer is yes. If narrow spectral lines appear in the data, and their amplitude has high statistical significance, then we can conclude that the sedimentation rate was constant, or at least was modulated at the same frequency as the data.

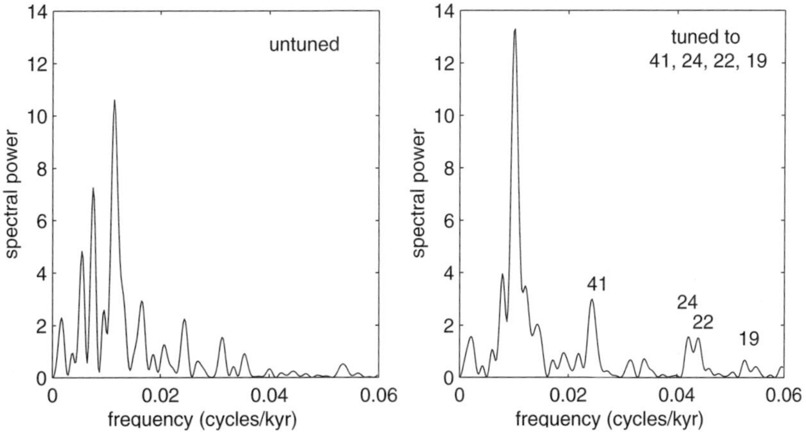

Fig. 8.14. Spectrum of Site 607, using untuned and tuned time scales.

This is not circular reasoning, but an evaluation of probabilities. Is it possible that the true $\delta^{18}O$ spectrum is actually a good match to that of eccentricity, but that variations in sedimentation rate transformed it into a single, narrow 100 kyr peak? It is possible, but it is extremely unlikely unless the noise level is high. If the statistical significance of the single narrow peak is at the 99.9% confidence level or above, then the odds of accidentally creating this peak are correspondingly small.

For the Site 607 $\delta^{18}O$ data shown on the right-hand side of Fig. 8.14, a tuned time scale was published by Ruddiman et al. (1989). The model that they used as the target for their data contained only the frequencies of precession and obliquity; thus their target included no power in the 100 kyr region. That is why we consider this time scale to be "minimally tuned". It should not be considered minimally-tuned for purposes of analysing the strength of the 41 or 23 kyr peaks, but it is legitimate to consider it an acceptable time scale as far as the 100 kyr power is concerned.

The same data, tuned to another time scale, which were based on a target that shows a strong eccentricity signal, is shown in Fig. 8.15. (This time scale was provided to us by M. Raymo.) The strongest peaks are near the eccentricity periods of 95 and 125 kyr, and there is even a small peak near 400. However, note how much the statistical significance of the peak has dropped. In the minimally-tuned data, the peak reaches over six times the mean, implying a statistical significance greater than $1-\exp(-6) = 99.8\%$. (The statistical significance would be higher if we calculated the mean without including the peak itself. The method used is much more conservative.) However, in the data tuned to the target curve target, the statistical significance is $1-\exp(-4.2) = 98.5\%$, substantially lower. In addition, the level of the background has increased. The presence of the secondary peak near 0.014 also suggests overtuning, according to the arguments of Section 5.6.2.

The fact that many data sets show a narrow peak even when the time scale is untuned allows us to draw two conclusions: the sedimentation rate was relatively

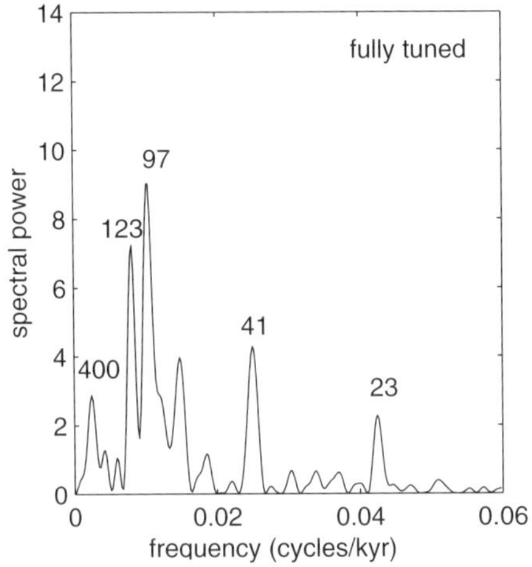

Fig. 8.15. Spectrum of Site 607, tuned to target containing eccentricity and precesssion.

constant in these locations, and the data are driven by a force that contains a single well-defined frequency.

Is it possible that a split peak, as seen in the highly-tuned data of Fig. 8.15, is more accurate than the minimally-tuned spectrum on the right side of Fig. 8.14? Could uncompensated variations in the time scale have accidentally merged the true, multiple peak into a single narrow peak? It is possible, but very unlikely.

8.5 THE ABSENCE OF THE 400 KYR CYCLE

If the 100 kyr peak is attributed to eccentricity, either directly or through the non-linear ice model, then a 400 kyr cycle should also be seen in the spectrum. This can be seen in the first column of Fig. 8.7 in which, for both the linear eccentricity model and the nonlinear ice model, the 400 kyr peak is the largest, and leaves the plot. (To see the full spectrum of eccentricity, see Fig. 2.9; for the full spectrum of the non-linear ice model, see Fig. 6.11.) Note that the 400 kyr peak is even stronger in the nonlinear ice model than it is in the linear eccentricity model.

Yet this peak is essentially absent in the spectra of the global ice proxy $\delta^{18}O$, as can be seen in the spectra of the data shown in Fig. 8.7. This absence has long been recognised as a problem for the insolation theory, and it has generally been accounted for by an *ad hoc* explanation: the period is so long, that the ice does not respond to it; see, for example, Imbrie et al. (1993a).

Yet this cycle does appear in other proxies. An outstanding example is the proxy for lake bed depths (Olsen and Kent, 1996). Their "Lake Bed Depth Rank" proxy simply assigns a number between 0 and 5 to each layer of rock, based on its texture, presence of cracks, and other physical features that suggest the depth of water at which the lake bed was formed. They show a spectral analysis of these depth ranks vs. time over ten million years during the Triassic which show a consistent pattern of eccentricity, with all three eccentricity periods present: 95, 125 and 400 kyr. A sample spectrum for their data, covering a one million year period, is shown in Fig. 8.16. The figure shows the spectrum with the usual frequency scale in A, and with an expanded scale in B which emphasises the 100 to 400 kyr region. Three strong peaks are seen in the low frequency region. Each of these is, within statistics, consistent with the periods expected for eccentricity: 400 kyr, 125 kyr and 95 kyr. The three peaks persist throughout the entire record of the lake beds.

It is also interesting to look at the power in the precession band, near 0.04 cycles/kyr. The highest power is at period 25.6 kyr; this does not agree with the peak period for present-day precession, which occurs at 23.7 kyr. Olsen believes the discrepancy may be due to long term changes in the orbit of the Earth, resulting from a combination of tidal drag and the chaotic behaviour of the Solar System.

The presence of strong spectral power in the 400 kyr region is also seen in another record in which the split eccentricity peak is clear, that of the coarse component (the large forams) for Site 806 in the western Pacific Ocean; see Section 8.4.1. In this sample, the frequency 0.0025 (period 400 kyr) is just outside of the

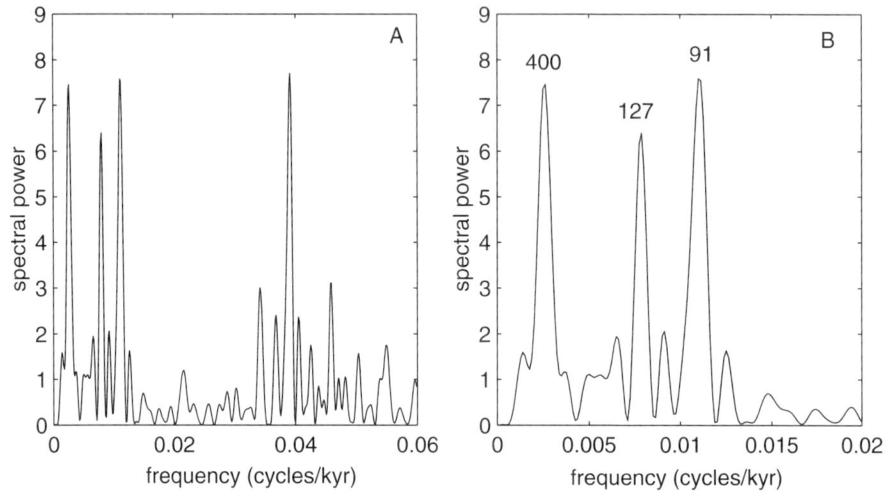

Fig. 8.16. Spectral power of Triassic lake bed depth ranks.

full-width at half-maximum of the maximum spectral power. This could be due to a statistical fluctuation (the frequency estimate will deviate from the full-width at half-maximum about as often as errors lie outside one standard deviation, i.e. about 32% of the time). It could also be due to interference from inclination; see Section 8.6.

The fact that the 400 kyr peak appears in other proxies, but does not appear in the global ice proxy, does not necessarily imply that the *ad hoc* explanation excusing its absence is incorrect. Clearly, different climate proxies can respond to different forcing functions. The lake bed depth ranks, for example, might respond to a pattern of rainfall that affected the precipitation in that area, and that could carry the 400 kyr cycle along with it. Simultaneously, the ice could be responding to a different aspect of eccentricity, and the response could suppress the 400 kyr cycle. Nevertheless, we must be cautious about accumulating too many *ad hoc* excuses, particularly if each problem or counter-example to a theory requires a different reason.

The orbital inclination model predicts no 400 kyr cycle, so if this is what is driving the ice ages, its absence in the sea floor $\delta^{18}O$ record presents no problem.

8.6 THE BISPECTRUM PROBLEM

The bispectrum tests aspects of coherence in the data that are not tested in the standard spectrum or coherence plots. The bispectrum of the $\delta^{18}O$ data from SPECMAP were discussed in Section 3.10. The bispectrum of SPECMAP (Fig. 3.28) shows peaks in areas that do not match the peaks from eccentricity-based models,

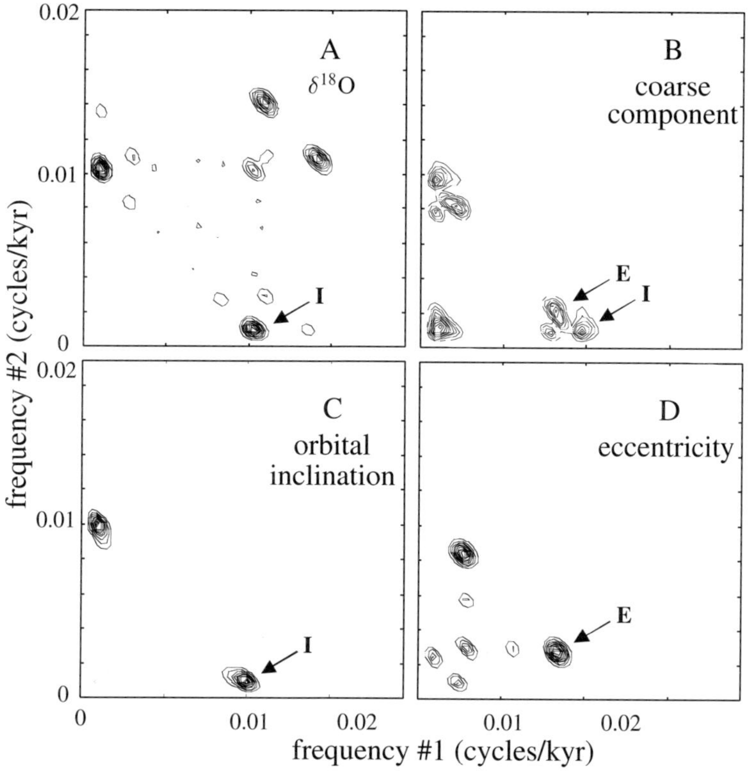

Fig. 8.17. Bispectra of δ^{18}O from Site 806 and models. Note that the bispectrum is symmetric about the diagonal.

but do match the prediction from the dust/orbital inclination model, shown in Fig. 3.29.

We now show the results of a similar analysis for Site 806 (Muller and MacDonald, 1997b). We do the analysis both for the δ^{18}O global ice proxy, and for the sand, also known as the coarse component, which we hypothesise is a proxy for local climate. The bispectra are shown in Fig. 8.17. The major signature of orbital inclination is the peak marked I at $(f_1, f_2) = (0.01, 0.001)$. The major signature of eccentricity is the peak marked E at $(f_1, f_2) = (0.008, 0.003)$. Note that the bispectrum of δ^{18}O shows only inclination. The bispectrum of the coarse component shows contributions from both eccentricity and inclination.

We had concluded, based on the spectra, that δ^{18}O was a closer match to inclination, and that the coarse component was a closer match to eccentricity. In Fig. 8.17 we see that this conclusion is strengthened by the bispectra. But the bispectra show that orbital inclination is also contributing to the local climate. This combination of effects may have led to the slight disagreement seen between the coarse component spectrum and that of eccentricity in Fig. 8.12.

8.7 THE PRECESSION PROBLEM

According to the Croll/Milankovitch concept, the ice ages are driven by changes in insolation, and the primary component of insolation comes from the 24/22/19 kyr cycles of the precessional parameter. The nonlinear ice model shows strong precession peaks (see Fig. 6.11). The precession signal is indeed strong, even dominant in many climate proxies, such as the terrigenous component of sediment, shown in Fig. 4.23.

Yet in the global ice proxy, $\delta^{18}O$ in sea floor cores, the precession signal is weak and sometimes absent. In the spectrum for Site 659 (Fig. 2.2), and the spectrum of $\delta^{18}O$ for Site 806 (Fig. 8.11) its amplitude is negligible (even if not statistically zero). In the $\delta^{18}O$ sea floor proxies, the strongest precession signal is seen in the spectrum of the uncorrected SPECMAP stack (Fig. 4.5, although much of this power disappears once the time scale is corrected, as shown in Fig. 5.19) and in the core for Site MD900963, used in the Low Latitude Stack (Fig. 5.42).

Precession is also a strong component of the Vostok atmospheric $\delta^{18}O$ but, as discussed in the section on the Dole effect (Section 4.7), that signal bears little similarity to the otherwise universal pattern seen in sea floor cores. If the atmospheric $\delta^{18}O$ is a valid global ice proxy, then the sea floor core signal is not. We will assume that the Vostok oxygen isotopic record represents an aspect of climate, but not global ice.

We illustrate the weakness of precession in the global ice proxy with yet another record, $\delta^{18}O$ from Site 552 (Shackleton and Hall, 1984). In Fig. 8.18 we show the spectrum from this site for the period 0 to 600 kyr. Next to it we show the spectrum of July 65N insolation.

The spectra have peaks at the same locations (except for the 100 kyr peak), but the amplitudes are dramatically different. In insolation, the precession peak is dominant; in the ice proxy, it is weak.

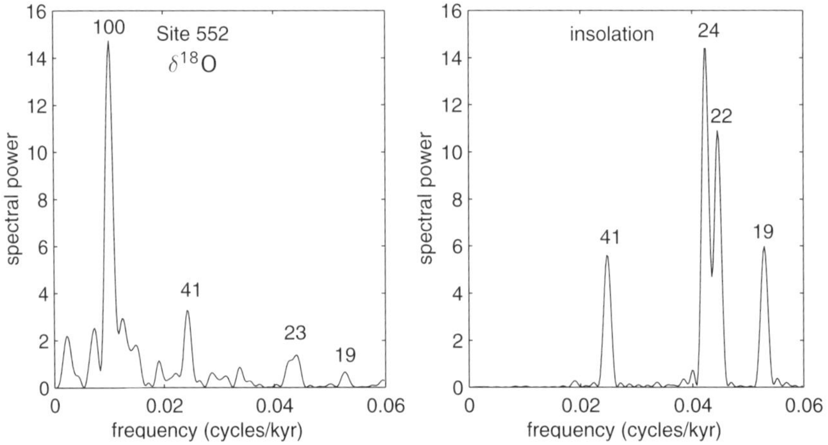

Fig. 8.18. Spectrum of $\delta^{18}O$ for Site 552, and that of July 65N insolation.

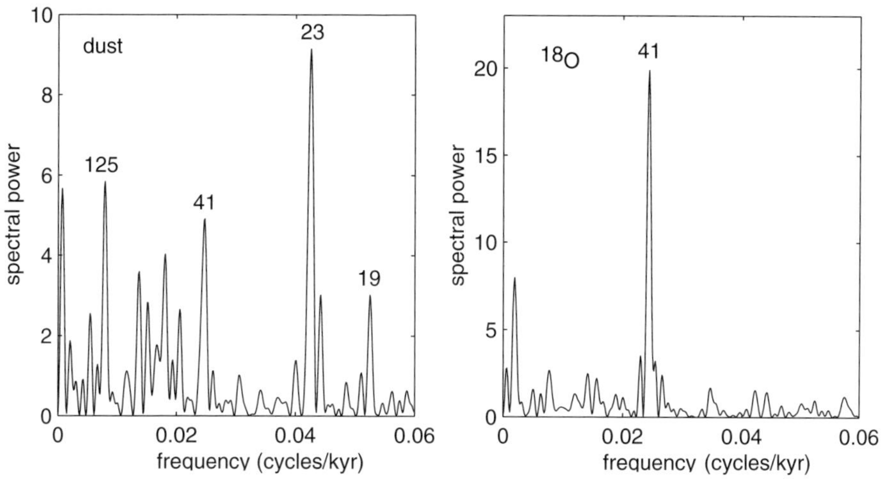

Fig. 8.19. Site 659 spectra, dust and $\delta^{18}O$, 1.5 to 2.5 Myr.

This low level of precession is a serious problem for the insolation-based theories, since it is the precession signal that is indirectly responsible for the dominant 100 kyr cycle. Is it possible that the signal is artificially suppressed, for example, by variations in the sedimentation rate? If the time scale has uncompensated fluctuations ("chatter"—see Section 5.1) as small as 10 kyr, then this could remove the coherence from the precession signal and make it disappear in the spectral plot, while leaving the 100 kyr cycle essentially unaffected. It is also possible that at some sites, low sedimentation rates coupled with bioturbation (the churning of the sediment by bottom-dwelling creatures) prevent precession from being recorded.

A strong argument against these possibilities is demonstrated by spectra from Site 659, discussed by Karner and Muller (2000a). The simultaneous spectra of dust and $\delta^{18}O$ variations are shown in Fig. 8.19 for Site 659 (Tiedemann et al., 1994) for the time interval 1.5 to 2.5 Ma. The time scale was adjusted to match the previously tuned time scale of Site 677 (Shackleton et al., 1990). Since the spectra come from the same sections of a single core, any mistakes in the time scales should affect both spectra equally. As can be seen in the figure, the dust signal shows a very strong precession signal, with peaks at 23 and 19 kyr.

These spectra are superimposed in Fig. 8.20, along with the spectrum of July 65N insolation. The dust signal shows a precession triplet with periods 24, 22 and 19 kyr; most of the spectral power is at period 24 kyr, but all the peaks are significant. There is also a significant obliquity peak at 41 kyr peak (0.024 cycles/kyr). This peak is also strong in the $\delta^{18}O$ data for this time period. The dust in the low frequency range (below 0.024) seems to have a rising background level, characteristic of red noise. Such noise could easily hide a peak in this range. We marked a peak at period 125 kyr. We could not be certain that there is no 95 kyr peak, since the high level of background could easily cancel it.

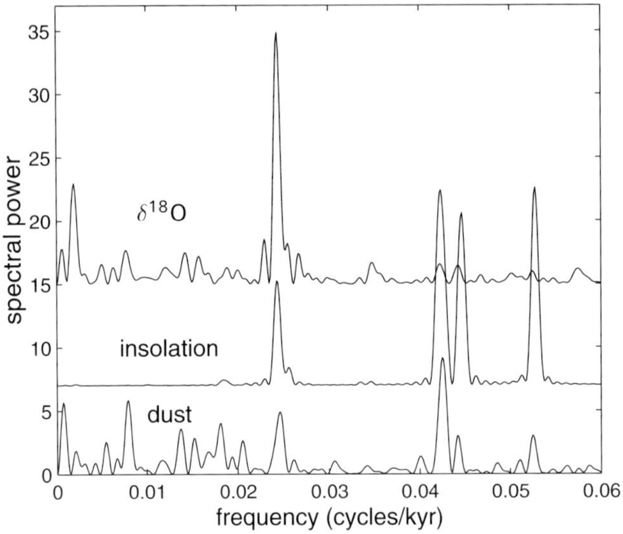

Fig. 8.20. Simultaneous spectra of Site 659, interval 1.5 to 2.5 Myr. The $\delta^{18}O$ and insolation spectral power have been offset from zero.

In the centre of the plot we show the spectral power for insolation during this time period. Since it is based on a theoretical calculation, the background is negligible, and it can be used to identify the important peaks in the other plots.

The top plot is the spectrum of $\delta^{18}O$ for the same period. It is very similar in character to the plot for Site 607, shown in Fig. 4.34. There is an extremely strong 41 kyr peak. But note the near absence of $\delta^{18}O$ spectral power in the precession region between 0.04 and 0.06 cycles/kyr. There is some there: small bumps coincident with the insolation spectra, but they are not even the highest points in this region.

We show the spectra of dust and $\delta^{18}O$ from this site for the period 0 to 600 kyr in Fig. 8.21. The spectrum for $\delta^{18}O$ in this time interval shows a small precession peak near period 23 kyr, similar to that seen in the corrected SPECMAP spectrum. The spectrum for the dust shows a stronger precession peak. From these spectra, we conclude that the precession peak truly is small in $\delta^{18}O$, and not an artifact of poor tuning or other experimental effects.

The precession problem does not exist in the orbital inclination model. In it, the total ice volume proxy $\delta^{18}O$ is postulated to be a sum of the dust contribution (which gives the 100 kyr peak) and a small contribution from insolation. In this model, there is no precession "problem", since the relative contributions of the insolation and dust components are not predictable.

8.8 THE 41 TO 100 KYR TRANSITION

About one million years ago (probably between 800 and 1200 kyr), the dominant cycle of glaciation changed from 41 kyr to 100 kyr. This is readily apparent in the

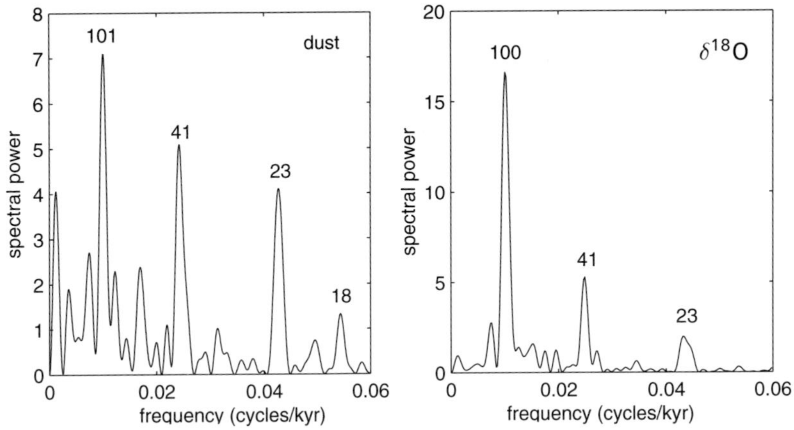

Fig. 8.21. Spectral power of Site 659 in dust component and in δ^{18}O. Note the strong precession signal in the dust and weak precession in the δ^{18}O proxy.

δ^{18}O data from Site 607, shown in Fig. 1.6, and it is reflected in the vast differences seen in the spectra for the periods on either side of this transition (e.g. in the spectrum for Site 607, shown in Fig. 8.14 and Fig. 4.34). Why did this occur? Some people argue that any glacial theory *must* account of this transition before it can be taken seriously. This position was taken by an anonymous referee for one of our papers. We disagree—despite the fact that the orbital inclination theory is the only one that offers a natural explanation for the shift. The shift was a singular event, and could reflect an unusual occurrence. In the insolation models, just such an *ad hoc* event is postulated, such as a change in the oceanic conveyor belt. Such a change can be documented in sea floor proxies, but that doesn't imply that the change drove the transition; it might be the other way. (But the fact that this explanation is *ad hoc* is not a problem. It is a *feature* of the theory.)

Nevertheless, if a separate explanation for the transition is forthcoming, and can be verified, then that is a strength. This is the case for the orbital inclination model. When the referee demanded an explanation for the transition, we supplied one: we postulated that the solar system dust underwent a sudden increase at that same time. Such an increase was not known, but it was plausible. The dust ring is continually replenished by collisions in the asteroid belt. If an unusually large collision took place at about 1 million years ago, the dust would have increased, and that would have made the amplitude of the dust-driven peak (with a 100 kyr period) to increase suddenly. This was the explanation that we added to the paper (Muller, 1994) to satisfy the referee. The paper was rejected anyway, and appeared only as a Lawrence Berkeley Laboratory Report.

About the same time that the orbital inclination theory was finally being published (Muller and MacDonald, 1994a, 1995), a new measurement of accreting dust was published (Farley, 1995). Farley used the light isotope of helium, ^{3}He, as a proxy for dust. ^{3}He is rare in the Earth's atmosphere because its light mass allows it to reach a high enough velocity to escape. However, ^{3}He is abundant in the solar wind,

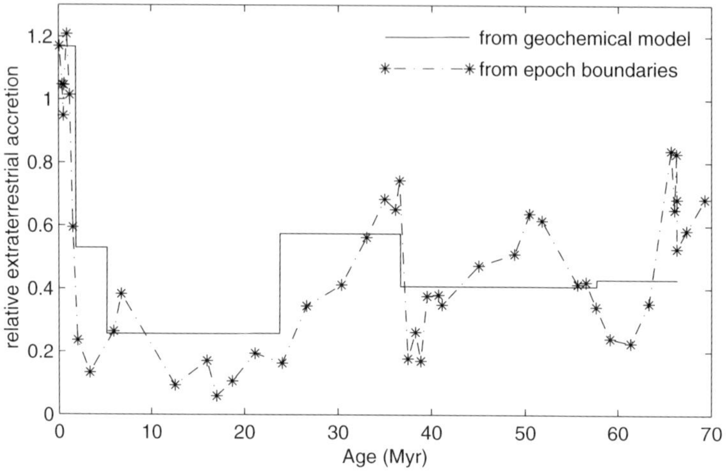

Fig. 8.22. Dust accretion estimated by sea floor ^3He measurements by Farley (1995). The sharp increase takes place between 2 and 1 Myr.

and it collides with and buries itself into interplanetary dust particles. Such particles, when they reach the ocean intact, become the primary source of ^3He on the sea floor.

Farley had measured the ^3He vs. time, and observed a sudden increase at about 1 Myr, as shown in Fig. 8.22. This implied that the level of accreting dust had undergone a similar increase. According to these measurements, the accreting dust suddenly shot up to a higher level than it had been at for the preceding 60 million years.

The increase in dust was certainly an unusual event. We don't know for certain why the increase took place. It could have been due to a particularly energetic collision in the asteroid belt. But the cause is less important, for the inclination theory, than the fact that this dramatic increase took place. The inclination theory depends on dust to act as a linking mechanism for the 100 kyr cycle. A sudden increase in dust provides a natural explanaiton for the strong surge in this cycle that took place at about the same time. Thus, the inclination model requires no *ad hoc* explanation for the transition. It relates it to the observed result of the dust increase.

8.9 PREDICTION VERIFIED: THE 100 KYR ACCRETION CYCLE

The orbital inclination model made a very explicit prediction: the accreting dust must show the same 100 kyr oscillation that is seen in the $\delta^{18}O$ record (Muller and MacDonald, 1995). The prediction was verified shortly afterwards by measurements of ^3He made at both Site 607 (Farley and Patterson, 1995) and Site 806 (Farley et al., 1996). The cycle was independently verified by Marcantonio (1995), although

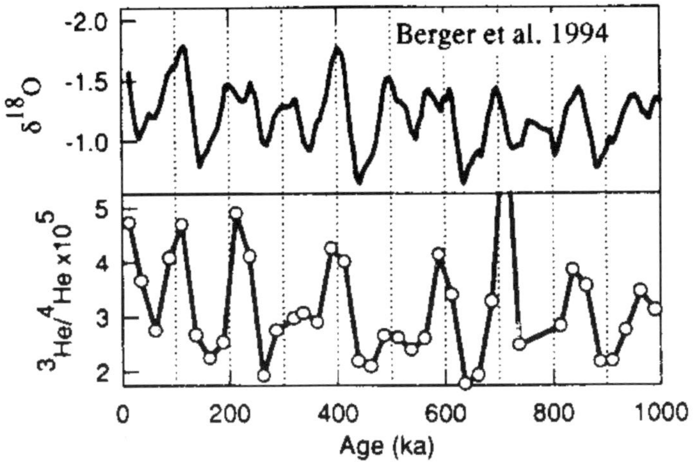

Fig. 8.23. 100 kyr cycles of dust, compared with $\delta^{18}O$ (Farley, 1996).

Marcantonio argued that the cycle could have been created by focusing by oceanic currents. We consider it unlikely that focusing would be observed at the location of Site 806, the Ontang Java Plateau; moreover, it would be doubly surprising to find similar focusing taking place at two locations as different as Site 806 in the equatorial Pacific, and Site 607 in the North Atlantic. The cycles reported by Farley are shown in Fig. 8.23.

Of course, the observations of cycles of dust simply verify the astronomical predictions—that dust should accrete with the cycles of orbital inclination. That observation does not verify that the cycles of climate are being driven by such dust. There are other issues. The observed cycles in Fig. 8.23 appear to suggest that the interglacials (high points in the $\delta^{18}O$ curve) are associated with periods of high accretion (high ^3He levels). The dust theory is unable to specify the sign of the effect, since according to the model (Chapter 7) dust could trigger clouds, or suppress them (e.g. through its effect on the atmospheric conductivity), and clouds can cause cooling (if low) or warming (if high). One must be cautious, however, before assuming that the ^3He is a direct measure of the bulk of the accreted dust. As discussed in Chapter 7, the time of transport of dust from its source in the asteroid belt to the Earth depends very strongly on the size of the particles. The ^3He signal gives a measurement of the dust accretion in the 7–20 micron size range (Farley et al., 1997). There are several effects that contribute to this. Very small particles do not have high surface area, and so they do not absorb as much ^3He from the solar wind. In addition, Poynting–Robertson drag (see Section 7.2) is more important for such particles, so they spend less time in the solar system absorbing ^3He before reaching the Earth. Large particles absorb a lot of ^3He, but they are severely heated upon entry to the atmosphere, so most of their ^3He is outgassed or released when the particle vaporises. Since the transport pattern to the Earth may be significantly different for the larger particles than for the particles that carry the ^3He, we should not yet assume that we know the phase relationship between the bulk of the accretion and the glacial cycles. The bulk accretion can be measured, in principle, through iridium or osmium abundances. It is very difficult to observe cycles as short as 100 kyr with these elements owing to contamination from terrestrial sources.

8.10 THE 41 KYR PROBLEM

For the period between 1 and 3 million years ago, the global ice proxy, $\delta^{18}O$, is dominated by the 41 kyr obliquity cycle, as shown by the spectra from many sites, including those of Site 607 and Site 677. In this time interval, the 100 kyr cycle is virtually absent, as are the cycles associated with precession. The standard Milankovitch model, based on northern hemisphere summer insolation, does not seem to work for this period. The dominance of the 41 kyr cycle is unexplained. This is a problem for all models.

There are several possible solutions to this problem. One is to postulate that the signal is driven by insolation, but that, during this period, the response of the climate system suppressed the high frequency precession period. This is not entirely satisfying, since the top precession period of 24 kyr is only a factor of 1.7 less than 41. They are not really very far apart. And the precession frequencies do show up in other proxies at this same time period.

Another possibility is that the cycle is driven by a different aspect of insolation. For example, northern latitude winter insolation is dominated by a 41 kyr cycle, so perhaps this is what is driving the cycles (Muller, 1999). The spectral power for January 65N insolation is shown in Figure 2.15. This spectrum looks very much like those of $\delta^{18}O$ during the 1 to 3 Myr time interval. Although the level of insolation is amall, the peak-to-peak variation is large: about 10 W/m^2. This can be seen in the plot of insolation in Fig. 8.24.

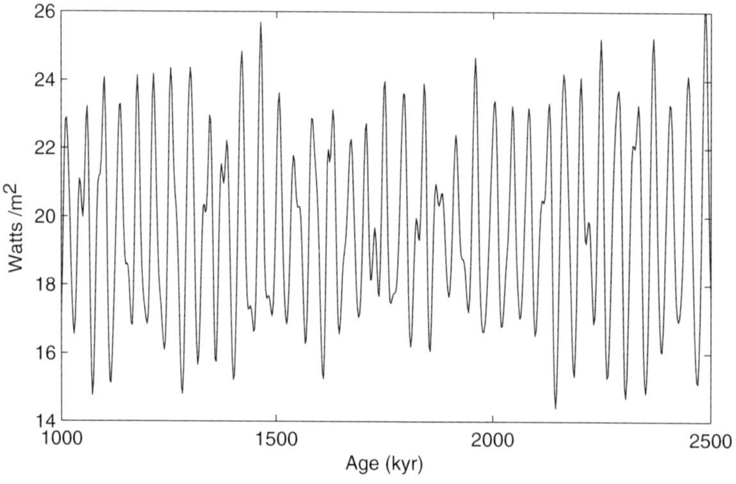

Fig. 8.24. January 65N insolation.

Note that the vertical scale shows a range from 14 to 26 Watts/m^2. This is a factor of 20 less than the insolation for the same latitude in July. Does it make sense to have this weak winter signal dominate over the summer signal? Possibly. The variation is larger than the equivalent radiative forcing that is expected, for example, from a doubling of the atmospheric CO_2, which is about 4 W/m^2. The amount of land north of latitude 65 is small, as is shown in Fig. 6.3. The formation of permanent glaciers appears to require land-based roots in the far north, shown by the examples of Greenland and Antarctica (the only places on Earth where extensive ice-age ice remains). The amount of land that remains dark during winter days is a strong function of obliquity, and that is what is driving the insolation curve of Fig. 8.24. Of course, this is all wildly speculative, and not supported by independent evidence.

Perhaps the most we can do now is to recognise that the dominance of the 41 kyr peak for a two million year period presents both a problem requiring a solution, and a clue giving us hints about the behaviour of climate.

8.11 THE SUDDEN TERMINATION PROBLEM

To illustrate the sudden nature of the terminations of the ice ages, we have plotted the $\delta^{18}O$ signal for Site 806 in Figure 8.25. We chose this site because there is strong evidence that the sedimentation rate is constant, so we could plot the data without tuning. The glacial terminations are normally indicated by the symbol "T" followed by a Roman numeral. The most dramatic terminations are indicated in the figure: TI, TII, TV, TVII and TIX. Although the ice ages are often described as having a "sawtooth" shape, from Fig. 8.25 we see that the main characteristic is that the terminations are usually more abrupt than the onset of glaciation.

Why do the ice ages often end so abruptly? In the insolation theories, this is attributed to the accidental coincidence of three cycles: eccentricity, obliquity and

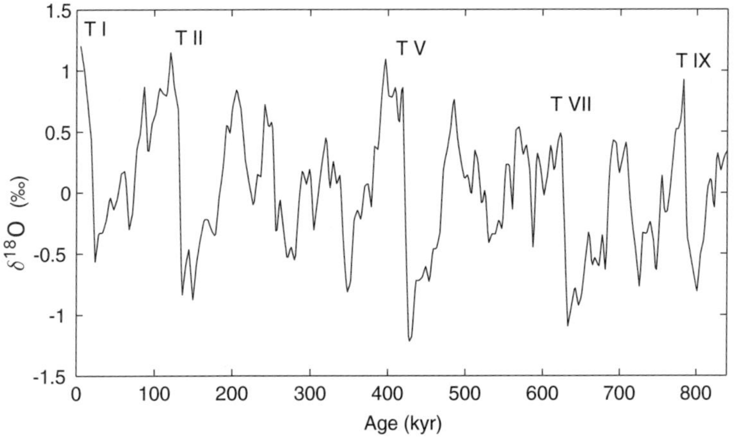

Fig. 8.25. $\delta^{18}O$ for Site 806. The sharpest glacial terminations are labelled.

precession. But only the precession cycle is sharp enough (going maximum to minimum in about 10 kyr) to give the observed sharpness. And the recent measurements of the termination ages, made by dating corals (see Section 8.3) show that these terminations occur at the wrong time to be ascribed to precession. The orbital inclination theory fares no better. A 100 kyr oscillation does not naturally account for an abrupt change in sea level.

A possible explanation for the sudden terminations is that they are not astronomically triggered. Rising insolation, driven either by eccentricity change or by reduced clouds (a possibility in the dust model), has limited effect on the glacial ice volume until, finally, the ice can no longer be maintained. As we discussed earlier, a similar phenomena is observed when river ice suddenly cracks in the spring. The water begins to flow, and the increased circulation provides a feedback that eliminates most of the ice within a few days. But the day that this occurs is not the same every year; it has a random component. For a simulation of such terminations, see Section 3.2.5.

8.12 THE "UNEXPLAINED" 70 KYR PEAK

MacDonald (1990) applied the maximum likelihood method of Section 3.7 in an attempt to see if all the frequencies present in the $\delta^{18}O$ data (from the SPECMAP stack) could be accounted for from known astronomical frequencies. The results of his search are shown in Fig. 8.26, adapted from his paper. Each white bar represents an expected frequency from eccentricity, obliquity or precession. The maximum amplitude for each of these is separately normalised to 1. Thus, for example, the peak near 0.0025 (period 400 kyr) has amplitude 1, and the other eccentricity peaks are smaller, but the obliquity peak near 0.024 (41 kyr period) is also normalised to amplitude 1.

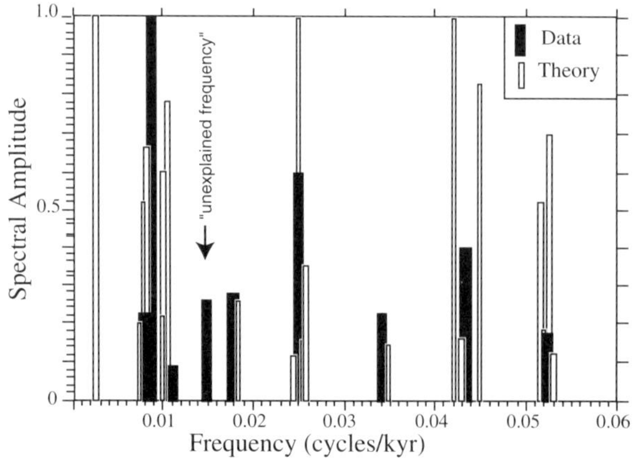

Fig. 8.26. Match between expected peaks and data.

MacDonald determined the significant frequencies in the data by the procedure described in Section 3.7. The first frequency was identified by its maximum power in the Lomb/Scargle periodogram. It was then removed from the data by a least-square fitting procedure, in effect, whitening the remaining data. Then the periodogram was recalculated, and a second frequency detected, and then removed. This procedure was followed until there were no longer any statistically-significant peaks in the final periodogram.

MacDonald found a remarkable overlap between orbital frequencies and the observed $\delta^{18}O$ frequencies, as can be seen in the figure. At that time, MacDonald attributed the spectral power near 0.009 to eccentricity, although we now prefer to identify it with orbital inclination. He concluded that with only one exception, *all* the frequencies matched known orbital frequencies. The one unexplained frequency is evident in Fig. 8.26 as the black bar near 0.0148 cycles/kyr, corresponding to period 68 kyr. He stated that this same frequency also appears in $\delta^{18}O$ data from core V19-30 from the Eastern Pacific. This was the only frequency present that was "not of orbital origin".

We now know that this frequency corresponds to the oscillation of the orbital parameter Ω as described in Section 2.2.6. Ω is the precession of the orbital inclination vector. It is shown as a function of time in Fig. 2.19 and Fig. 2.22. We show the spectra of Ω in Fig. 8.27, analysed in both the ecliptic frame and then in the invariable frame. In both frames, the dominant frequency is at 0.0144 cycles/kyr, corresponding to period 70 kyr. Thus, we have found an orbital parameter for MacDonald's one unexplained frequency.

Note that the 70 kyr peak that appears in spectra should not always be attributed to Ω. As we showed in Section 5.6.2, there is another possible origin for peaks in this region: the mistuning of the 100 kyr cycle to an eccentricity based model. In the example in that section, the overtuning gave rise to a 77 kyr peak. Thus, overtuning was the likely source of the 77 kyr peak seen in the spectrum of the low latitude stack of Bassinot et al. (Fig. 5.43), and of the 77 kyr peak seen in the spectrum of Site 849 (Fig. 5.46).

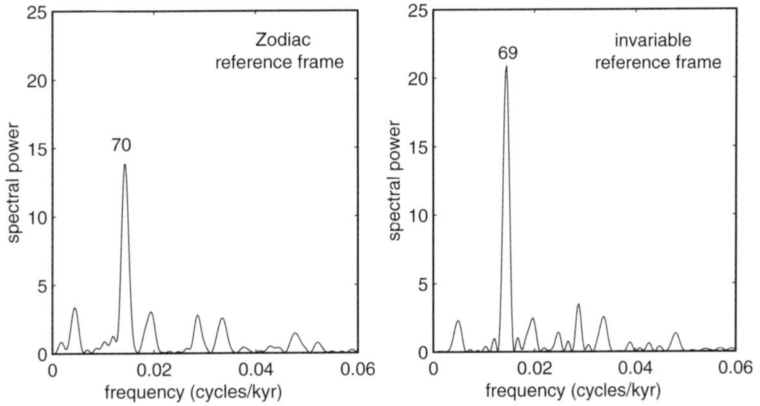

Fig. 8.27. Spectra of Ω.

8.13 VOSTOK

In this data we will review the data from the Vostok ice core, but we will use the minimally-tuned time scale described in Section 5.2.1. This time scale enhances several features in the spectra, in particular, those relating to insolation. However, as we described in that section, we believe that this time scale is more accurate than the untuned scale and therefore the proper one to use.

The spectrum of atmospheric $\delta^{18}O$ is shown in Fig. 8.28, along with the spectrum of July 65N insolation. The similarities are striking, particularly the agreement of the frequencies of the dominant peaks. The ratio of the 23 and 19 kyr peaks are a reasonable match to insolation. The $\delta^{18}O$, however, has a strong narrow peak at 105 kyr, and excess power at 41 kyr.

The spectrum of the deuterium record is shown in Fig. 8.29. It is plausible to interpret the spectrum as representative of temperature. There is a strong narrow peak at 104 and a second strong peak at 41. The peak at 104 is compatible with either an eccentricity interpretation (shown) or an orbital inclination interpretation. However, in the 420 kyr span of the data the absence of a 400 kyr peak (expected from eccentricity) is significant and this argues in favour of inclination as the cause.

In the remainder of the plots in this section, we show the spectrum of Vostok methane, CO_2, dust, and sodium (sea salt). The same peaks recur, but in different ratios. The most straightforward interpretation of these signals is that we are seeing a combination of insolation and orbital inclination. The 100 kyr peak could also be due to eccentricity, if some mechanism is suppressing the 400 kyr cycle. The strength of the 41 kyr peak is somewhat mysterious. Some people might be tempted to identify the 29 kyr peak as one of the "known" secondary peaks of obliquity (see Fig. 2.13) but the amplitude is far too large, and some *ad hoc* mechanism would have to be found to amplify it but not the 41.

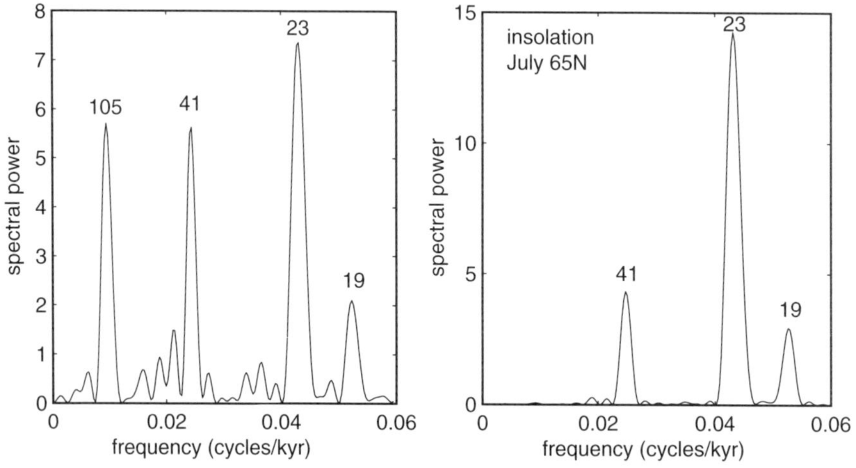

Fig. 8.28. Atmospheric $\delta^{18}O$ from Vostok, and July 65N insolation.

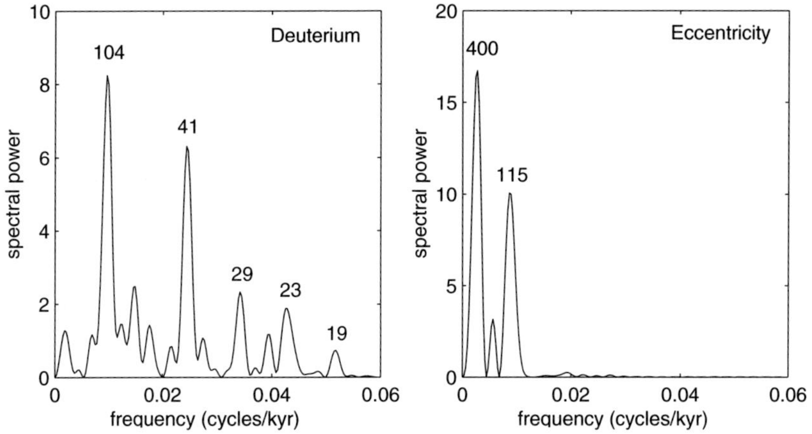

Fig. 8.29. D_2 spectrum from Vostok.

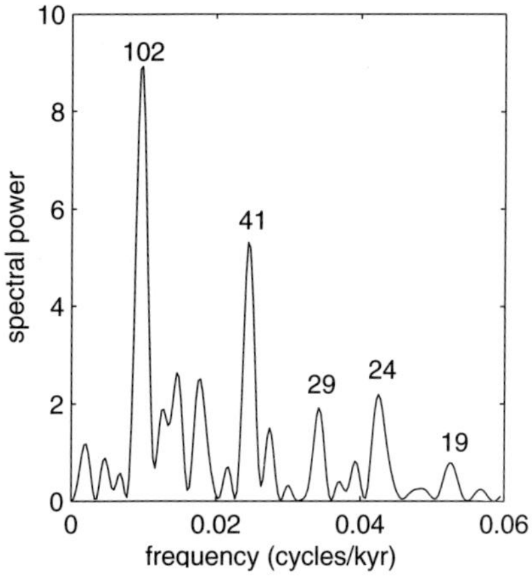

Fig. 8.30. Methane spectrum from Vostok.

8.14 MILLENNIAL SCALE OSCILLATIONS

"The aim of science is not to open the door to infinite wisdom, but to set a limit to infinite error."

Bertolt Brecht, *The Life of Galileo*, 1939

In addition to the prominent 100 kyr cycles, the detailed data available from the Greenland Ice Sheet Project, GISP2 (Grootes et al., 1993), the Greenland Ice Project, GRIP (Johnsen et al., 1992; Dansgaard et al., 1993) and some sea floor cores, have given us data with sufficiently high resolution that we can search for much higher frequency oscillations, with periods in the kiloyear range and shorter. In the

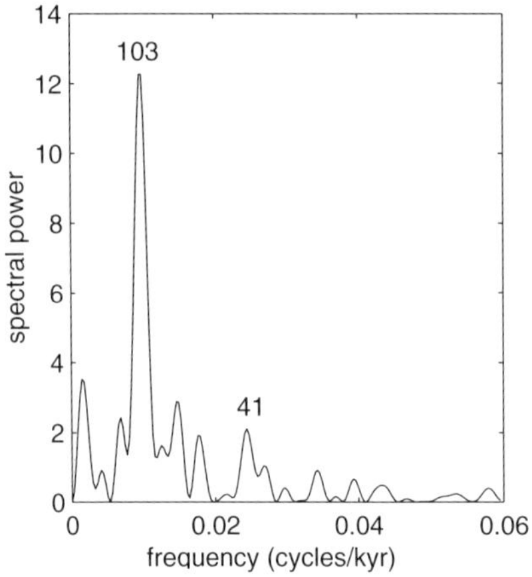

Fig. 8.31. CO_2 spectrum from Vostok.

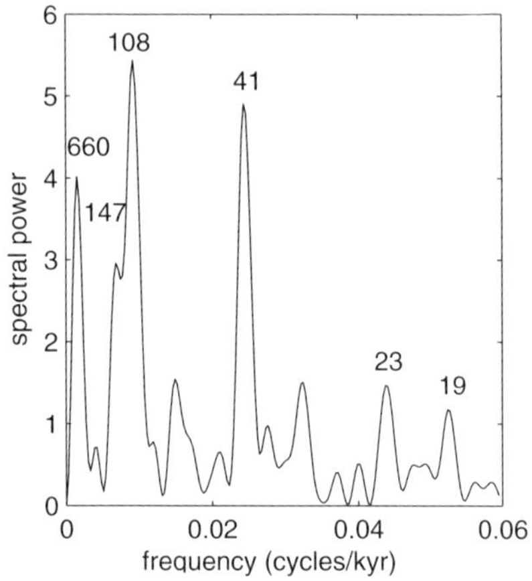

Fig. 8.32. Dust spectrum, Vostok.

literature, these have been referred to as "millennial scale" oscillations. There is certainly climate variability which occurs in time intervals of a thousand years. Variations that take place on this time scale are of particular interest to us because they are not far removed from the climate variations we experience during our lives. The Little Ice Age could be such an example. About the year 1300 we sank into a cold period from which we are only now emerging. An extensive discussion of millennial scale variability can be found in *Mechanisms of Global Climate Change at Millennial Time Scales* (Clark et al., 1999b).

Such variations can always be mathematically expressed as a sum of sines and cosines; that is the meaning of Fourier analysis. But we learn most from Fourier analysis when the resulting spectrum is dominated by a few strong lines. We will show that this is not the case for the millennial scale oscillations. There has been extensive discussion in the literature about the possibility of a 1.5 kyr oscillation in the data. We will show that there is an excess of spectral power in this region, but that the variations it represents is qualitatively different than we have seen in the longer time periods. There are no narrow peaks which can be easily attributed to astronomical causes. Our analysis will follow that of Karner et al. (Karner, 2000c).

We begin with the data which have the best time scale and the most detail: the $\delta^{18}O$ data from the Greenland Ice Sheet Project, GISP2 (Grootes et al., 1993). The $\delta^{18}O$ data for GISP2 are shown in Fig. 1.4. The spectrum for these data covering the last 120 kyr is shown in Fig. 8.34. Note that it is a log plot (to emphasise small peaks) and that the maximum frequency plotted is 2 cycles/kyr. We are looking at a much higher frequency range than when we discuss orbital cycles. The data show a generally red spectrum (peaking near zero), although the structure is not smooth. There is a slight excess of spectral power near the frequency 0.7 cycles/kyr. This excess has been extensively studied and, even though it is barely above the noise, we believe that it is statistically significant. But there are surprises in the data too, and these were missed in most of the early papers published about this excess.

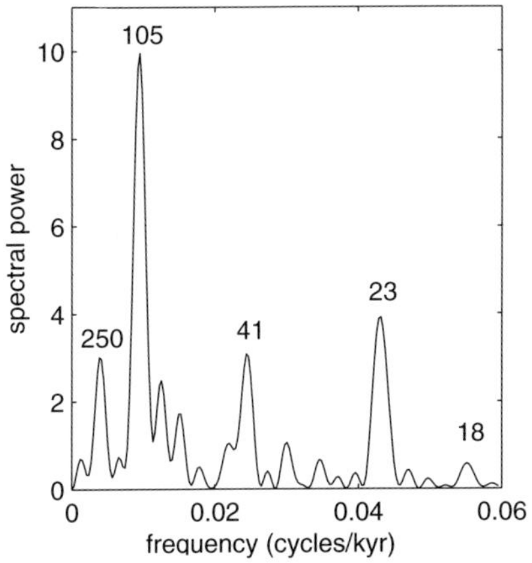

Fig. 8.33. Sodium spectrum, Vostok.

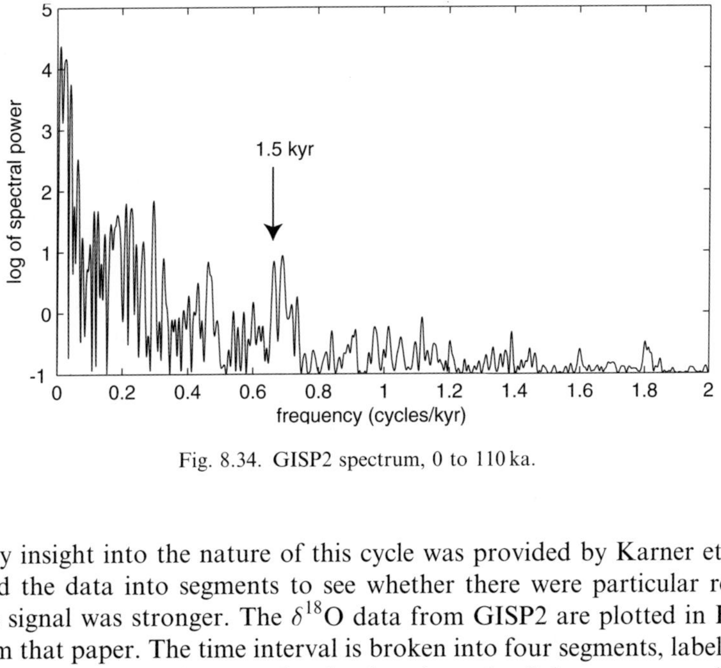

Fig. 8.34. GISP2 spectrum, 0 to 110 ka.

The key insight into the nature of this cycle was provided by Karner et al., who subdivided the data into segments to see whether there were particular regions in which the signal was stronger. The δ^{18}O data from GISP2 are plotted in Fig. 8.35, taken from that paper. The time interval is broken into four segments, labelled A, B, C and D, and the spectral power for the data in each of these segments is plotted below. The arrow in each of these spectra points to the expected location of a 1.5 kyr peak. Note that a significant peak occurs only in the time interval from 26 to 36 kyr. Nowhere else is the spectral power significant.

Karner et al. analysed this issue in another way, by performing a spectral analysis of the data passed through a 10 kyr sliding window. As the window moved, the maximum spectral power in three regions was measured: in the vicinity of period 1.5 kyr, and in windows on either side of that period. The values at the side

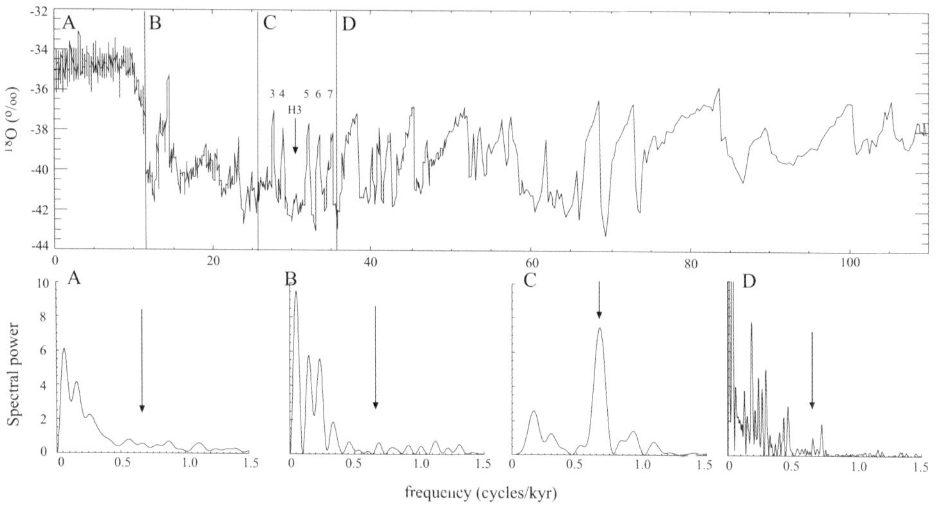

Fig. 8.35. GISP2 spectra for subdivided time intervals.

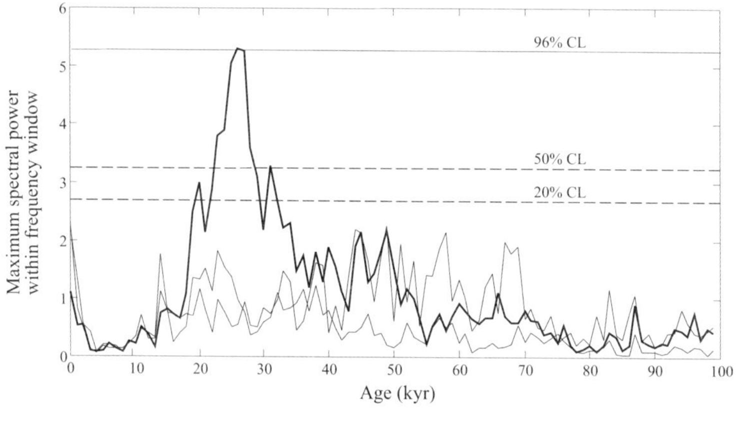

Fig. 8.36. Moving window analysis of GISP2.

frequencies were used to estimate the background levels, i.e. the levels at which spectral fluctuations appeared that were unrelated to the putative 1.5 kyr oscillation. The results are shown in Fig. 8.36. It is clear that there was statistically significant spectral power at the 1.5 kyr line only in the vicinity of the period 26 to 36 kyr. (The peak is slightly broader than this range, since a sliding window begins to pick up oscillations from this period when its centre is within 5 kyr of the region.)

Mayewski et al. (1997) reported a large number of statistically significant peaks in the millennial range occurring in the time interval 0–11 kyr (the Holocene). They used a combination signal called the "Polar Circulation Index". Karner et al. also did a spectral analysis of this interval, and the results are shown in Fig. 8.37. Although there is a "peak" at period 1.5 kyr (indicated by the vertical line), there

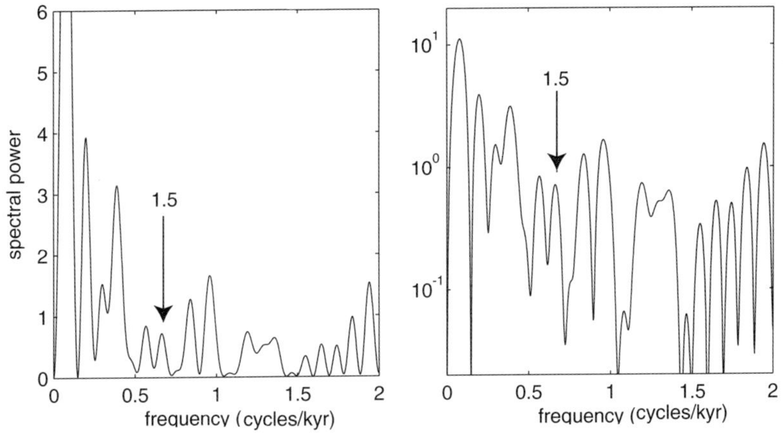

Fig. 8.37. Holocene PCI spectrum (linear and log scales).

are numerous other peaks with similar power in the adjacent frequency bins. If we use the methods explained in Section 3.11, we would estimate the statistical significance of the peak by comparing it with adjacent bins. The conclusion would be that none of the fluctuations in this region is statistically significant. Mayewski et al. regard the other peaks not as a background, but as other significant millennial-scale oscillations. Thus, in their analysis, nearly half of the spectrum shows peaks that they consider significant above the 99% confidence level. Of course, we cannot prove their estimate of background to be wrong. As the example of Penzias and Wilson illustrates,[1] "One man's background is another man's signal".

Nevertheless, it is clear that to the extent that millennial scale oscillations appear in these data, they do not appear as isolated narrow spectral lines, but as fluctuations that are indistinguishable (in a statistical sense) from continuous noise.

As a second example of millennial scale variability, we show a proxy called "hematite stained grains", analysed by Bond et al. (1999). These stains, found in sediment in the North Atlantic (Site 659), are believed to be a measure of ice-rafted debris, and thus a measure of the intensity of glaciation. The percentage of such grains in the sediment for the last 80 kyr is plotted in Fig. 8.38.

The spectrum of these data is shown in Fig. 8.39. An arrow points to the 1.5 kyr period. If we interpret this as a single narrow peak, then it has marginal statistical significance. More importantly, it has very little climate significance: the power under the small peak accounts for only 3% of the variance of the signal. This is not the sort of variation that could account for dominant phenomena, such as the Little Ice Age. However, the power excess in the region spanning 0.4 to 0.7 cycles per kyr (period 1.4 to 2.5 kyr) does rise higher than in the adjacent regions, and may not be red noise.

[1] They won the Nobel Prize in Physics for their discovery of the Cosmic Microwave Background. It was termed *background* because it interfered with radio communications. Dicke and his group immediately recognised that it was the remnant radiation left over from the Big Bang of cosmology.

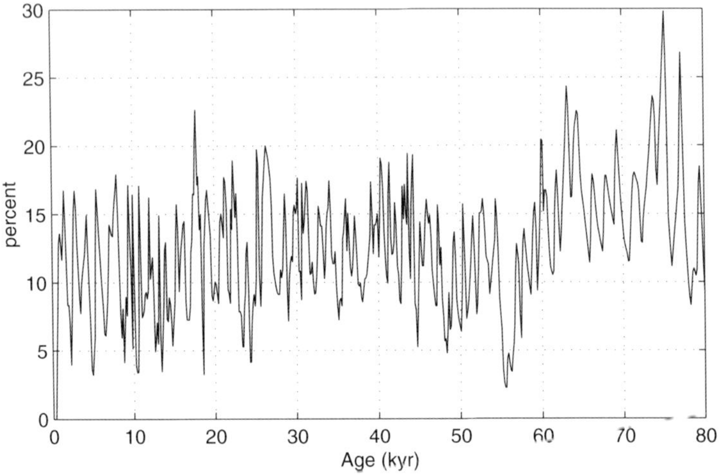

Fig. 8.38. Hematite-stained grains.

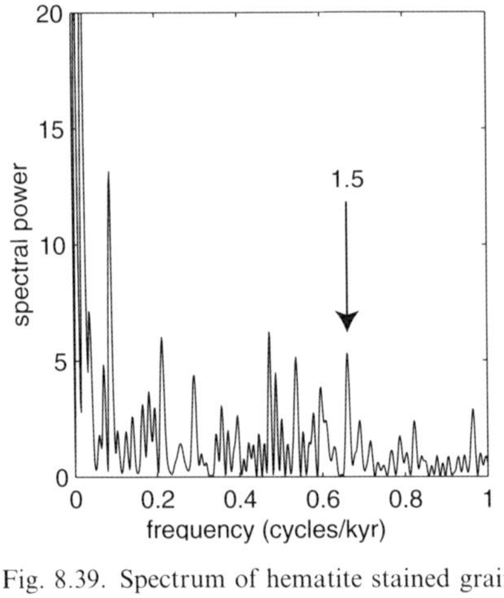

Fig. 8.39. Spectrum of hematite stained grains.

The power in this interval accounts for 16% of the variance. Indeed, the paper by Bond et al. (1999) does not claim that this is a narrow peak. If we include power from adjacent regions, the peak could be a broad one, broken by speckling. A broad peak could result from inaccuracy in the time scale (drift), and the possibility that the oscillation arises not from space but from a geophysical mechanism that we could classify as a relaxation oscillator.

It is certainly *possible* that the millennial scale oscillations are astronomically driven. However, the nature of the spectra is in stark contrast with the spectra we have for the glacials cycles. We argued that the 41 kyr cycle was clearly astronomical, and so was the 100 kyr cycle. The 23 kyr cycle was less evident, but its coincidence in period with the known insolation cycles of the Croll/Milankovitch theory is persuasive. The 41 and 23 kyr cycles are *predicted* by any reasonable facsimile of the insolation theory, and when they appear in the data, the obvious conclusion is that they are related to astronomy. But the situation for the millennial scale oscillations is quite different. No theory predicted an astronomical cycle of 1.5 kyr. We must avoid the temptation of identifying every cycle that we see with astronomy. The fact that astronomy has been so successful for the larger glacial cycles does not mean that it is a paradigm that should necessarily be adopted for all analyses of climate variability.

9

Status of our understanding

We shall not cease from exploration
And the end of all our exploring
Will be to arrive where we started
And know the place for the first time.

T.S. Eliot, *Four Quartets*, 1942

In this chapter we draw the conclusions that we think follow from the data and analysis described in this book. We encourage you, however, to examine the data yourself, to continue to examine new data as it becomes available, and to draw your own conclusions. The first eight chapters of this book stand on their own. They contain the data and techniques that must be understood and applied by anyone who is trying to understand glacial cycles. This chapter, describing our conclusions, is the least important chapter in the book. It should certainly not be among the first chapters that you read.

The glacial cycles are driven by astronomical forcing. This is a consequence of both the match in frequencies and the narrow width of their spectral features, i.e. their phase coherence. Models that attribute the cycles to natural oscillations of the Earth's climate dynamics do not account for these fundamental coincidences. The reasoning that lead to this conclusion is described in Section 2.1.

We do not fully understand what drives the 100 kyr cycle of the ice ages. Many models have been devised to try to explain this cycle, but they all involve enough adjustable parameters to make the interpretation suspect. There is no compelling theory that allows us to do first principle computations. Even the best versions of the Milankovitch theory (the mathematical nonlinear models; see especially Sections 6.2.1 and 6.4.8) simply assume that the global ice is proportional to the driving force, and use adjustable parameters to relate the two. This is not surprising, given the fact that even computations of present-day climate require numerous adjustable parameters to make the theory fit the data. How could we hope to do better with paleoclimate data?

Models that predict phenomena which disagree with the observations either should be ruled out or, at the minimum, should be put on the defensive. It is always possible that an apparently minor modification of a theory will salvage it—but until that modification is found, the theory must be considered untenable. Most of the published models have such serious shortcomings.

Insolation, the canon of Milankovitch (Milankovitch, 1941), certainly affects climate—but it is not clear that it plays any role in driving the 100 kyr glacial cycle. Among the various proxies, the influence of insolation is most clearly seen in the isotopic composition of air entrapped in the Vostok ice, where the major frequencies of obliquity and precession are strong (see especially Fig. 4.37 and Fig. 4.40). Insolation is also seen clearly in the terrigenous component of sea floor cores, where the same frequencies dominate (Fig. 4.22). Small contributions from insolation are present in $\delta^{18}O$ from sea floor cores, but they are inconsistent from site to site. They may be more closely linked to local climate rather than to global ice and, in some cases, they may have been artificially enhanced by overtuning. Insolation appears to be strong in the lake bed depth ranks (Fig. 8.16). All of these proxies point to insolation playing a major role in climate, a result that surprises almost nobody, and which would surely bring a smile to the souls of Croll and Milankovitch.

The dominant proxy for the glacial ice volume, $\delta^{18}O$ from sea floor cores, tells a very different story. More than any other proxy, it shows consistent behaviour in samples taken from many latitudes and longitudes around the world (see Fig. 4.2). The overall behaviour of this proxy, at least in benthic forams from sea floor cores, is widely regarded to be a measure of global ice. The pattern it shows is clear. In the last million years, it is dominated by a strong narrow single 100 kyr peak; see, in particular, Fig. 8.7 and Fig. 8.8. It is most dramatically seen in the $\delta^{18}O$ data from Site 659, shown in Fig. 2.2. It is also seen in other proxies that relate directly to the glacial cycles, including the carbonate isopleths (which respond to sea level, see Fig. 4.26, and the loess, see Fig. 4.29). The next strongest peak in the $\delta^{18}O$ data is the 41 kyr peak of obliquity, and the weakest peaks are those of precession.

The narrow shape of the 100 kyr peak has particular significance. It was first noticed in 1994 (Muller and MacDonald, 1994b). At that time, its presence contradicted every published theory of the 100 kyr cycle (see Table 6.1). Subsequently, this narrow peak has not been easily forced into the procrustean bed of Milankovitch. It is not the 400, 125, 95 kyr triplet predicted by most insolation-based theories. The only insolation-based model that has successfully reproduced it is Berger's memory model (Section 6.4.8) but that model appears to have a fatal causality problem (see Section 8.3). The reason that it is so difficult to make the theory fit the data, is that although the same frequencies are present, the spectra of the data from sea floor $\delta^{18}O$ don't even resemble the spectra of insolation, even after extensive tuning. The precession frequencies, with periods of 24, 22 and 19 kyr, are weak, and in many records apparently absent. This is true not only for the last million years when the 100 kyr cycle was dominant, but in the preceding two million years when obliquity dominated. This weakness would not be so dramatic, if not for the fact that other proxies show such strong precession cycles. At Site 659 (Fig. 8.19), precession is seen strongly in the terrigenous component while simultaneously being weak in global ice proxy $\delta^{18}O$. This shows that the absence is not due to chatter or drift in the time scale. The precession peaks are relatively weak (climatically, not statistically) in most sea floor records of $\delta^{18}O$. They appear to be strongest in equatorial samples, and this suggests that they respond to local climate and not global ice.

In most of the $\delta^{18}O$ data sets, a 41 kyr signal is also present, and this corresponds to the period of obliquity. Between three and one million years ago, the 41 kyr period totally dominates. It is seen even in untuned $\delta^{18}O$ data; an example in an Atlantic core is shown in Fig. 4.34, and in a Pacific core in Fig. 6.7. The narrowness of these peaks implies that the sedimentation rate was remarkably constant, at least at these sites. Yet, the 41 kyr oscillation is not easily explained by the standard insolation theories, linear or nonlinear. The theories which attribute the ice cycles to variations in northern hemisphere summer insolation should all be dominated by precession, not obliquity (see Fig. 2.14). Yet the precession is weak or absent in most sea floor $\delta^{18}O$ records. We know from other proxies (e.g. terrigenous components, atmospheric oxygen) what precession looks like when it is strong, and in the global ice proxies, it isn't. One possibility is that the 41 kyr oscillation is driven by winter insolation, not summer. This has a dominant 41 kyr peak (again, see Fig. 2.14). Another possibility is that the oscillation is driven by a symmetric response to northern plus southern hemisphere insolation. However, this latter model requires an almost perfect cancellation of the precession components, and we consider this unlikely. For lack of a better model, we will accept the 41 kyr cycle is probably insolation driven. (But that doesn't mean that you should accept this.)

In the published literature, the 100 kyr cycle has a standard interpretation: it is due to eccentricity, perhaps indirectly through the envelope of the precession parameter. Many papers even refer to the 100 kyr cycle as the *eccentricity cycle*, as if this origin were an established fact. Eccentricity is also clearly seen in climate proxies, particularly in the coarse foraminifera component of Site 806 (Fig. 8.11), as well as in the Triassic depth ranks (Fig. 8.16). Its presence (and absence, in the $\delta^{18}O$ proxy) in Site 806 is clear in both the spectrum and in the bispectrum (Fig. 8.17), analyses that are independent. The weakness of the precession in these records presents a problem. In nonlinear models, the precession derives its strength from the envelope of precession, but if precession is absent, then where does eccentricity come from? The nonlinear model of W. Berger (Section 6.4.8) addresses this issue, using an hypothesised "memory effect" that tends to wash out low frequency signals, while allowing ice to accumulate. Another possibility is *total* insolation over the entire Earth. The insolation, integrated over an entire year, does not depend on obliquity or precession, but only on eccentricity. But this theory is farfetched, since the variations of total insolation due to eccentricity are tiny, only a part in a thousand. The dominance of eccentricity in the coarse component of Site 806 is just another mystery to be solved. But at least it shows what eccentricity should look like, when it does show up.

The attribution of the 100 kyr cycle to eccentricity, either directly or through nonlinear models, has serious problems. These were discussed in Chapter 8. It is worthwhile reiterating these problems. They include the following:

1. *The Stage-11 problem.* At an age of about 400 kyr, we have one of the greatest terminations that has ever taken place in glacial records. And yet the eccentricity is undergoing only very small oscillations at this time.
2. *Causality problem.* Accurately dated climate series from Devils Hole in Nevada show a strong termination that is virtually complete by 135 kyr. Data from coral reefs (Section 8.3) substantiate this date. Yet insolation at that time had not yet risen significantly. The termination precedes the rising insolation; the cause appears to follow the effect. This is an extremely serious problem for any theory that attributes the terminations to precession cycles.
3. *Unsplit peak problem.* The insolation-based theories predict that the power in the 100 kyr vicinity should be split into a 95/125 kyr doublet. Rather than a doublet, the $\delta^{18}O$ global ice proxy data show a single narrow peak near 100 kyr. Moreover, this narrow peak is evident in $\delta^{18}O$ from Site 806, a record that does show the split peak appearing simultaneously in a different proxy. This simultaneous presence indicates that the split peak is not missed because of tuning or analysis error.
4. *400 kyr peak problem.* Eccentricity should have a 400 kyr cycle in addition to the 100 kyr power. The absence of this cycle has been attributed to *ad hoc* mechanisms. Yet in those proxies in which eccentricity is clearly seen (based on other criteria), such as the Site 806 coarse component, and the lake bed depth ranks, the 400 kyr cycle is strong and clearly seen.
5. *Stage-1 problem.* This is completely analogous to the Stage-11 problem, but it happened just 12,000 years ago, and so it is particularly well documented. Insolation underwent only very minor oscillation, and yet the glacial oscillation was one of the strongest ones in the record.
6. *Bispectrum problem.* The bispectrum of $\delta^{18}O$ does not match the bispectrum of eccentricity. This problem is made more evident by the fact that in one of the cores for which this is observed (Site 806), the bispectrum of the coarse component does match that of eccentricity.

These problems are not completely independent. The split peak in eccentricity gives rise to a cancellation in the time domain that is responsible for the Stage-1 and Stage-11 problems. Nevertheless, they need to be treated as separate issues, since some records show the problems in one form and not the other. For example, climate records that span only the last 100 kyr (such as the GISP2 ice core) can clearly reflect the Stage-1 problem, even though its time duration is not long enough to be able to evaluate whether the 100 kyr power is split or single. Even if we were to take the point of view that no one of these problems by itself should be considered as fatal to the insolation theory, the combination is serious enough that we should be searching hard for an alternative explanation.

As an alternative, we have described in some detail (Chapter 7) a model in which the 100 kyr cycle is driven by changes in orbital inclination. The inclination does not exert its effect through insolation, but rather through the effect it has in modulating the accretion of interplanetary dust. The dust, in turn, can affect climate through its effect on cloud cover and ozone. Even with the simplifying approximation that the response is linear, and only two parameters (an amplitude and a delay), the fit to the data is excellent.

The inclination model has none of the six problems listed above for insolation. In particular:

1. *No Stage-11 problem.* The orbital inclination is an excellent fit to the data, going back as far as 900 kyr.
2. *No causality problem.* The observed $\delta^{18}O$ signal lags the orbital inclination by 33 kyr, a reasonable delay (a little more than 1/4 cycle; see Section 4.4).
3. *No split peak problem.* The spectrum of orbital inclination is an excellent match to the spectra of the data.
4. *No 400 kyr peak is predicted,* in agreement with observation.
5. *No Stage-1 problem.*
6. *No bispectrum problem.* The bispectrum of the data has all the frequencies present in the bispectrum of the theory.

In addition, the orbital inclination theory has several independent "successes". The first of these was a published prediction that was subsequently verified. They are:

1. *Interplanetary dust cycles.* The inclination theory predicted that measurements of accreted dust would show a 100 kyr modulation (Muller and MacDonald, 1995). This modulation was subsequently discovered in both Atlantic and Pacific Ocean sites; see Section 8.9.
2. *41/100 kyr transition.* The inclination theory relates the 100 kyr cycle (but not the 41 kyr cycle) to the presence of interplanetary dust. Such dust has been observed to undergo a major increase at about 1 million years ago, a result that was not known when the theory was first published. This increase in dust (probably due to a collision in the asteroid belt), naturally accounts for the sudden appearance of the 100 kyr cycle at that time.
3. Noctilucent clouds. The cloud formation peaks during the passage through the invariable plane, the plane of symmetry of orbital inclination.

The inclination theory has two deficiencies. It does not account for the presence of the 41 and 23 kyr peaks, so we must add a role for insolation in local climate. Secondly, it does not readily account for the sudden terminations. It postulates they are due to instabilities, but this is an *ad hoc* reason.

The literature is full of *ad hoc* arguments. If Occam's razor makes any sense for science, it should lead us to distrust theories which have many such explanations and numerous adjustable parameters. New theories, when published, should be suspect unless they list (and perhaps even number) their *ad hoc* assumption and parameter lists. We believe that the orbital inclination theory ranks near the top of the list, in that it has fewer such assumptions than virtually any other theory.

Several papers (see, for example, Rial, 1999 and Berger, 1999) argue that the Milankovitch theory can be patched up, thereby rendering the orbital inclination theory "unnecessary". But the fact that the insolation model came first does not give it a higher priority or scientific status. There is no "grandfather clause" that gives older theories the right to avoid having their spelling checked. Models must be evaluated on their merits, not on their seniority.

It seems clear that climate is complex, but potentially comprehensible. The presence of narrow spectral lines means that the cycles of glaciation are not driven completely chaotic by butterflies flapping their wings. The clues tell us that there are a few simple phenomena which dominate. Many proxies give clean signals that match known astronomical phenomena. Insolation dominates the Vostok atmospheric $\delta^{18}O$, but is only a weak contributor to glacial ice. Terrestrial dust, both in Vostok ice records and in the sea floor terrigenous component, appears to have an eccentricity cycle. Yet the wind-blown dust that led to the formation of the loess shows a single narrow spectral peak, reminiscent of orbital inclination. Glacial ice volume is an excellent match to orbital inclination. There are many other clues. What are they telling us? We have the feeling that they can be untangled, if only we could look at them in the right way.

At this point, we return to a question that we raised in the Preface. Suppose we were to look at the data with fresh eyes, as if for the first time. Which model would we consider the most capable of explaining the glacial cycles? The answer is important, because if the dust model is correct, then major phenomena are taking place in the atmosphere that are not being included in current global climate models. If we refuse to learn from history, we may be doomed to relive it.

Appendix 1

Bispectral analysis

In this section we give some technical background and examples of the bispectrum. The bispectrum helps to identify nonlinear processes in which the oscillations at two frequencies generate a third frequency. Bispectral analysis consists of the search for three simultaneous signals, in three different frequency bands, which obey the equation $f_1 + f_2 = f_3$.

Bispectra can behave as a fingerprint; a process that contains a bispectral pattern will typically imprint that same pattern on processes that it drives. Thus, when we compare bispectra of $\delta^{18}O$ to those of eccentricity or orbital inclination, we are looking to see which is responsible. Bispectra are often sensitive to the presence of frequencies that are otherwise difficult to detect, so they yield information that can be statistically independent from the results of the spectra.

The first empirical bispectrum was calculated by Hasselman et al. (1963) using ocean height data. They observed a strong interaction of a dominant spectral component with itself (harmonic generation) and appreciable interaction with other spectral components. These interactions resulted in the peaking of the crests of the shallow water wave as the ocean waves moved onshore. Since this example was published, hundreds of applications have been described in the literature in such varied fields as turbulent fluid flow, electrical brain activity, plasma interactions, dynamical astronomy and astrophysics (Tyron, 1981; Nikias and Petropulu, 1993).

We begin by defining the bispectrum and the steps involved in calculating the bispectrum from a digital time series. We consider a zero mean time series $x(t)$ that may be represented by its Fourier components X_k

$$x(t) = \sum_{k=-\infty}^{\infty} X_k e^{-2\pi i f_k t}$$

The power spectrum $P(k)$ and the bispectrum $B(k, l)$ are then defined by

$$P(k) = E[X_k X_k^*]$$

$$B(k, l) = E[X_k X_l X_{k+l}^*]$$

The operator $E[\ ]$ is the expectation value or averaging operator. X_k^* is the complex conjugate of X_k, $X_k^* = X_{-k}$.

The power spectrum describes the contribution to the variance of $x(t)$, $(E[x^2(t)])$ from the spectral component at f_k of width $\Delta f (= 1/T)$ where T is the length of the record. The bispectrum represents the contribution to the skewness $(E[x^3(t)])$ of those spectral components at f_k and f_l of area $(\Delta f)^2$. With this interpretation we have

$$E[x^2(t)] = \sum_k P_k$$

$$E[x^3(t)] = \sum_{k,l} B(k, l)$$

The definition of the bispectrum clearly shows that the bispectrum $B(k, l)$ measures the statistical dependence between three frequency bands centred at f_k, f_l, and f_{k+l}. If there is energy at these bands, as there would be in any noisy record, then each band is characterised by an amplitude and a phase. If the phases are statistically independent, the sum or difference phases will be randomly distributed over $(-\pi, \pi)$. When a statistical averaging, denoted by the operator $E[\ ]$, is carried out, the bispectrum will vanish as a result of random phase mixing. Alternatively, if the three frequency bands are coupled to each other, the total phase will not be random at all, although the phase of each band may be randomly changing. This situation is one of the phase coherence among the three bands. In this case, the statistical averaging will not lead to a vanishing bispectrum. Phase coherence among various bands in the spectrum of a time series is evidence that the series was produced by a nonlinear process (MacDonald, 1989).

A simple mechanical example illustrates the concept of the bispectrum. Consider a machine in which three wheels are interconnected through new teeth gears which initially operate without chatter. Suppose that their angular frequencies are ω_1, ω_2, and $\omega_1 + \omega_2$. The machine will radiate acoustic energy at the corresponding frequencies f_1, f_2, and $f_1 + f_2$. A power spectrum of the acoustic field radiated by the machine would show peaks at f_1, f_2, and $f_1 + f_2$ and possibly at their harmonics, $2f_1$, $2f_2$, etc., with a continuum of background noise. The bispectrum of the acoustic signal would have a peak at (f_1, f_2) and, if the harmonics are generated, at (f_1, f_1), (f_2, f_2), etc. As the gears wear, perfect phase lock is lost, and there is a certain chatter as the gears slip randomly. Frequency coupling is maintained, with the power spectrum showing peaks at f_1, f_2, and $f_1 + f_2$, but phase coupling is lost. As phase coupling is lost, the bispectral peak diminishes in amplitude. Indeed, gear wear in machinery is operationally monitored by continuously calculating the instantaneous bispectrum (Sato et al., 1976) and noting the amplitude of the bispectrum.

We note from its definition that the bispectrum is highly symmetric:

$$B(k, l) = B(l, k) = B^*(-k, -l) = B(-k - l, l) = B(k, -k - l)$$

Further, the bispectrum detects only the quadratic (symmetric) part of a nonlinearity. If a linear signal $x(t)$ passes through a cubic device, $y(t) = x^3(t)$, then the bispectrum of $y(t)$ vanishes.

COMPUTATION OF THE BISPECTRUM

Numerous schemes have been advanced over the years to estimate the complex bispectrum of a time series. Hasselman et al. (1963), in the pre-FFT days, used narrowband pass filters to estimate X_k, X_l, and X_{k+l}. Alternatively, in analogy with Blackman–Tukey, the triple correlation function r_3 (τ_1, τ_2) is calculated

$$r_3(\tau_1, \tau_2) = \frac{1}{N} \sum_{k=J}^{J+N-1} x(k)x(k+\tau_1)x(k+\tau_2)$$

and the double Fourier transform is then taken with respect to τ_1 and τ_2 (Lohmann and Wirnitzer, 1984).

Our procedure follows current practice (Nikias and Petropulu, 1993). We assume that the sampling interval Δt is sufficiently small that the Nyquist frequency $f_N = 1/(2\Delta t)$ is larger than any spectral component of interest present in $x(t)$. We further assume that the record length, $T = N\Delta t$, is large enough to assure adequate frequency resolution. If frequency bands f_k and f_l contain significant power, then we desire that the elementary bandwidth $\Delta f = 1/T$ be much smaller than $|f_k - f_l|$. This consideration determines how we carry out the phase averaging. One method would break the record up into M separate records and then average the computed bispectrum over the M records. This procedure reduces the effective resolution by a factor of M. Instead, from the full record we compute a moving average of the nearest neighbour estimates of the bispectra in the (f_1, f_2) plane, making use of the full symmetry of the bispectrum. This procedure smudges the resulting peaks in the bispectrum but retains the full resolution for estimates of individual points of the bispectrum.

In summary, the computational procedure involves:

1. subtracting the mean value for the record;
2. windowing with a Hanning window to reduce leakage (Harris, 1978);
3. computing the Fourier amplitudes and phases using the FFT techniques; and
4. estimating the bispectrum by averaging in the (f_1, f_2) plane.

A Matlab routine for calculating the bispectrum is given in Appendix 2.

Appendix 2

Matlab programs

In this Appendix we include several simple programs that facilitate spectral analysis along the lines recommended in this book. We use the program Matlab, which is widely available for many platforms, and which is particularly simple to program and read because of its ability to handle vectors. This is not the place to give a general tutorial on Matlab. However, some of the functions in Matlab are very powerful and very well matched to paleoclimate work, so it is useful to show how they can be used. The programs were written in Matlab 5.2.

NOTE ON VECTOR AND MATRIX PROGRAMMING

A vector, in Matlab, is a set of numbers represented by one symbol. For example, the vector `t = [1:1000]` represents all the numbers between 1 and 1000. We can operate on a vector with a function to get a new vector. The Matlab equation `y = sin(t)` creates a new vector y which has 1000 values, the sine of each number from 1 to 1000. The use of vectors allows Matlab to run very quickly. In addition, once you get used to it, the vector notation results in code that is remarkably clear and easy to read. It does not have as many "for loops" or "do loops" cluttering up the program.

Matlab also allows the use of matrices in place of vectors. So, for example, if f is a vector of frequencies, then `sin(6.28*f'*t)` is a matrix of numbers containing the sine for every combination of frequency and time. Technical note: the f in the equation was followed by a prime symbol; this is necessary to turn f into a column vector so that the matrix multiplication would be rectangular. In the code that follows, we do not make full use of this matrix feature, for two reasons. First, we found that it does not make the code easier to understand; just the opposite. Second, full use of matrices does not make the code run faster; again, just the opposite! This has to do with the way matrix operations are compiled.

READING DATA

Suppose that we have a text document "isotope_meas.dat" with two columns, separated by spaces. The first column contains depths of the samples, and the second column contains the isotope measurements h. The document looks like this:

```
%   depth    isotope value
    0        14.7
    1.2      19.2
    1.7      15.2
    2.5      27.1
    3.1      10.9
```

The % symbol prior to the words "depth" and "isotope value" makes Matlab skip these lines when it reads the file. The file can be read by the following statements:

```
load isotope_meas.dat
d = isotope_meas(:,1);
h = isotope_meas(:,2);
```

These statements put the values into vectors d and h that look like this:

```
d = [0      1.2   1.7   2.5   3.1 ]'
h = [14.7 19.2  15.2  27.1  10.9 ]'
```

SIMULATED DATA

We give a short code that will generate a set of depths and simulated data. We will use the results of this code to illustrate the other examples given in this Appendix. The code will generate a signal o18 that oscillates with two periods 95 and 125 kyr, and add a random background signal. We will also add a small random number to each age point to simulate the fact that the data are not equally spaced in time.

```
% create simulated Monte Carlo data
% create evenly spaced age points between 0 and 600 kyr
      age = [0:600]';
% sets random number generator to fixed starting point
      rand('state',0);
% add a small random number to each age (except the end points)
      ager = age + 0.3* rand(size(age))-.15;
            ager(1) = age(1); ager(601) = age(601);
      depth = age/10;   % creates depth between 0 and 60
% generate random background
bkg = interp1([0:10:600],rand(61,1),ager);
% create signal at frequencies f1 and f2
      f1 = 1/95; f2 = 1/125;
```

```
      sig = cos(2*pi*f1*ager) + cos(2*pi*f2*ager +pi);
% add the signals together, putting the noise at 10%
      o18 = sig + 0.1*bkg;
% plot the results
      figure; plot(ager,o18);
```

The resulting plot is shown below:

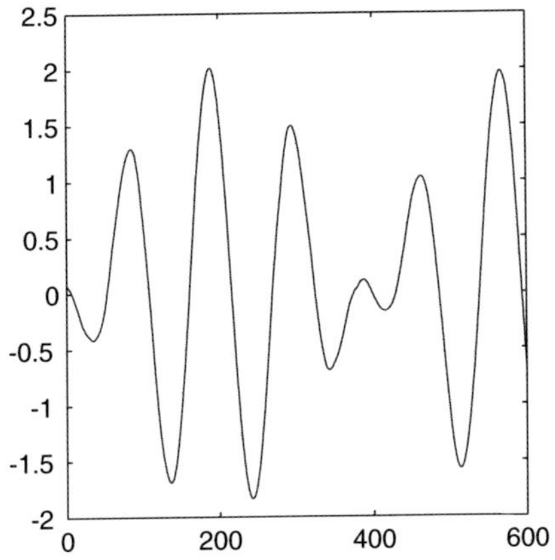

ADJUSTING TO AGE TIE POINTS

Suppose that, as a result of tuning or radiometric measurements, we know that certain depths D have ages T, where

```
D = [0 15 25 35 45 60]';
T = [0 20 27 32 40 60]';
```

Let us use the same vector "depth" that was generated by the previous program. (We assume it is still stored in the computer memory.) Then the ages for all the data points (with a linear interpolation) can be calculated by the single Matlab statement:

```
newage = interp1(D,T,depth);
```

This gives a vector "new age" with the interpolated values of time for each data point. Alternative interpolation procedures could be used by substituting the statements:

```
newage = interp1(D,T,depth,'spline');
newage = interp1(D,T,depth,'cubic');
```

The new data can now be plotted by the single statement:

```
plot(newage, o18);
```

Here is a program that illustrates the procedure. First run the simulated data code. Then run this:

```
D = [0 15 25 35 45 60]';
T = [0 20 27 32 40 60]';
newage = interp1(D,T,depth);
plot(newage, o18);
```

The resulting plot is shown below.

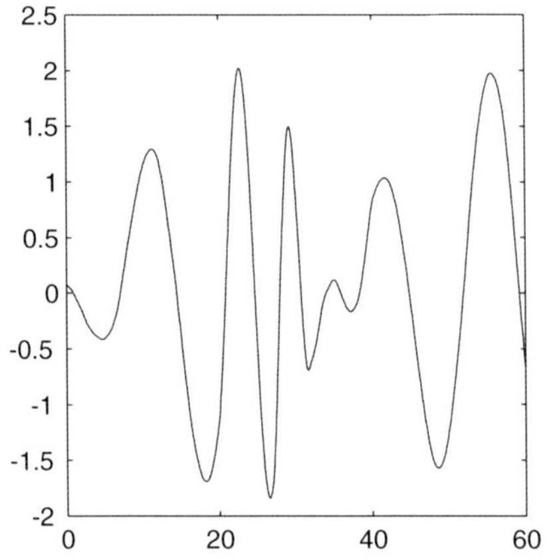

Obviously it is distorted by the fact that we have "adjusted" a time scale that was already correct, to new tie points that were not correct (we just made them up). This is the sort of distortion that you get if you overtune a time scale.

INTERPOLATING TO EVENLY-SPACED TIME POINTS

The ages "ager" are not uniformly spaced. Suppose that we need evenly spaced ages, to facilitate making a Fast Fourier Transform (FFT). We begin by creating a set of evenly spaced ages by picking the maximum and minimum ages, the number of points we want, and then using the Matlab program linspace. Then we use the function interp1 to interpolate the values of o18 to the new ages. First run the simulated data code. Then append the following statements:

```
% generate 512 evenly-spaced data points
      amin = 0; amax = 600; n = 512;
      a = linspace(amin,amax,n);
% interpolate the o18 data to the new ages
      o18_interp = interp1(ager,o18,a);
% plot the results
      plot(a, o18_interp);
```

The resulting plot is shown below.

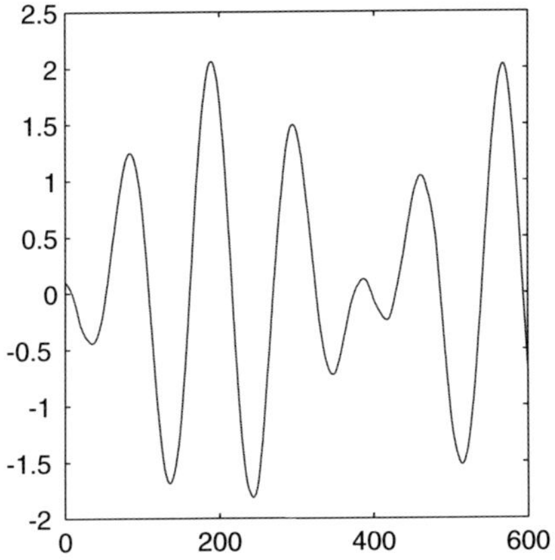

FAST FOURIER TRANSFORM

To calculate the Fourier transform, first interpolate the data to uniform spacing in age, as described above. Then subtract the average of the data:

```
o = o18_interp - mean(o18_interp);
```

Alternatively, you can remove both the mean and the trend (average slope) by the statement:

```
o = detrend(o18_interp);
```

For zero padding, pick a power of two that is about ten times larger than your data set. So, for example, if you have 500 data points, then we will pad the result to npt $= 2^{13} = 8192$. This is done automatically in Matlab with the call to FFT:

```
npt = 8192;
amp = FFT(o, npt);
```

(For most paleoclimate data, npt $= 8192$ will work; if your data array is much larger, or if you prefer a smoother plot, consider using npt $= 2^{14} = 16384$ or npt $= 2^{15} = 32768$.)

The variable `amp` consists of the complex Fourier amplitudes, which contain both real amplitude and phase. To calculate the Fourier power, use:

```
P = amp .*conj(amp);
```

Don't forget to put the period before the multiply sign! This period tells Matlab to do the multiplication component by component. The phase can be calculated by taking the arctangent of the imaginary over the real parts:

```
phase = angle(amp);
```

Finally, we need to calculate the frequencies corresponding to these power levels. The frequency resolution is determined by the Nyquist frequency. The first npt/2 elements contain the power for the frequencies from f = 0 to the Nyquist frequency, given by Nyquist = $1/(2\Delta t)$, where Δt is the time between adjacent points. (f = 0 represents the DC level, and this should be close to zero since we subtracted the mean.) We can calculate the frequencies from the Matlab statements:

```
Nyquist = 0.5/(a(2)-a(1));
f = linspace(0,Nyquist, npt/2);
```

The power spectrum should be normalised. Our usual convention is to normalise it to a mean of one unit. This can be done by the statement,

```
P = P/mean(P(1:npt/2));
```

You can now look at your spectrum by plotting power vs. frequency. For paleoclimate data, much of the interesting power is in the lower part of the spectrum. So, for example, suppose you are working with 1 Myr of data, initially interpolated to 1 kyr spacing. Suppose the maximum frequency you wish to plot is fmax. Typically, fmax might be 0.06 cycles/kyr. The number of points you want to plot is given by:

```
num = (npt/2)*fmax/Nyquist;
```

This should be rounded to make it an integer:

```
num = round(num);
```

And you can now make the power spectrum plot:

```
plot(f(1:num),P(1:num));
```

We put all this together in the following example. First run the simulated data code. Then the interpolating to evenly space time points code. Then run the FFT code. We do that all in the following script. (Most of the comment statements have been omitted.)

```
% simulated data code
      age = [0:600]';
      rand('state',0);
      ager = age + 0.3* rand(size(age))-.15;
            ager(1) = age(1); ager(601) = age(601);
            depth = age/10;   % creates depth between 0 and 60
      bkg = interp1([0:10:600],rand(61,1),ager);
            f1 = 1/95; f2 = 1/125;
            sig = cos(2*pi*f1*ager) + cos(2*pi*f2*ager +pi);
            o18 = sig + 0.1*bkg;
% interpolating to evenly-spaced data code
      amin = 0; amax = 600; n = 512;
      a = linspace(amin,amax,n);
      o18_interp = interp1(ager,o18,a);
```

```
% FFT code
          o = o18_interp - mean(o18_interp);
          npt = 8192;
     amp = fft(o, npt);
     P = amp .*conj(amp);
     Nyquist = 0.5/(a(2)-a(1));
     f = linspace(0,Nyquist, npt/2);
     P = P/mean(P(1:npt/2));
     figure; plot(f(1:200),P(1:200))
     xlabel('frequency (cycles/kyr)');
     ylabel('spectral power');
```

The resulting plot is shown below:

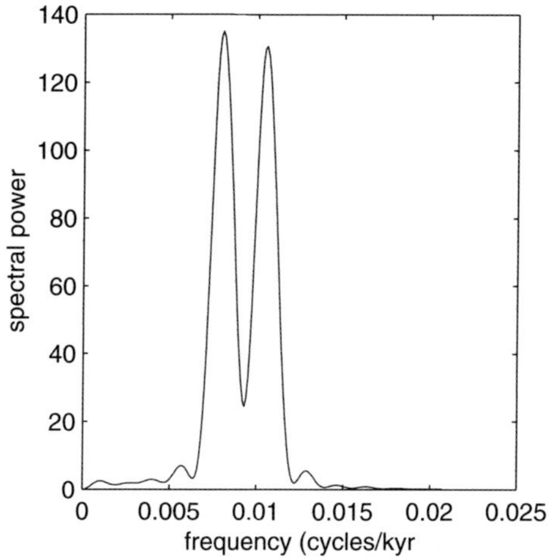

TAPER

To apply a taper to the data, you simply multiply the data by the taper function prior to taking the FFT. For example, the Hanning taper (as described in the main text) is given by

```
g = 0.5*(1 - cos(2*pi*a/amax));
```

To apply the taper, just multiply the data, point by point, by it:

```
o = g.*o;
```

Adding these two lines to the example program yields the following plot of spectral power:

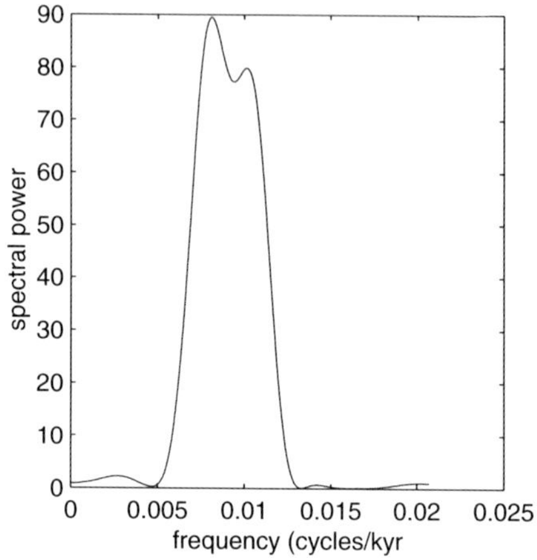

The sidelobes have been reduced, but at a cost: the resolution has been reduced to the point that the two peaks are no longer resolved (at least by the standard Rayleigh criterion). To get a better idea of the sidelobe reduction, we save o as o1 before we apply the taper, and then plot both functions together using the Matlab statement

```
semilogy(f,o,f,o1);
```

The results of this are shown below. Your results may differ, if your random number generator gives a different background.

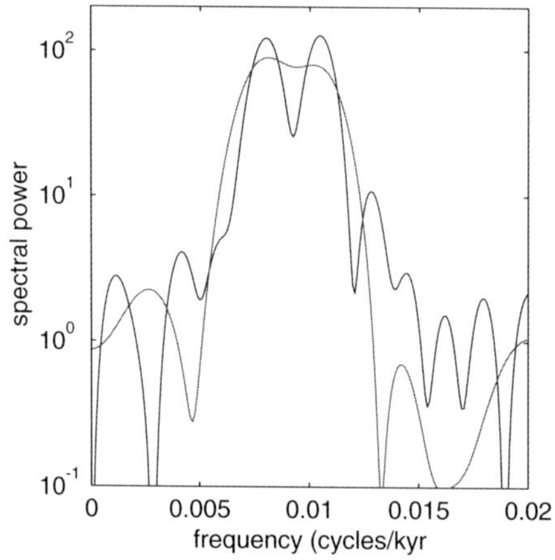

BACKGROUND SIMULATION

There are many ways to create a "random" background. Each will have advantages and disadvantages. We suggest that you try some of the methods described below, and adjust the parameters, until you find one that produces background that "looks similar" to the real background in both the time domain and in the frequency domain. Then use that one for your Monte Carlo simulations.

RANDOM POINTS WITH INTERPOLATION

This routine generates random values between 0 and 1, places them at times spaced every 10 kyr, and then does a linear interpolation for points in between. This is the code we have used in the previous examples of this appendix.

```
% random numbers every 10 kyr
        t = [0:1000]';
        t2 = [0:10:1000]'; rand('state',0); % sets random
                                            %     generator
        r = rand(size(t2));
        C = interp1(t2,r,t); figure; plot(t,C);
```

This code results in the following figure:

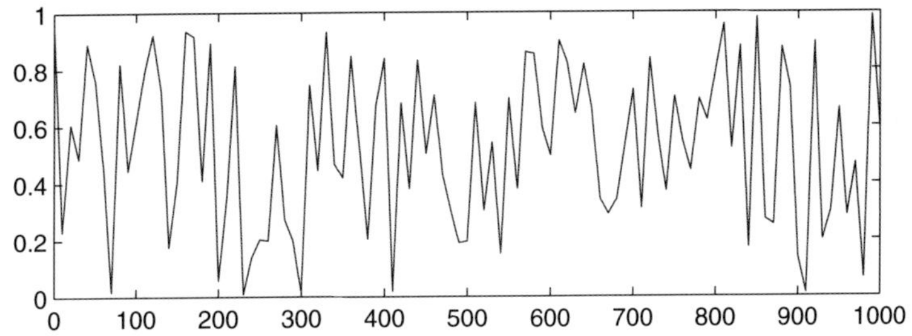

For a less jagged data set, the final interpolation can be done with a spline instead of linear interpolation, by substituting for the last line the statement:

```
C = interp1(t2, r, t, 'spline');
```

Using the same random points with the spline function gives the less jagged plot below:

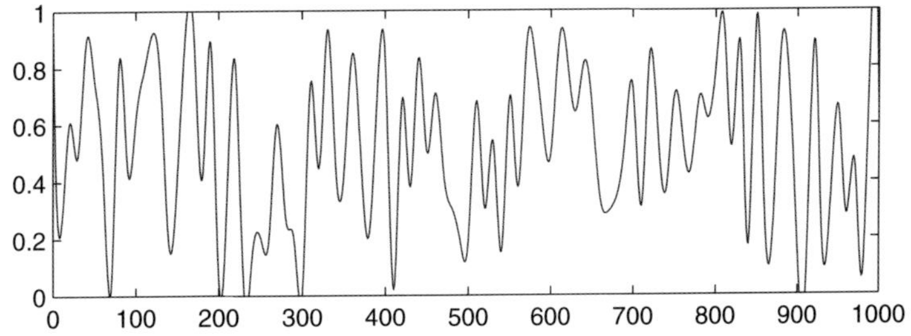

RANDOM POINTS AVERAGED

We create random numbers, and then perform a running average to obtain the simulated data. We do the running average by performing a convolution with a constant function of length SM. The plot generated is shown after the code.

```
% create a random set
      age = [0:1000]';
      rand('state',0);
      r = rand(size(age));
% smoothes the vector A by averaging nearest SM point
      SM = 15; % averages over 15 adjacent points
      B = ones(SM,1); % B is the smoothing function.
% smooth by performing the convolution
      otemp = conv(r,B);
% Vector otemp longer than age, so shorten it.
      k = fix((length(B)-1)/2);
      o = ctemp(k:k+length(A)-1);
      figure; plot(age,o);
```

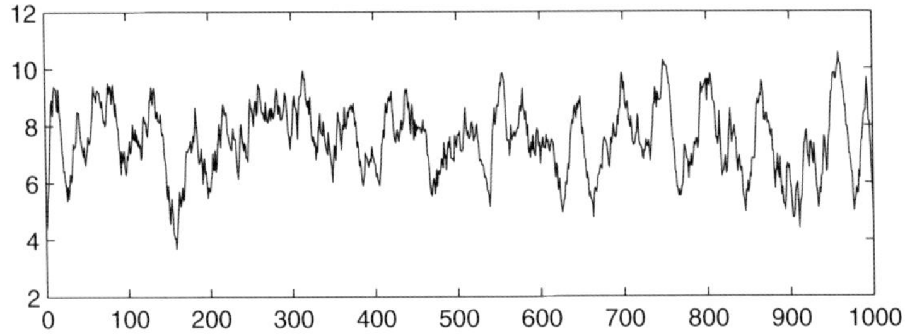

RANDOM POINTS CONVOLVED WITH GAUSSIAN

Instead of using a flat function to perform the smoothing, the convolution can be done with a Gaussian. This means that the running average emphasises nearby points more than distant ones.

```
age = [0:1000]';
      rand('state',0);
      r = rand(size(age));
% pick the RMS of the Gaussian function for smoothing, in bins
      sigma = 3;
% create Gaussian over range of 5 sigma
      B = exp(-([0:5*sigma]-2.5*sigma).^2/(2*sigma.^2));
% smooth and shorten smoothed vector C to same length as A
      otemp = conv(B,r); % performs convolution.
```

```
k = fix((length(B)-1)/2);
o = otemp(k:k+length(A)-1
figure; plot(age,o)
```

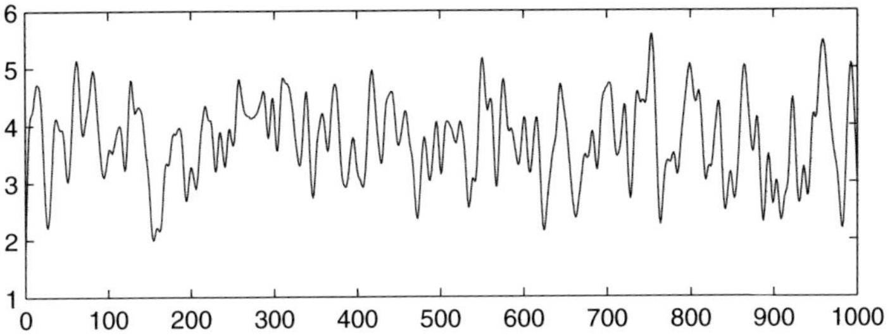

LOMB/SCARGLE PERIODOGRAM

The following function calculates the Lomb/Scargle periodogram. Input to the function is a vector y containing the data points (e.g. δ^{18}O values), a vector t containing the age of each data point, and a vector freq containing the frequencies at which the periodogram is to be calculated. The function should be written and saved as a separate file titled "lomb.m". Below is the code for lomb.m, and a separate program "test_lomb.m" that can be used to test it. To illustrate the fact that the Lomb periodgram does not require evenly-spaced data, it uses the unevenly spaced data points

```
% program test_lomb.m
      age = [0:600]';
      rand('state',0);
      ager = age + 0.3* rand(size(age))-.15;
      ager(1) = age(1); ager(601) = age(601);
      depth = age/10; % creates depth between 0 and 60
      bkg = interp1([0:10:600],rand(61,1),ager);
            f1 = 1/95; f2 = 1/125;
            sig = cos(2*pi*f1*ager) + cos(2*pi*f2*ager+pi);
            o18 = sig + bkg;
      % pick frequencies to evaluate spectral power
            freq = [0:0.0001:0.02]';
            power = lomb(age,o18,freq);
            power(1) = 0;
      % normalize to average 1
            power = power/std(power);
      % plot the results
            figure; plot(freq,power)
            xlabel('frequency (cycles/kyr)');
            ylabel('spectral power');
```

```
function power = lomb(t,y,freq)
      % this program should be saved with the name lomb.m
      % Calculates Lomb periodogram.
      % set constants
            nfreq = length(freq);
            fmax = freq(nfreq);
            fmin = freq(1);
            power = zeros(nfreq,1);
            f4pi = freq*4.*pi; pi2 = pi*2.;
            n = length(y);
            cosarg = zeros(n,1);
            sinarg = zeros(n,1);
            argu = zeros(n,1);
            var = cov(y); % variance
      % subtract mean
            yn = y - mean(y);
      % do one Lomb loop (the t loop is vectorized)
            for fi = 1:nfreq;
                  sinsum = sum(sin(f4pi(fi)*t));
                  cossum = sum(cos(f4pi(fi)*t));
                  tau = atan2(sinsum,cossum);
                  argu = pi2*freq(fi)*(t-tau);
                  cosarg = cos(argu);
                  cfi = sum(yn.*cosarg);
                  cosnorm = sum(cosarg.*cosarg);
                  sinarg = sin(argu);
                  sfi = sum(yn.*sinarg);
                  sinnorm = sum(sinarg.*sinarg);
                  power(fi)=(cfi*cfi/cosnorm+sfi*sfi/
                                                sinnorm)/(2*var);
            end
```

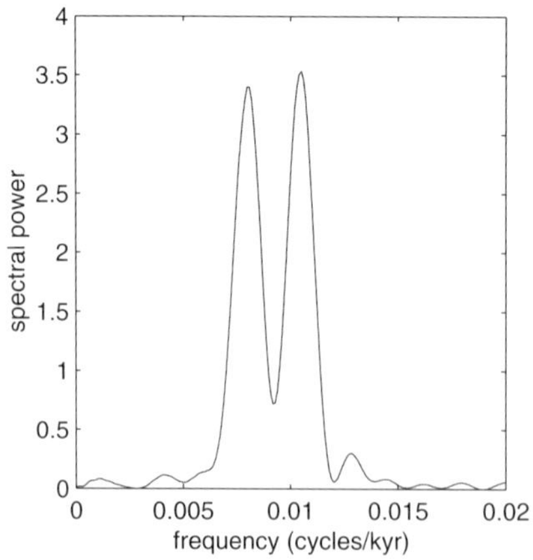

BLACKMAN–TUKEY METHOD

```
% code to calculate frequencies using the Blackman-Tukey method
% first set up data with three periods (100, 41, and 23)
% to demonstrate the method
      t = [1:600];
      y = sin(2*pi*t/100)+sin(2*pi*t/41) + sin(2*pi*t/23);
% choose lags
      lag = 1/3; % you should try other values!
      lagno = round(lag*length(y)); % "number" of lags
% calculate autocorrelation function
      if lag == 1;
            c = xcorr(o);
      else
            c = xcorr(o,o,lagno);
      end
% shift the function to place the zero offset point in
% first location of the vector
      center = ceil(length(c)/2);
      c1 = c(center:length(c));
% do the FFT and take the real part
      p = real(fft(c1)); % p is the spectral power.
% normalize to mean of 1
      p = p/std(p);
% plot the results
      fNyquist = 0.5/(t(2)-t(1));
      f = linspace(0,fNyquist,floor(lagno/2+1));
      fig; plot(f,p(1:length(f)));
% set the maximum frequency on the plot to 0.06
      ax = axis;
      axis([ax(1) 0.06 ax(3) ax(4)])
      ylabel('spectral power');
      xlabel('frequency (cycles/kyr)');
```

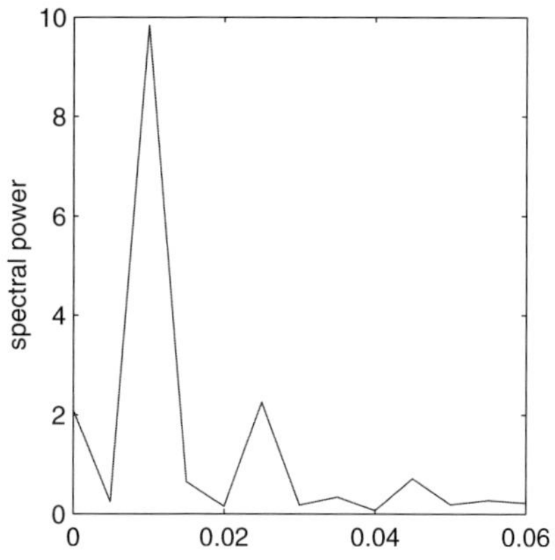

MAXIMUM ENTROPY METHOD

In the code below, we generate an artificial data set, and then use the maximum entropy method to estimate the spectrum. We invoke the program "pyulear", which is available in the optional "Signal Toolbox" of Matlab.

An interesting exercise to try is the following: add a random background to the data, using the routines on page 288, with various amplitudes. Try increasing the order of the maximum entropy method (make it greater than 100) and see how this affects the ability to see the three frequencies present in the simulation.

```
% create artificial data to test the method
        t = [1:600];
        y = sin(2*pi*t/100)+sin(2*pi*t/41)+sin(2*pi*t/23);
% choose Maximum Entropy parameters
        npts = 2^13;
        df = 1/(t(2)-t(1)); % sampling frequency
        order = 100; % order of the approximation;
% invoke the Matlab program pyulear
[p,f] = pyulear(y,order,npts,df);
p = p/std(p);
        % plot the result
figure;
plot(f,p); % or substitute semilogy(f,p);
% set the maximum frequency on the plot to 0.06
        ax = axis;
        axis([ax(1) 0.06 ax(3) ax(4)])
        ylabel('spectral power');
        xlabel('frequency (cycles/kyr)');
```

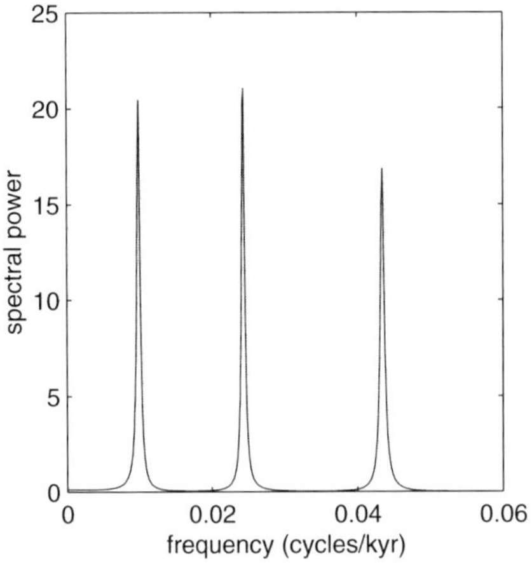

BISPECTRUM

Given a vector of ages t, and a vector of data y, this routine will calculate and plot the bispectrum. The maximum frequency used is set at fmax = 0.03; the user can change this.

When calculating the bispectrum, it is easy to create spurious peaks if the spectrum has strong sidelobes. To minimise this danger, we multiply the data by a Hanning filter prior to calculating the bispectrum.

The data vector y should be a real data set (such as the $\delta^{18}O$ record from SPECMAP) and the ages of the data points should be in a vector t with the same length.

```
% y and t must be previously read into the program
% remove the mean
      y = y - mean(y);
% multiply data vector "y" by a Hanning window
      n = length(y);
      w = .5*(1 - cos(2*pi*(1:n)'/(n+1)));% the Hanning window
      yw = y.*w;
% do Fourier transform of the windowed data
      npts = 2^13;             % padded length
      amp = fft(yw,npts);      % the complex Fourier amplitude
% find the frequencies
      Nyquist = 0.5/(t(2)-t(1));  % the highest frequency
      fmin = 2*Nyquist/npts;      % the lowest nonzero
                                              frequency
      f = [0:fmin:Nyquist];       % the frequency vector
      maxf = 0.02;                % max frequency for
                                         Bispectrum plot
      nfreq = fix(maxf/fmin);
% calculate bispectrum
      B = zeros(nfreq,nfreq);
      for i = 1:nfreq
            for j = 1:nfreq
                B(i,j) = amp(i).*amp(j).*conj(amp(i+j-1));
            end
      end
% Smooth the bispectrum by convolving with
% a small filter array
            filt = [0 .0 0.5 0.0; 0.5 1.0 0.5; 0.0 0.5 0.0];
            BC = abs(conv2(filt,B)).^2;
% shift Bispectrum (to undo shifting caused by convolution)
            Bispec = BC(2:nfreq,2:nfreq);
            freq = f(1:nfreq-1);
% Plot results
            figure('Position', [100 100 400 400]);
            contour(freq,freq,Bispec,10)
```

BAND PASS FILTER

This function will take data from the time domain, remove all frequencies except those within a band between fa and fb, and then return the filtered data. The input for this function is the column vector "y" containing the data, the column vector "age" containing the ages, and the two frequencies fa and fb. The data points must be equally spaced in time. The program sets up a test, with two frequencies present. The function bandpassfilter should appear as a separate file, called bandpassfilter.m. The plots the program produces are shown at the end.

```
% set up data with two frequencies, periods 100 and 41
        age = [1:.1:1000];
        y = sin(2*pi*age/100)+sin(2*pi*age/41);
% pick frequency limits that will pass only the 41
        fa = 1/80;
        fb = 1/20;
        ynew = bandpassfilter(age,y,fa,fb);
% plot the results
        figure; plot(age,y); title('unfiltered');
        figure; plot(age,ynew);
                title('filtered');setymax(2);setymin(-2);

% the following should be a separate file called
                                                % bandpassfilter.m

function ynew = bandpassfilter(age,y,fa,fb)
        f1 = min([fa fb];
        f2 = max([fa fb]);
        n = length(age);
        dt = age(2)-age(1);   % sampling interval
        fNyq = 1/(2*dt);              % Nyquist frequency
        y = y - mean(y);              % remove mean
        ft = fft(y);          % fast fourier transform
        fre = linspace(0,2*fNyq,n)'; % frequency vector
        [m,k1] = min(abs(fre - f1));  % find k1 corresponding
                                      % to freq f1
        [m,k2] = min(abs(fre - f2));  % find k2
                              % corresponding to freq f2
        k3 = n-k2+2;          % conjugate points
        k4 = n-k1+2;
% zero all but the desired band.
```

```
% Do for negative frequencies too.
      ft(1:k1) = zeros(k1,1);
      ft(k2:k3) = zeros(k3-k2+1,1);
      ft(k4:n) = zeros(k1-1,1);
% take inverse FFT; remove the
% imaginary part (from roundoff errors)
      ynew = real(ifft(ft));
```

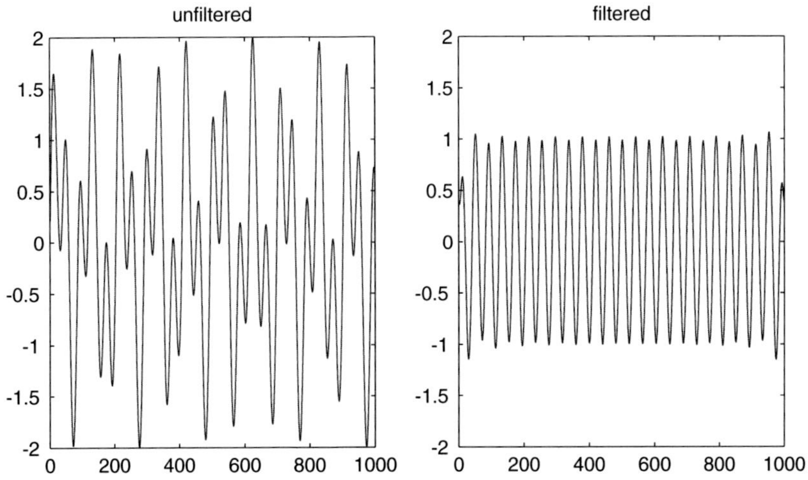

Appendix 3

Maximum Entropy Plots

In this appendix we present the spectra of several of the important data sets using the maximum entropy method for the spectral estimate. The number of poles used in all cases was 250.

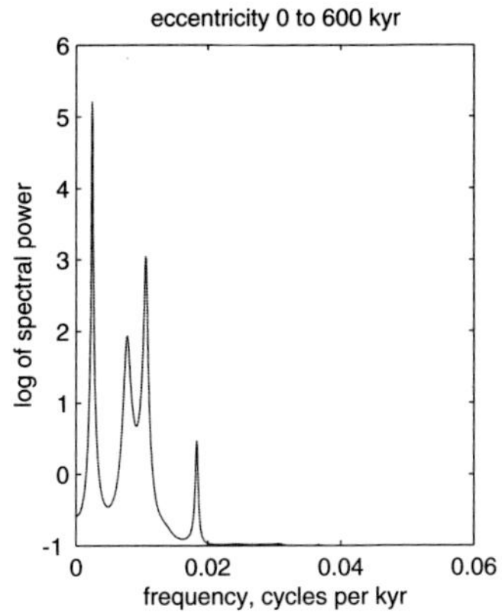

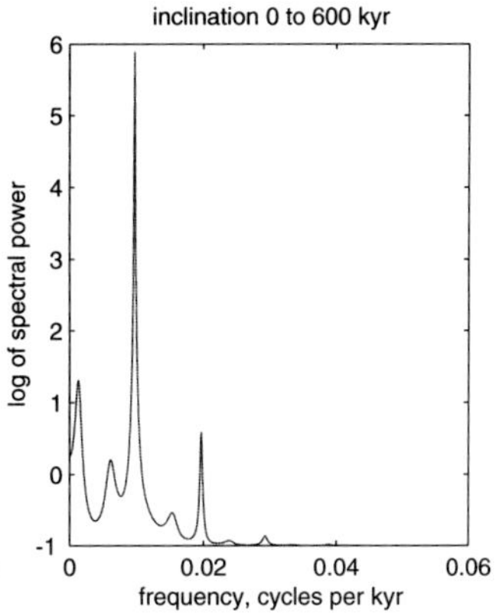

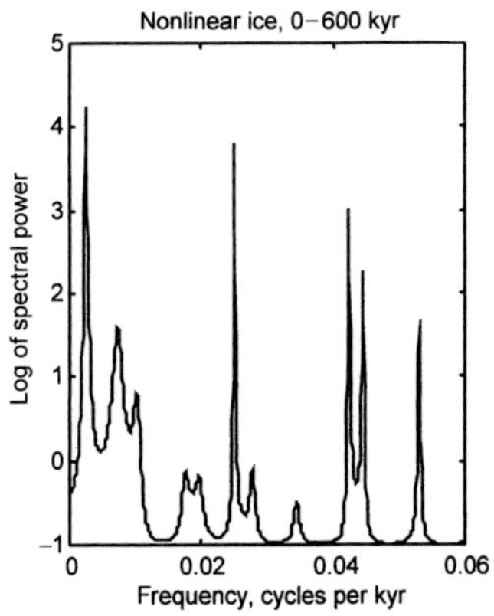

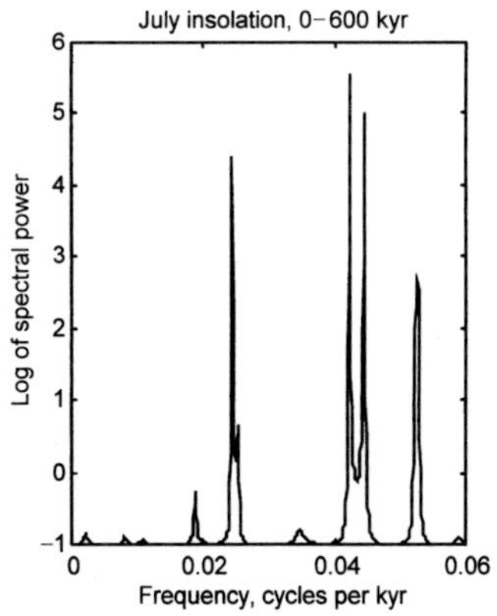

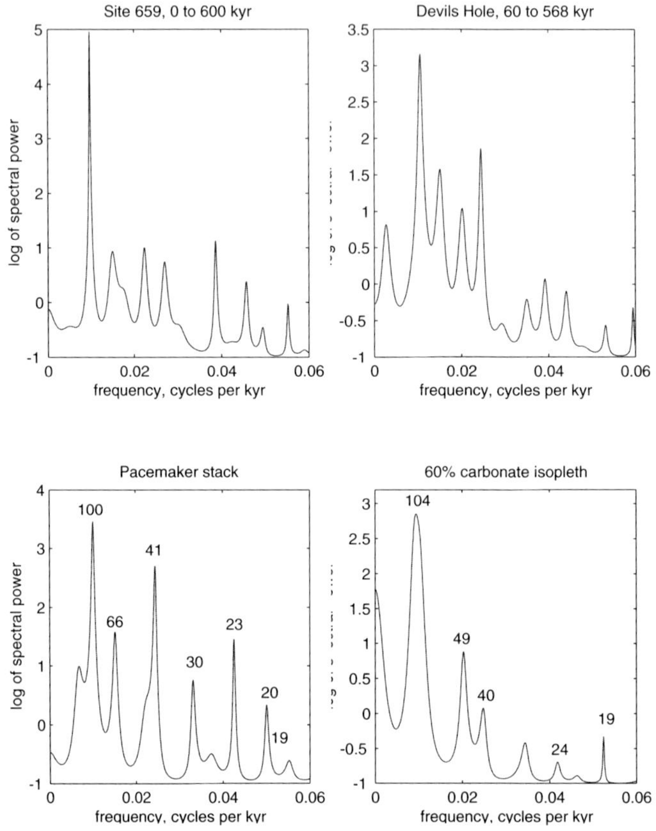
Site 659, 0 to 600 kyr
log of spectral power
frequency, cycles per kyr
Devils Hole, 60 to 568 kyr
frequency, cycles per kyr
Pacemaker stack
100
41
66
23
30
20
19
log of spectral power
frequency, cycles per kyr
60% carbonate isopleth
104
49
40
19
24
frequency, cycles per kyr

Appendix 4

Websites for data

Note: if your browser requires it, put http:// before each address, except for the ftp site listed last.

for Greenland ice data (GRIP, GISP, etc)
`www.gisp2.sr.unh.edu/GISP2/`

NOAA Paleoclimatology Program (World Data Center A for Paleoclimatology)
`www.ngdc.noaa.gov/paleo/paleo.htm`
`medias.meteo.fr/paleo/www/anglais/activities/donnees/`

Temperature records
`www.ncdc.noaa.gov/ol/climate/research/1998/anomalies/land_o_C.all`

Climate System Research Center, U. Massachusetts
`www.geo.umass.edu/climate/html`

American Quaternary Association
`pc53.chm.nau.edu/amqua/`

International Geosphere-Biosphere Programme
`www.pages.unibe.ch/pages.html`

International Union for Quaternary Research
`inqua.nlh.no/`

Quaternary Research Association
`gra.org.uk`

Orbital inclination data
`muller.lbl.gov`

FTP sites: ftp://ftp.ngdc.noaa.gov/paleo/paleocean/

References

Aiken, A.C. and McPeters, R.D. Meteoric material and the behaviour of upper stratospheric polar ozone. *Geophysical Research Letters (Supplement)* 1986; 13: 1300–1303.

Baksi, A.K., Hougton, B., McWilliams M., Tanaka, H. and Turner, G. What is the age of the Brunhes-Matuyama polarity transition? *EOS Trans. Am. Geophys. Union* 1991; 72: 135.

Baksi, A., Hsu, V., McWilliams, M. and Farrar, E. $^{40}Ar/^{39}Ar$ dating of the Brunhes/Matuyama geomagnetic field reversal. *Science* 1992; 256: 356–357.

Bard, E., Hamelin, B. and Fairbanks, R.G. Comparison between radiocarbon and uranium series ages on glacial age Barbados corals. *Nature* 1990a; 345: 405–409.

Bard, E., Hamelin, B. and Fairbanks, R.G. Additional uranium series ages for Barbados corals by the mass spectrometry methods. *Nature*, 1990b; 346: 456–458.

Bard, E., Hamelin, B., Fairbanks, R.G. and Zindler, A. Calibration of the ^{14}C time scale over the past 30,000 years using mass spectrometric U-Th ages from Barbados corals. *Nature* 1990c; 345: 405–409.

Barker, J.L. and Anders, E. Accretion rate of cosmic matter from iridium and osmium contents of deep-sea sediments. *Geochimica et Cosmochimica Acta* 1968; 32: 627–645.

Bassinot, F.C., Labeyrie, L.D., Vincent, E., Quidelleur, X., Shackleton, N.J. and Lancelot, Y. The astronomical theory of climate and the age of the Brunhes-Matuyama magnetic reversal. *Earth and Planetary Science Letters* 1994; 126: 91–108.

Bender, M., Sowers, T., Dickson, M-L., Orchardo, J., Grootes, P., Mayewski, P.A. and Meese, D.A. Climate correlations between Greenland and Antarctica during the past 100,000 years. *Nature* 1994; 372: 663–666.

Berger, A. Long-term variation of daily insolation and Quaternary climatic changes. *J. Atmos. Sc.* 1978; 35: 2362–2367.

Berger, A. and Loutre, M. Astronomical solutions for paleoclimate studies over the last 3 million years. *Earth and Planetary Science Letters* 1992; 111: 369–382.

Berger, W.H. The 100-kyr ice-age cycle: Internal oscillation or inclinational forcing? *Int. Journal Earth Sciences* 1999; 88: 305–316.

Berger, W.H., Bickert, T., Schmidt, H. and Wefer, G. Quaternary oxygen isotope record of pelagic foraminifers: Site 908, Ontong Java Plateau. In: *Proceedings of the Ocean Drilling Program, Scientific Results, Vol. 130*. Ocean Drilling Program, College Station, TX, 1993; 381–395.

Berger, W.H., Bickert, T., Wefer, G. and Yasuda, M.K. Brunhes-Matuyama boundary: 790 k.y. data consistent with ODP Leg 130 oxygen isotope records based on fit to Milankovitch template. *Geophysical Research Letters* 1995; 22: 1525–1528.

Berger, W.H., Yasuda, M., Bickert, T., Wefer, G. and Takayama, E. Quaternary time scale for the Ontong Java Plateau: Milankovitch template for the Ocean Drilling Program, Site 806. *Geology* 1994; 22: 463–467.

Birchfield, G.E. and Grumbine, R.W. "Slow" physics of large continental ice sheets and underlying bedrock, and its relation to the Pleistocene ice ages. *J. Geophys. Res.* 1985; 83: 4123–4125.

Birchfield, G.E. and Weertman, J. A note on the spectral response of a model continental ice sheet. *J. Geophys. Res.* 1978; 83: 4123–4125.

Blackman, R.B. and Tukey, J.W. *The Measurement of Power Spectra*. Dover, NY, 1958.

Bond, G.C., Showers, W., Elliot, M., Evans, M., Lotti, R., Hajdas, I., Bonani, G. and Johnson, W. The North Atlantic's 1–2 kyr climate rhythm: Relation to Heinrich events, Dansgaard/Oeschger cycles and the little ice age. In: Clark, P.U., Webb, R.S. and Keigwin, L.D. (Eds) *Mechanisms of Global Climate Change at Millennial Time Scales, Vol. 112*. American Geophysical Union, Washington, DC, 1999; 35–58.

Bradley, R.W. *Paleoclimatology: Reconstructing Climates of the Quaternary*. International Geophysics Series, Vol. 68. Academic Press, NY, 1999.

Broecker, W.S. *The Glacial World According to Wally*. Eldigio Press, NY, 1995.

Broecker, W.S. and Oversby, V.M. *Chemical Equilibria in the Earth*. McGraw-Hill, NY, 1971.

Broecker, W.S., Thurber, D.L., Goddard, J., Ku, T., Matthews, R.K. and Mesolella, J. Milankovitch hypothesis supported by precise dating of coral reefs and deep-sea sediments. *Science* 1968; 159: 1–4.

Broecker, W.S. and van Donk, J. Insolation changes, ice volumes, and the ^{18}O record in deep-sea cores. *Rev. Geophys, Space Phys.* 1970; 8: 169–197.

Brownlee, D.E. Cosmic dust: Collection and research. *Ann. Rev. of Earth Planet. Sci.* 1985; Vol. 13; 147–173.

Calder, N. Arithmetic of ice ages. *Nature* 1974; 252: 216–218.

Carlson, S. A double-blind test of astrology. *Nature* 1985; 318: 419–425.

Clark, P.U., Alley, R.B. and Pollard, D. Northern hemisphere ice-sheet influence on global climate change. *Science* 1999a; 286: 1104–1111.

Clark, P.U., Webb, R.S. and Keigwin, L.D. *Mechanisms of Global Climate Change at Millennial Time Scales*. Geophysical Monographs, Vol. 112. American Geophysical Union, Washington, DC, 1999b.

Craig, H. Isotopic variations in meteoric waters. *Science* 1961; 133: 1702–1703.

Craig, H. and Gordon, L.I. Deuterium and oxygen-18 variations in the ocean and the marine atmosphere. In: *Stable Isotopes in Oceanographic Studies and Paleotemperatures*. Consiglio Nazionale delle Ricerche, Laboratorio di Geologica Nucleare, Pisa, 1965; 1–122.

Crowley, T.J. Potential reconciliation of Devils Hole and deep-sea Pleistocene chronologies. *Paleoceanography* 1994; 9: 1–5.

Crowley, T.J. and Kim, K.-Y. Milankovitch forcing of the last interglacial sea level. *Science* 1994: 265 (9 September): 1566–1570.

Crowley, T.J., Kim, K.-Y., Mengel, J.G.T. and Short, D.A. Modeling 100,000-year climate fluctuations in pre-Pleistocene time series. *Science* 1992; 255: 705–707.

Cuffey, K.M. and Clow, G.D. Temperature, accumulation, and ice sheet elevation in central Greenland through the last deglaciation. *Journal of Geophysical Research* 1997; 102: 26,383–26,396.

Danby, J.M.A. *Fundamentals of Celestial Mechanics*. Willmann-Bell, Richmond, VA, 1992.

Dansgaard, W. Stable isotopes in precipitation. *Tellus* 1964; 16: 436–468.

Dansgaard, W., Johnsen, S.J., Clausen, H.B., Dahl-Jensen, D., Gundestrup, N.S., Hammer, C.U., Hvidberg, C.S., Steffensen, J.P., Sveinbjornsdottir, A.E., Jouzel, J. and Bond, G. Evidence for general instability of past climate from a 250-kyr ice-core record. *Nature* 1993; 364 (15 July): 218–220.

DeBlonde, G. and Peltier, W.R. A one-dimensional model of continental ice volume fluctuations through the Pleistocene: Implications for the origin of the mid-Pleistocene climate transition. *J. Clim.* 1991; 4: 318–344.

DeMenocal, P.B. Plio-Pleistocene African climate. *Science* 1995; 270: 53–59.

Dermott, S.F., Grogan, K., Gustafson, B.A.S., Jayaraman, S., Kortenkamp, S.J. and Xu, Y.L. Sources of interplanetary dust. In: Gustafson, B.A.S. and Hanner, M.S. (Eds) *Physics, Chemistry and Dynamics of Interplanetary Dust, Vol. 104*. Astronomical Society of the Pacific, 1996; 143–153.

Dermott, S.F., Jayaraman, S., Xu, Y.L., Gustafson, B.A. and Liou, J.C. A circumsolar ring of asteroid dust in resonant lock with the Earth. *Nature* 1994; 369: 719–723.

Dermott, S.F., Nicholson, P.D., Burns, J.A. and Houck, J.R. Origin of the Solar System dust bands discovered by IRAS. *Nature* 1984; 312: 505–509.

Dole, M., Lane, G.A., Rudd, D.P. and Zaukelies, D.A. Isotopic composition of atmospheric oxygen and nitrogen. *Geochim. Cosmochim. Acta* 1954; 6: 65.

Eddy, J.A. The Maunder Minimum. *Science* 1976; 192: 1189.

Eddy, J.A. Climate and the changing sun. *Climatic Change* 1977; 1: 173.

Edwards, R.L., Cheng, H., Murrell, M.T. and Goldstein, S.J. Protactinium-231 dating of carbonates by thermal ionization mass spectrometry: Implications for Quaternary climate change. *Science* 1997; 276: 782–788.

Edwards, R.L. and Gallup, C.D. Dating of the Devils Hole calcite vein. *Science* 1993; 259: 1626.

Emiliani, C. Paleotemperature analysis of Caribbean cores P-6304-8 and P-6304-9 and a generalized temperature curve for the past 425,000 years. *J. Geol.* 1966; 74: 109–124.

Esat, T.M., McCulloch, M.T., Chappell, J., Pillans, B. and Omura A. Rapid fluctuations in sea level recorded at Huon Peninsula during the penultimate deglaciation. *Science* 1999; 283: 197–201.

Esser, B.K. and Turekian, K.K. Accretion rate of extraterrestrial particles determined from osmium isotope systematics of Pacific pelagic clay and manganese nodules. *Geochimica et Cosmochimica Acta* 1988; 52: 1382–1388.

Farley, K. A 70 million year record of extraterrestrial helium fallout to the deep sea. *Nature* 1995; 378: 153–156.

Farley, K. and Patterson, D.B. A 100 kyr periodicity in the flux of extraterrestrial He-3 to the seafloor. *Nature* 1995; 378: 600–603.

Farley, K.A., Love, S.G. and Patterson, D.B. Atmospheric entry heating and helium retentivity of interplanetary dust particles. *Geochimica et Cosmochimica Acta* 1997; 61: 2309–2316.

Farley, K.A., Patterson, D.B. and Love, S.G. Helium in seafloor sediments: Size of He-3 bearing cosmic dust grains and observations of 100 ka cyclicity in He-3 and He-4 at ODP Site 806. *Eos* 1996; 77: F415–F416.

Farrell, J.W. and Prell, W.L. Pacific $CaCO_3$ preservation and ^{18}O since 4 Ma: Paleoceanic and paleoclimatic implications. *Paleoceanography* 1991; 6: 319–331.

Faure, G. *Principles of Isotope Geology*. John Wiley & Sons, NY, 1986.

Fiocco, G. and Grams, G. On the origin of noctilucent clouds: Extraterrestrial dust and trapped water molecules. *Journal of Atmospheric and Terrestrial Physics* 1971; 33: 815–824.

Flynn, G. Interplanetary dust particles collected from the stratosphere: Physical, chemical, and mineralogical properties and implications for their sources. *Planet. Space Sci.* 1994; 42: 1151–1161.

Flynn, G.J., Sutton, S.R., Bajt, S., Klock, W., Thomas, K.L. and Keller, L.P. The volatile content of anhydrous interplanetary dust. *Meteoritics* 1993; 28(3): 349.

Foyle, B. and Haurwitz, B. Noctilucent clouds. *Space Science Reviews* 1966; 6: 279–340.

Friis-Christensen, E. and Lassen, K. Length of the solar cycle: An indicator of solar activity closely associated with climate. *Science* 1991; 254: 698.

Gallèe, H., van Ypersele, J.P., Fichefet, T., Marsiat, I., Tricot, C. and Berger, A. Simulation of the last glacial cycle by a coupled, sectorially averaged climate-ice sheet model, 2, Response to insolation and CO_2 variation. *J. Geophys. Res.* 1992; 97: 15,713–15,740.

Gallée, H. van Ypersele, J.P., Fichefet, T., Tricot, C. and Berger, A. Simulation of the last glacial cycle by a coupled, sectorially averaged climate-ice sheet model, 1, The climate model. *J. Geophys. Res.* 1991; 96: 13,139–13,161.

Gallup, C.D., Cutler, K.B., Cheng, H., Edwards, H.L., Speed, R., Taylor, F.W., Adkine, J.F. and Burr, G.S. Sea level record for the last 200,000 years from concordant U-Th and U-Pa dates of fossil corals: Implications for deep-sea temperature changes and for early sea level rise during Termination II. *EOS Transactions* 1999; 80: F581.

Ganapathy, R. The Tunguska explosion of 1908: Discovery of meteoritic debris near the explosion site and at the South Pole. *Science* 1988: 220: 1158–1161.

Gold, T. Resonant orbits of grains and the formation of satellites. *Icarus* 1975; 25: 489–491.

GRIP members. Climate instability during the last interglacial period recorded in the GRIP ice core. *Nature* 1993; 364: 203–207.

Grootes, P.M., Stuiver, M., White, J.W.C., Johnsen, S. and Jouzel, J. Comparison of oxygen isotope records from the GISP2 and GRIP Greenland ice cores. *Nature* 1993; 366: 552–554.

Grün, E., Zook, H.A., Gechtig, H. and Giese, R.H. Collisional balance of the meteoritic complex. *Icarus* 1985; 62: 244–272.

Gustafson, B.A.S. and Hanner, M.S. *Physics, Chemistry, and Dynamics of Interplanetary Dust*. Proceedings of IAU Colloquium 150, Astronomical Society of the Pacific, 1996.

Harris, F. On the use of windows for harmonic analysis with discrete Fourier transform. *Proceedings IEEE* 1978: 66: 51–56.

Hartmann, D.L. Radiative effects of clouds on Earth's climate. In: Hobbs, P.V. (Ed.) *Aerosol-Cloud Climate Interactions, Vol. 54*. Academic Press, 1993.

Hasselman, K., Munk, W. and MacDonald, G. Bispectra of ocean waves. In: Rosenblatt, M. (Ed.) *Time Series Analysis*. John Wiley & Sons, NY, 1963; 125–139.

Hays, J., Imbrie, J. and Shackleton, N. Variations in the Earth's orbit: Pacemaker of the ice ages. *Science* 1976; 194: 1121–1132.

Henderson, G.M. and Slowey, N.C. Direct U-Th isochron dating of the penultimate deglaciation and the implications for the mechanisms linking insolation to climate. *EOS Transactions* 1999; 80 (AGU Fall Meeting): F580–F581.

Herman, J.R. and Goldberg, R.A. *Sun, Weather and Climate*. Dover, NY, 1978.

Hirayama, K. Group of asteroids probably of common origin. *Astr. J.* 1918; 31: 185–188.

Horowitz, P. and Hill, W. *The Art of Electronics*. Cambridge University Press, 1991.

Hughes, D.W. Meteors. In: McDonnell, J.A.M. (Ed). *Cosmic Dust*. John Wiley, New York, 1978; 123–185.

Hunten, D.M., Turco, R.P. and Toon, O.B. Smoke and dust particles of meteoric origin in the mesosphere and stratosphere. *Journal of the Atmospheric Sciences* 1980; 37, 1342–1357.

Imbrie, J. and Imbrie, K.P. *Ice Ages: Solving the Mystery*. Harvard University Press, MA, 1979.

Imbrie, J., Berger, A., Boyle, E., Clemens, S., Duffy, A., Howard, W., Kukla, G., Kutzbach, J., Martinson, D., McIntyre, A., Mir, A., Molfino, B., Morley, J., Peterson, L., Pisias, N., Prell, W., Raymo, M., Shackleton, N. and Toggweiller, J. On the structure and origin of major glaciation cycles: 2. The 100,000 year cycle. *Paleoceanography* 1993a; 699–735.

Imbrie, J., Hays, J., Martinson, D., McIntyre, A., Mix, A., Morley, J., Pisias, N., Prell, W. and Shackleton, N. The orbital theory of Pleistocene climate: Support from a revised chronology of the marine $\delta^{18}O$ record. In: Berger, A. et al. (Eds) *Milankovitch and Climate, Part 1*. D. Riedel, Dordrecht, 1984; 269–305.

Imbrie, J. and Imbrie, J.Z. Modeling the climate response to orbital variations. *Science*, 1980; 207: 943–952.

Imbrie, J., Mix, A.C. and Martinson, D.G. Milankovitch theory viewed from Devils Hole. *Nature* 1993b; 363: 531–533.

IPCC. *Climate Change 1995*. Cambridge University Press, 1996.

Izett, G. and Obradovich, J. Dating of the Matumaya–Brunhes boundary based on $^{40}Ar/^{39}Ar$ ages of the Bishop Tuff and Cerro San Luis rhyolite. *Geol. Soc. Am. Abstracts with Programs* 1991; 23(6): A106.

Jackson, A.A. and Zook, H.A. A Solar System dust ring with the Earth as its shepherd. *Nature* 1989; 337: 629–637.

Janecek, T.R. High-resolution carbonate and bulk grain-size data for Sites 803–806 (0–2 MA). *Proceedings of the Ocean Drilling Program, Scientific Results* 1993; 130: 761–773.

Johnson, G.R. Brunhes–Matuyama magnetic reversal dated at 790,000 yr B.P. by marine-astronomical correlations. *Quat. Res.* 1982; 17: 135–147.

Johnson, R.G. and Wright, H.E. Great Basin calcite vein and the Pleistocene time scale: Comment. *Science* 1989; 246: 262.

Johnsen, S.J., Clausen, H.B., Dansgaard, W., Fuhrer, K., Gundestrup, N., Hammer, C.U., Iversen, P., Jouzel, J., Stauffer, B. and Steffensenn, J.P. Irregular glacial interstadials recorded in a new Greenland ice core. *Nature* 1992; 359: 311–316.

Jones, P.D., New, M., Parker, D.E., Martin, S. and Rigor, I.G. Surface air temperature and its changes over the past 150 years. *Rev. Geophys.*, 1999; 37: 173.

Jouzel, J. and Merlivat, L. Deuterium and oxygen-18 in precipitation: Modeling of the isotopic effects during snow formation. *Jour. Geophysical Research* 1984; 89D: 11,749–11,757.

Karner, D.B. and Muller, R.A. *Weakness of Precession in Global Ice Volume Records*, 2000b, Lawrence Berkeley Laboratory Report LBNL-44777.

Karner, D.B. and Muller, R.A. A causality problem for Milankovitch. *Science* 2000a; 288: 2143–2144.

Karner, D.B., Medeiros, B., Muller, R.A. *Dansgaard-Oeschger Events and the 1.5-kyr Cycle*, 2000c, Lawrence Berkeley Laboratory Report LBNL-44529.

Katz, M.E., Pak, D.E., Dickens, G.R. and Miller, K.G. The source and fate of massive carbon input during the latest Paleocene thermal maximum. *Science* 1999; 286: 1531–1533.

Keesee, R.G. Nucleation and particle formation in the upper atmosphere. *Journal of Geophysical Research* 1989; 94: 14,683–14,692.

Kelsall, T., Weiland, J.L., Franz, B.A., Reach, W.T., Arendt, R.G., Dwek, E., Freudenreich, H.T., Hauser, M.G., Moseley, S.H., Odegard, N.P., Silverberg, R.F. and Wright, E.L. The COBE diffuse infrared background experiment search for the cosmic infrared background. II. Model of the interplanetary dust cloud. *Astrophysical Journal* 1998; 508: 44–73.

Kernthaler, S.C., Toumi, R. and Haigh. J.D. Some doubts concerning a link between cosmic ray fluxes and global cloudiness. *Geophysical Research Letters* 1999; 26: 863–865.

Kirkby, J. *Beam Measurements of a CLOUD (Cosmics Leaving Outdoor Droplets) Chamber*. CERN, Geneva, 1998.

Kominz, M.A., Heath, G.R., Ku, T.L. and Pisias, N.G.E. Bruhes time scales and the interpretation of climage change. *Earth Planet. Sci. Lett.* 1979; 45: 394–310.

Kominz, M.A. and Pisias, N.G. Pleistocene climate: Deterministic or stochastic? *Science* 1979; 204: 171–173.

Kortenkamp, S.J. and Dermott, S.F. Accretion of interplanetary dust particles by the Earth. *Icarus* 1998; 135: 469–495.

Kukla, G. Comment on Cesare Emiliani "The Pleistocene epoch and the evolution of man". *Current Anthropology* 1968; 9: 37–39.

Kukla, G. Correlation between loesses and deep-sea sediments. *Geol. Fören. Stockhom Förh.* 1970; 92: 148–180.

Kukla, G., Heller, F., Ming, L.X., Chun, X.T., Sheng, L.T. and Sheng, A.Z. Pleistocene climates in China dated by magnetic susceptibility. *Geology* 1988; 16: 811–814.

Kyte, F.T. and Wasson, J.T. Accretion rate of extraterrestrial matter—Iridium deposited 33 to 67 million years ago. *Science* 1986: 232: 1225–1229.

Labitzke, K. and van Loon, H. The signal of the 11-year sunspot cycle in the upper troposphere–lower stratosphere. *Space Science Reviews* 1997; 80: 393–410.

Lamb, H.H. The early Medieval warm epoch and its sequel. *Palaeogeography, Palaeoclimatology, Palaeoecology* 1965; 1: 13–37.

Landau, L.D. and Lifshitz, E.M. *Mechanics, Third Edition*. Pergamon Press, 1976.

Laskar J., Joutel, F. and Boudin, F. Orbital, precessional, and insolation quantities for the Earth from −20 Myr to +10 Myr. *Astron. Astrophys.* 1993; 270: 522–533.

Lee, R.B., Gibson, M.A., Wilson, R.S. and Thomas, S. Long-term total solar irradiance variability during sunspot cycle 22. *Journal of Geophysical Research* 1995; 100: 1667–1675.

Lohmann, A. and Wirnitzer, B. Triple correlations. *Proceedings IEEE* 1984; 72: 889–890.

Lomb, N. Least squares frequency analysis of unequally spaced data. *Astrophys. Space Sci.* 1976; 39: 447–462.

Love, S.G. and Brownlee, D.E. A direct measurement of the terrestrial mass accretion rate of cosmic dust. *Science* 1993; 262: 550–553.

Low, F.G., Beitema, D.A., Gautier, T.N., Gillett, F.C., Beichman, C.A., Neugebauer, G., Young, E.Y., Aumann, H.H., Boggess, N., Emerson, J.P. and Habing, H.J. et al. Infrared cirrus: New components of the extended infrared emission. *Astrophysical Journal* 1984; 278: L19–L22.

Ludwig, K.R., Simmons, K.R., Szabo, G., Winograd, I.J., Landwehr, J.M., Riggs, A.C. and Hoffman, R.J. Mass spectrometric ^{230}Th-^{234}U-^{238}U dating of the Devils Hole calcite vein. *Science* 1992; 258: 284–287.

Ludwig, K.R., Simmons, K.R., Winograd, I.J., Szabo, B.J., Landwehr, J.M. and Riggs, A.C. Last interglacial in Devils Hole, by N.J. Shackleton, a reply. *Nature*, 1993a; 362: 596.

Ludwig, K.R., Simmons, K.R., Winograd, I.J., Szabo, B.J. and Riggs, A.C. Dating of the Devils Hole calcite vein by R. Lawrence Edwards and Christina D. Gallup, a response. *Science* 1993b; 259: 1626–1627.

Lyons, W.B., Mayewski, P.A., Spencer, M.J., Twickler, N.M.S. and Graedel, T.E. A northern hemispheric volcanic chemistry (1869–1984) record and climatic implications using a south Greenland ice core. *Annals of Glaciology* 1990; 14: 176–182.

MacDonald, G.J. Spectral analysis of time series generated by nonlinear processes. *Reviews of Geophysics* 1989; 27: 449–469.

MacDonald, G.J. Global climate change. In: MacDonald, G.J. and Sertorio, L. (Eds) *Global Climate and Ecosystem Change*. Plenum Press, NY, 1990; 1–95.

Mankinen, E.A. and Dalrymple, G.B. Revised geomagnetic polarity time scale for the interval 0–5 m.y. B.P. *J. Geophys. Res.* 1979; 84: 615–626.

Marcantonio, F. A comparative study of accumulation rates derived by He and Th isotope analysis of marine sediments. *Earth and Planetary Science Letters* 1995; 133: 549–555.

Marion, J.B. and Thornton, S.T. *Classical Dynamics of Particles and Systems, Fourth Edition*. Saunders College Publishing, Fort Worth, 1995, p. 126.

Maurette, M., Jehanno, C., Robin, E. and Hammer, C. Characteristics and mass distribution of extraterrestrial dust from the Greenland ice cap. *Nature* 1987; 328 (20 August): 699–702.

Mayewski, P.A., Meeker, L.D., Twickler, M.S., Whitlow, S., Yang, Q., Lyons, W.B. and Prentice, M. Major features and forcing of high-latitude northern hemisphere atmospheric circulation using a 110,000-year-long glaciochemical series. *Journal of Geophysical Research* 1997; 102: 26,345–26,366.

Mayewski, P.A., Lyons, W.B., Spencer, M.J., Twickler, M.S., Buck, C.F. and Whitlow, S. An ice core record of atmospheric response to anthropogenic sulphate and nitrate. *Nature* 1990; 346: 554–556.

Milankovitch, M. *Canon of Insolation and the Ice-Age Problem*. Royal Serbian Academy, Belgrade, 1941.

Mix, A.C., Pisias, N.G., Rugh, W., Wilson, J., Morey, A. and Hagelberg, T.K. Benthic foraminifer stable isotope record from Site 849 (0–5 Ma); Local and global climate changes. *Proceedings of the Ocean Drilling Program Scientific Results* 1991; 138: 371–412.

Morely, J.J. and Hays, J.D. Towards a high-resolution, global, deep-sea chronology for the last 750,000 years. *Earth Planet. Sci. Lett.* 1981; 53: 279–295.

Muller, R.A. Glacial cycles and extraterrestrial accretion. *Lawrence Berkeley Laboratory Report LBL-35665*, Berkeley, 1994.

Muller, R.A. *Solar Forcing and Glaciation, Proceedings of the Stage-11 Workshop*. USGS Open File Report 99-312, R. Poore et al., eds, pp. 45–50, 1999.

Muller, R.A. and MacDonald, G.J. Glacial cycles: Orbital nutation dominated for the last 900,000 years. *Lawrence Berkeley Laboratory Report LBL-35667*, Berkeley, 1994a.

Muller, R.A. and MacDonald, G.J. High-resolution spectral analysis of glacial cycles: Eccentricity is ruled out. *Lawrence Berkeley Laboratory Report LBL-35666*, Berkeley, 1994b.

Muller, R.A and MacDonald, G.J. Glacial cycles and orbital inclination. *Nature* 1995; 377: 107–108.

Muller, R.A. and MacDonald, G.J. Glacial cycles and astronomical forcing. *Science* 1997a; 277: 215–218.

Muller, R.A. and MacDonald, G.J. Simultaneous presence of orbital inclination and eccentricity in proxy climate records from Ocean Drilling Program Site 806. *Geology* 1997b; 25: 3–6.

Muller, R.A. and MacDonald, G.J. Spectrum of 100 kyr glacial cycle: Orbital inclination, not eccentricity. *Proc. Nat. Acad. Sci.* 1997c; 94: 8329–8334.

Neeman, B. Orbital tuning of Paleoclimate Records: A Reassessment. *Lawrence Berkeley Laboratory Report LBNL-39572*, 1993.

Neugebauer, G., Beichman, C.A., Soifer, B.T., Aumann, H.H., Chester, T.J., Gautier, T.N., Gillett, F.C., Hauser, M.G., Houck, J.R., Lonsdale, C.J., Low, F.J. and Young, E.T. Early results from the Infrared Astronomical Satellite. *Science* 1984; 224: 14–21.

Nikias, C. and Petropulu, A. *Higher-order Spectra Analysis*. Prentice-Hall, NJ, 1993.

Oerlemans, J. Model experiments of the 100,000-yr glacial cycle. *Nature* 1980; 287: 430–432.

Olsen, P.E. and Kent, D.V. Milankovitch climate forcing in the tropics of Pangaea during the Late Triassic. *Palaeogeography, Palaeoclimatology, Palaeoecology* 1996; 122: 1–26.

Paillard, C. The timing of Pleistocene glaciations from a simple multiple-state climate model. *Nature* 1998; 391: 378–381.

Party, S.S. In: Kroenke, L.W., Berger, W.H., Janecek, T.R. et al. (Eds) *Proceedings of the Ocean Drilling Project, Initial Reports, Site 806, Vol. 130*, 1991; 291–355.

Paul, A. and Berger, W.H. Climate cycles and climate transitions as a response to astronomical and CO_2 forcings. In: Harff, J., Lemke, W. and Stattegger, K. (Eds) *Computerized Modeling of Sedimentary Systems*. Springer, Heidelberg, 1999; 223–245.

Percival, D.B. and Walden, A.T. *Spectral Analysis for Physical Applications: Multitaper and Conventional Univariate Techniques*. Cambridge University Press, 1993.

Petit, J.R., Jouzel, J., Raynaud, D., Barkov, N., Barnola, J.M. Basile, I., Bender, M., Chappellaz, J., Davis, J., Delaygue, G., Delmotte, M., Kotlyako, V., Legrand, M., Lipenkov, V., Lorius, C., Pépin, L., Ritz, C., Saltzman, E. and Stievenard, M. Climate and atmospheric history of the past 420,000 years from the Vostok ice core, Antarctica. *Nature* 1999; 399: 429–436.

Peucker-Ehrenbrink, B. Accretion of extraterrestrial matter during the last 80 million years and its effect on the marine osmium isotope record. *Geochimica et Cosmochimica Acta* 1996; 60(17): 3187–3196.

Pisias, N.G. Geologic time series from deep-sea sediments: Time scales and distortion by bioturbation. *Marine Geology* 1983; 51: 99–113.

Pisias, N.G. and Shackleton, N.J. Modeling the global climate response to orbital forcing and atmospheric carbon dioxide changes. *Nature* 1984; 310: 757–759.

Pollard, D. A coupled climate-ice sheet model applied to the Quaternary ice ages. *J. Clim.* 1983; 1: 965–997.

Poynting, J.H. Radiation in the solar system: Its effect on temperature and its pressure on small bodies. *Phil. Tran. Roy. Soc. London* 1904; A202: 525.

Prather, M.J. and Rodriguez J.M. Antarctic ozone—meteoric control of HN03. *Geophysical Research Letters* 1988; 15: 1–4.

Prell, W.L. Oxygen and carbon isotope stratigraphy for the Quaternary of Hole 502B; evidence for two modes of isotopic variability. In: Prell, W.L. and Gardner, J.V. (Eds) *Initial Reports of the Deep Sea Drilling Project, Vol. LXVIII.* US Government Printing Office 1982; 455–464.

Press, W.H., Flannery, B., Teukolsky, S. and Vetterling, W. *Numerical Recipes.* Cambridge University Press, 1993.

Priestley M.B. *Spectral Analysis and Time Series, Vol. 1.* Academic Press, NY, 1981.

Quayle, R.G., Peterson, T.C., Basist, A.N. and Godfrey, C.S. An operational near real time global temperature index. *Geophys. Res. Lett.* 1999; 26: 333–336.

Quinn, T.R., Tremaine, S. and Duncan, M. A three million year integration of the Earth's orbit. *Astronomical Journal* 1991; 101: 2287–2305.

Rau, A.V., Karner, D. and Muller, R. Sedimentation rate modelling and the absence of precession in recent climate records. In preparation.

Raymo, M.E. Global climate change: A three million year perspective. In: Kukla, G.J. and Went, E. (Eds) *Start of a Glacial, Vol. I 3.* Springer-Verlag, Heidelberg, 1992; 207–223.

Raymo, M.E. The timing of major climate terminations. *Paleoceanography* 1997; 12: 577–585.

Raymo, M.E. Glacial puzzles. *Science* 1998; 281: 1467–1468.

Raymo, M.E., Oppo, D.W. and Curry, W. The mid Pleistocene climate transition: A deep sea carbon isotope perspective. *Paleoceanography* in revision, 1997.

Raymo, M.E., Ruddiman, W.F., Backman, J., Clement, B.M. and Martinson, D.G. Late Pliocene variation in northern hemisphere ice sheets and North Atlantic deep water circulation. *Paleoceanography* 1989; 4: 413–446.

Rial, J.A. On the origin of the long period sawtooth shape of the late Pleistocene paleoclimate records: The first derivative of the earth's orbital eccentricity. *Geophysical Research Letters* 1995; 22f: 1997–2000.

Rial, J.A., Pacemaking the ice ages by frequency modulation of the earth's orbital eccentricity. *Science* 1999; 285: 564–568.

Rife, D.C. and Boorstyn, R.R. Single-tone parameter estimation from discrete-time observations. *IEEE Trans. Information Theory* 1974; IT-20: 591–598.

Robertson, H.P. Dynamical effects of radiation in the solar system. *Mon. Not. Roy. Astron. Soc.* 1937; 97: 423.

Rocchia, R., Bonté, P., Jéhanno, C.I., Robin, E., de Angelis, M. and Boclet, D. Search for the Tunguska event relics in the Antarctic snow and new estimation of the cosmic iridium accretion rate. In: Sharpton, V.L. and Ward, P.D. (Eds) *Geological Society of America Special Paper 247.* Geological Society of America, CO 1990; 189–193.

Rozanski, K., Araguas-Araguas, L. and Gonfiantini, R. Isotopic patterns in modern global precipitation. In: Swart, P.K., Lohmannn, K.C., McKenzie, J. and Savin, S. (Eds) *Climate Change in Continental Isotopic Records*. American Geophysical Union, Washington, DC, 1993; 1–36.

Ruddiman, W.F., Raymo, M. and McIntyre, A. Matuyama 41,000 cycles: North Atlantic Ocean and northern hemisphere ice sheets. *Earth and Planetary Science Letters* 1986; 80: 117–129.

Ruddiman, W.F., Raymo, M.E., Martijnson, D.G., Clement, B.M. and Backman, J. Pleistocene evolution: Northern hemisphere ice sheets and North Atlantic Ocean. *Palaeoceanography* 1989; 4: 353–412.

Salby, M.L. *Fundamentals of Atmospheric Physics*. Academic Press, NY, 1996.

Saltzman B. and Verbitsky, M.Y. Multiple instabilities and modes of glacial rhythmicity in the Plio-Pleistocene: A general theory of late Cenozoic climate change. *Climate Dynamics* 1993; 9: 1–15.

Saltzman, B. and Verbitsky M.Y. Predicting the Vostok CO_2 curve. *Nature* 1995; 377: 690.

Santer, B.D., Wigley, T.M.L., Gaffen, D.J., Bengtsson, L., Doutriaux, C., Boyle, J.S., Esch, M., Hnilo, J.J., Jones, P.D., Meehl, G.A., Roeckner, E., Taylor, K.E. and Wehner, M.F. Interpreting differential temperature trends at the surface and in the lower troposphere. *Science* 2000; 287: 1227–1232.

Sarnthein, M. and Tiedemann, R. Toward a high-resolution stable isotope stratigraphy of the last 3.4 million years: Sites 658 and 659 off Northwest Africa. *Proc. Ocean, Drilling Program, Scientific Results* 1989; 108: 167.

Sato, T., Sasaki, K. and Nakamura, Y. Real-time bispectral analysis of gear noise and its application to contactless diagnosis. *J. Acoust. Soc. Am.* 1976; 62: 382–387.

Scargle, J. Studies in astronomical time series analysis, II, Statistical aspects of spectral analysis of unevenly spaced data. *Astrophys. J.* 1982; 263: 835–853.

Seibold, E. and Berger, W.H. *The Sea Floor: An Introduction to Marine Geology*. Springer-Verlag, NY, 1996.

Self, S., Rampino, M. and Barbera, J.J. The possible effects of large 19th and 20th century volcanic eruptions on zonal and hemispheric surface temperatures. *Journal of Volcanology and Geothermal Research* 1981; 11: 41–60.

Shackleton, J.J., Berger, A.L. and Peltier, W.R. An alternative astronomical calibration of the lower Pleistocene timescale based on ODP Site 677. *Transactions of the Royal Society of Edinburgh, Earth Sciences* 1990; 81: 251–261.

Shackleton, N. Carbon-13 in Uvigerina: Tropical rainforest history and the equatorial Pacific carbonate dissolution cycles. In: Andersen, N. and Malahoff, A. (Eds) *The Fate of Fossil Fuel Carbon Dioxide in the Oceans, Vol. 6*. Plenum Press, NY, 1977.

Shackleton, N.J. Last interglacial in Devils Hole, Nevada. *Nature* 1993; 362: 596.

Shackleton, N.J. and Hall, M.A. Oxygen and carbon isotope stratigraphy of Deep Sea Drilling Project hole 552A: Plio-Pleistocene glacial history. *Initial Reports of the Deep Sea Drilling Project, Vol. 81*. US Government Printing Office, Washington, DC, 1984; 599–609.

Shackleton, N.J. and Hall, M.A. Stable isotope history of the Pleistocene at ODP site 677. *Proceedings of the Ocean Drilling Program, Scientific Results, Vol. 111*. Ocean Drilling Program, College Station, TX, 1989; 295–316.

Shackleton, N.J. and Opdyke, N.D. Oxygen isotope and paleomagnetic stratigraphy of equatorial Pacific core V28-238: Oxygen isotope temperatures and ice volumes on a 10^5-year time scale. *Quaternary Res.* 1973; 3: 39–55.

Short, D.A., Menge, J.G., Crowley, T.J., Hyde, W.T. and North, G.R. Filtering of Milankovitch cycles by Earth's geography. *Quat. Res.* 1991; 35: 157–173.

Shu, F.H. *The Physical Universe: An Introduction to Astronomy*. University Science Books, Mill Valley, 1982.

Sowers, T. and Bender, M. Climate records covering the last deglaciation. *Science* 1995; 269: 210–214.

Spell, T.L. and McDougall, I. Revisions to the age of the Brunhes-Matuyama boundary and the Pleistocene geomagnetic polarity timescale. *Geophysical Research Letters* 1992; 19: 1181–1184.

Sussman, G.J. and Wisdom, J. Chaotic evolution of the Solar System. *Science* 1992; 257: 56–62.

Svensmark, H. Influence of cosmic rays on earth's climate. *Phys. Rev. Lett.* 1998; 81: 5027–5030.

Svensmark H. and Friis-Christensen, E. Variation of cosmic ray flux and global cloud coverage—a missing link in solar-climate relationships. *Journal of Atmospheric and Solar-Terrestrial Physics* 1997; 59: 1225–1232.

Sykes, M.V., Greenberg, R., Dermott, S., Nicholson, P., Burns, J. and Gautier, T. Dust bands in the asteroid belt. In: Binzel, T.G.R. and Matthews, M. (Eds) *Asteroids II*. University of Arizona Press, Tucson, AZ, 1989; 336–367.

Takahashi, H., Yokoyama, Y., Fireman, E.L. and Lorius, L. Iridium content of polar ice and accretion rate of cosmic matter. *Lunar and Planetary Science* 1979; 9: 1131–1133.

Tauxe, L., Deino, A., Behrensmeyer, A. and Potts, R. Pinning down the Brunhes-Matuyama and upper Jaramillo Boundaries: A reconciliation of orbital and isotopic time scales. *Earth and Planetary Science Letters* 1992; 109: 561–572.

Taylor, K.C., Lamorey, G.W., Doyle, G.A., Alley, R.B., Grootes, P.M., Mayewski, P.A., White, J.W.C. and Barlow, L.K. The "flickering switch" of late Pleistocene climate change. *Nature* 1993; 361: 432–436.

Thierstein, H.R., Geitzenauer, K.R., Molfino, B. and Shackleton, N.J. Global synchroneity of late Quaternary coccolith datum levels: Validation by oxygen isotopes. *Geology* 1977; 5: 400–414.

Thomas, G.E. Mesospheric clouds and the physics of the Mesopause region. *Reviews of Geophysics* 1991; 29: 553–575.

Thomas, G.E. and Olivero, J.J. Climatology of polar mesospheric clouds, 2, Further analysis of Solar Mesosphere Explorer data. *J. Geophys. Res.* 1989; 94: 14,673–14,681.

Thompson, L.G., Davis, M.E., Mosley-Thompson, E., Sowers, T.A., Henderson, K.A., Zagorodnov, V.S., Lin, P-N., Mikhalenko, V.N., Campen, R.K., Bolzan,

J.F., Cole-Dai, J. and Brancou, B. A 25,000-year tropical climate history from Bolivian ice cores. *Science* 1998, 282: 1858–1864.

Thompson, L.G., Yao, T., Davis, M.E., Henderson, K.A., Mosley-Thompson, E., Lin, P-N., Beer, J., Synal, H-A., Cole-Dai, J. and Bolzan, J.F. Tropical climate instability: The last glacial cycle from a Qinghai-Tibetan ice core. *Science* 1997; 276: 1821–1825.

Tiedemann, R., Sarnthein, M. and Shackleton, N.J. Astronomic timescale for the Pliocene Atlantic $\delta^{18}O$ and dust flux records of Ocean Drilling Program Site 659. *Paleoceanography* 1994; 9: 619–638.

Tinsley, B.A. Correlations of atmospheric dynamics with solar wind-induced changes of air-earth current density into cloud tops. *Journal of Geophysical Research* 1996; 101: 29,701–29,714.

Toon, O.B. and Farlow, N.H. Particles above the tropopause: Measurements and models of stratospheric aerosols, meteoric debris, nacreous clouds, and noctilucent clouds. *Ann. Rev. Earth Planet. Sci.* 1981; 9: 19–58.

Tuchman, B.W. *A Distant Mirror: The Calamitous 14th Century*. Alfred A. Knopf, NY, 1978.

Tukey. J. An introduction to the calculations of numerical spectrum analysis. In: Harris, B. (Ed.) *Advanced Seminar on Spectral Analysis of Time Series*. John Wiley & Sons, NY, 1967.

Tuncel, G. and Zoller, W.H. Atmospheric iridium at the South Pole as a measure of the meteoritic component. *Nature* 1987; 703–705.

Turco, R.P., Toon, O.B., Hamill, P. and Whitten, R.C. Effects of meteoric debris on stratospheric aerosols and gases. *Journal of Geophysical Research* 1981; 86: 1113–1128.

Turco, R.P., Toon, O.B., Whitten, R.C., Keesee, R.G. and Hollenback, D. Noctilucent clouds—Simulation studies of their genesis, properties and global influences. *Planetary and Space Science* 1982; 30: 1147–1181.

Tyron, P. *The Bispectrum and Higher Order Spectra: A Bibliography*. US Government Printing Office, Washington, DC, 1981.

van Trees, H.L. *Detection, Estimation and Modulation Theory, Part I*. John Wiley & Sons, NY, 1968.

Walden, T., McCoy, E. and Percival, D. The effective bandwidth of a multitaper spectral estimator. *Biometrika*, 1995; 82: 201–214.

Weertman, J. Rate of growth or shrinkage of nonequilibrium icesheets. *J. Glaciol.* 1964; 5: 145–158.

Whipple, F.L. and Hawkins, G.S. On meteors and rainfall. *Journal of Meteorology* 1956; 13: 236–240.

Whittle, P. The simultaneous estimation of a time series harmonic component and covariance structure. *Trab. Estadist* 1952; 3: 43–57.

Winograd, I.J. and Coplen, T.B. Reply to Johnson and Wright. *Science* 1989; 246: 263.

Winograd, I.J., Coplen, T.B., Landwehr, J.M., Riggs, A.C., Ludwig, K.R., Szabo, B.J., Kolesar, P.T. and Revesz, K.M. Continuous 500,000-year climate record from vein calcite in Devils Hole, Nevada. *Science* 1992; 258: 255–260.

Winograd, I.J. and Landwehr, J.M. *A Response to "Milankovitch Theory Viewed from Devils Hole" by J. Imbrie, A.C. Mix and D.G. Martinson*. US Geological Survey, Reston, VA, 1993.

Winograd, I.J., Landwehr, J.M., Ludwig, K.R., Coplen, T.B. and Riggs, A.C. Duration and structure of the past four interglaciations. *Quaternary Research* 1997: 48: 141–154.

Index